Studies in Classification, Data Analysis, and Knowledge Organization

More information about this series at http://www.springer.com/series/1564

Paolo Mariani · Mariangela Zenga
Editors

Data Science and Social Research II

Methods, Technologies and Applications

Springer

Editors
Paolo Mariani
Department of Economics
Management and Statistics
University of Milano-Bicocca
Milan, Italy

Mariangela Zenga
Department of Statistics
and Quantitative Methods
University of Milano-Bicocca
Milan, Italy

ISSN 1431-8814 ISSN 2198-3321 (electronic)
Studies in Classification, Data Analysis, and Knowledge Organization
ISBN 978-3-030-51221-7 ISBN 978-3-030-51222-4 (eBook)
https://doi.org/10.1007/978-3-030-51222-4

This Springer imprint is published by the registered company Springer Nature Switzerland AG
The registered company address is: Gewerbestrasse 11, 6330 Cham, Switzerland

Preface

As digital technologies, the Internet and social media become increasingly integrated into society, a proliferation of digital footprint of human and societal behaviours are generated in our daily lives. All these data provide opportunities to study complex social systems, by the empirical observation of patterns in large-scale data, quantitative modelling and experiments. The social data revolution has not only produced new business models, but has also provided policymakers with better instruments to support their decisions.

This book consists of a selection of the papers presented at the Second International Conference on Data Science and Social Research held in Milan in February 2019 (https://www.dssr2019.unimib.it). The conference aimed to stimulate the debate between scholars of different disciplines about the so-called data revolution in social research. Statisticians, computer scientists and experts on social research discussed the opportunities and challenges of the social data revolution to create a fertile ground for addressing new research problems.

The volume collects 30 contributions focused on the topics for complex social systems. Several papers deal in new methodological developments to extract social knowledge from large scale data sets and new social research about human behaviour and society with large datasets, either mined from various sources (e.g. social media, communication systems) or created via controlled experiments. Moreover, some contributions analysed integrated systems to take advantage of new social data sources; others discussed big data quality issues, both as a reformulation of traditional representativeness and validity and as emerging quality aspects such as access constraints, which may produce inequalities.

All contributions were subjected to peer-review and are listed in alphabetical order of the first author.

We would like to express our gratitude to the Scientific, Program and Local Committees that allowed the realisation of the conference. Moreover, we would thank the authors and the anonymous referees who made the creation of the volume possible. Our deep gratitude also goes to Laura Benedan that supported us in every organisational stage of this book.

Milan, Italy
February 2020

Paolo Mariani
Mariangela Zenga

Contents

Digital Methods and the Evolution of the Epistemology of Social Sciences

Enrica Amaturo and Biagio Aragona

Abstract After ten years that the debate on big data, computation and digital methods has been a contested epistemological terrain between some who were generally optimistic, and others who were generally critical, a large group of scholars, nowadays, supports an active commitment by social scientists to face the digital dimension of social inquiry. The progressive use of digital methods needs to be sustained by an abductive, intersubjective and plural epistemological framework that allows to profitably include big data and computation within the different paradigmatic traditions that coexist in our disciplines. In order to affirm this digital epistemology it is critical to adopt a methodological posture able to elaborate research designs with and against the digital, trying to exploit what digital techniques can give as added value, but going to test their reliability, alongside others techniques, including qualitative ones.

1 Introduction

Big data, computation, and digital methods have been a contested epistemological terrain for the last decade of social research. In this controversy, oversimplifying, we have had two groups, with different postures on the matter. On one side, those who were generally optimistic, on the other side, those who were generally critical (Salganik 2018).

Since the advent of big data, the epistemological debate has then developed between two opposites: revolution and involution. According to revolution, disruptive technical changes transform in better both the sciences and the methods once consolidated within the different scientific disciplines (Lazer 2009; Mayer-Schonberger and Cuckier 2012). On the contrary, involution states that digital methods and big data impoverish social sciences and their method (Boyd and Crawford 2012), becoming

E. Amaturo · B. Aragona (✉)
University Federico II of Naples, Naples, Italy
e-mail: aragona@unina.it

P. Mariani and M. Zenga (eds.), *Data Science and Social Research II*, Studies in Classification, Data Analysis, and Knowledge Organization,
https://doi.org/10.1007/978-3-030-51222-4_1

a threat to the empirical sociology based on surveys and interviewing (Savage and Burrows 2007).

These views developed around two main topics: the quality of digital data, and the role of technology in social research. "Revolutionary" scholars believe that more data automatically lead to better research. This actually is not sustained by facts. Loads of data may actually increase the level of "noise" in data, so we can hardly distinguish rumors form signals (Torabi Asr and Taboada 2019). Moreover, more data does not mean better data *tout court*, because the risk of *garbage in - garbage out* is higher (Kitchin and Lauriault 2015). For these reasons, scholars who maintain involution are critical with the quality of digital data sources, which are mainly secondary data repurposed with new objectives. In analog research most of the data that were used for social research were created with the purpose of research, while in digital social research huge amount of data are created by corporations and organizations with the intent of making profit, delivering services, or administering public affairs.

In a similar way, optimistics are enthusiast about the possibilities opened by digital technology, while critics contrast the adoption of technology. Scholars who sustain the revolution in social sciences believe that technology is the driver of innovation and of advances in knowledge. This technological determinism promotes an idea in which the scientific disciplines stands at the passive end (Marres 2017), with technology being a force of improvement for research. On the contrary, most critics believe that the reconfigurability of digital infrastructures and devices is contested and it needs continue demonstration and testing. Big data and digital methods are effective as far as we would be in the condition to inspect the theoretical assumptions within the data and the socio-technical processes that shape them.

After ten years that the debate on big data and digital methods has developed on these two opposite visions, a large group of scholars, nowadays, supports an active commitment by social scientists to face the digital dimension of social inquiry (Orton-Jhonson and Prior 2013; Lupton 2014; Daniels and Gregory 2016). On this basis, instead of opposing revolution and involution, it should be more correct to talk about an epistemic evolution. These scholars advocate that in order to effectively improve the unfolding of social phenomena, digital methods and big data have to be integrated with traditional data sources and methods already existing in social sciences. The works developed by these scholars focus on the definition of an epistemology of social research that adopts a critical posture on the role that digital technology must have in scientific research, but, at the same time, creative on the possibilities offered by technology to research (Marres 2017; Halford and Savage 2017).

In this article, we argue that one way to do that is concentrating more efforts on the construction of research designs for and against the digital. With the objective to avoid ideological positions about the role of digital methods in social research, we should then promote an active engagement in testing the different instruments of digital research, such as big data, machine learning, platforms analytics, search as research tools, and so on. Moreover, digital technologies carry the risk of flattening social research only on two phases: data analysis, and communication of results. The development of techniques for elaborating increasingly large databases—which often are not even fully understood by social scientists (Kitchin 2014a)—pushes to consider the process of research as an analytical process, effectively neglecting the other research phases. Likewise, data visualization and infographics, which are so fundamental to digital data communication, can produce the same risks (Halford and Savage 2017). One possible answer to these concerns is again to invest more efforts in developing digital research designs. By refocusing the attention on the research design we can restore importance to all the other research phases.

2 Revolution

The idea that digital methods, computation and big data are revolutionary for social sciences has developed upon three main epistemic features: objectivity, induction and computation.

The objectivity leaves from the fact that reality is considered independent from technology and sociality. Big data, platforms and digital ecosystems are seen as windows on the social reality. The scientific method is data driven, and—resting on the possibility to track human behavior with unprecedented fidelity and precision—exploring existing data may be more useful than building models on why the people behave the way they do. The objectivity of reality has the further consequence of repurposing the dualism between the researcher and the reality, between subject and object, that is typical of positivism. What is different with early positivism is the objective of research, which is more centered on quantification and description than on looking for causes of phenomena. Data are the core of this approach to social science, and computational methods are the required tools to learn from these data. This view on the digital is often espoused by scholars in quantitative computational social sciences (Lazer 2009) and data science.

The revolutionary idea has been also sustained by discussions about innovation in social research, which are more methodological than epistemological (i.e. if digital techniques should be considered new or not (Marres 2017). The new-vs-old dichotomy was firstly promoted by Rogers (2013), which distinguished between digitized and digital native techniques. The former are those that already existed in analog form, and that are "migrated" on the web (for example web surveys and digital ethnography), the latter are those "born" on the web, such as web scraping techniques, and search as research tools. The division of digital methods in these two groups has at least two weaknesses that may be envisaged. First of all, those who advocate the existence of digital native techniques propose the idea of methodological development as guided by technology. Technology = new; social sciences = old. Furthermore, this opposition implies that the development of research methods in social sciences should come from the "outside", from disciplines such as computer science and data science. But, all the techniques that are having great relevance in digital research: "have an inter-disciplinary origin … and can be qualified as" mixed "techniques, in the sense that they combine computational elements and sociological elements" (Marres 2017, p. 104). Halford (2013) believes that digital techniques are not at all "alien" to social sciences, but rather that the techniques incorporated in digital platforms and devices are built on consolidated and lasting methodological principles. By stressing the revolution of social sciences and their methods, scholars focus on ruptures instead than on connections, with the result that the continuity between traditions of social research and digital techniques will no longer be recognized.

In order to clarify this point, there is the example of the analytics that have been developed by *Google* to do research through searches made on its search engine. By using these analytics, such as *Google Trends*, Google carried out a famous study on the ability to predict the propagation of influenza before the research institute (CDC) that was responsible for measuring its spread. That work (Ginsberg et al. 2009) was used to state that digital native techniques enabled forms of analysis that could not be realized before (Mayer-Schonberger and Cuckier 2012; Rogers 2013). Abbott (2011), however, noted that these tools rested on very traditional forms of analysis. For example, *Trends* uses the analysis of temporal and territorial series to count how many times keywords have been searched on the search engine. That is, even if the technique is innovative with respect to the technological and computational aspects, the underlying methodological principle is very old. It would then make more sense, even at the methodological level, to appropriately examine how digital techniques lead/do not lead to social research method evolution, rather than focusing the attention on its revolution.

3 Evolution

It is not difficult to support the idea that big data, digital methods and computation contribute to an evolution of social research methods, and more generally of social sciences. The spreading of large databases from a variety of sources gives the possibility of doing research in many ways, and of improving techniques that were already used in the past (i.e., content analysis and network analysis). At the same time, critical data science research (Iliadis and Russo 2016) is emerging, with the aim to assess the social consequences of the processes of digitalization, and consequent datafication, of various sectors of society. The collection, analysis and processing of data, networks and relationships through digital methods therefore manages to create also new points of contact between digital and not digital social science (Orton-Jhonson and Prior 2013; Daniels and Gregory 2016).

This evolutionary posture starts with an active commitment by social scientists to confront with the technological dimension of social inquiry. Its main features are: intersubjectivity, abduction and mixed methods.

Intersubjectivity refers to the fact that social reality is dependent on the socio-technical activities that are made to grasp it. Data are not a simple reflection of a world that "is", but are thoroughly "produced". The separation between object and subject must be overcome. It acknowledges the role of platforms (Van Dijck et al. 2018) and «methodological dispositifs» (Ruppert 2013) in shaping reality.

Digital platforms are changing with a velocity that is not usual for social science data (Chandler et al. 2019). For example, with longitudinal survey data, breaks in the series are rare and very carefully implemented inside the overall longitudinal research design. On the contrary, platforms change all the time, and changes occur in at least three ways (Salganik 2018): they may change in who is using them, in how they are used, and in how the platform work. Some examples are: during 2012 US Presidential Election the proportions of tweets by women increased and decreased from day to day; *Facebook* in Italy started to be a social network to reconnect the school community, and now is used also as a form of advertising; in 2018 *Twitter* decided to double the number of digits in tweets from 140 to 280. All these kinds of changes may have an impact on research results.

Also the "methodological dispositifs" may impact on results. "Methodological dispositifs" are the material objects and ideas that configure the ways we do research. They are not simply research methods, but they are also the same objects of analysis. To understand the role played by these dispositifs a close inspection of data assemblages should be realized (Kitchin 2014b; Aragona 2017). Data assemblage is a complex socio-technical system composed of many elements and elements that are thoroughly entwined, whose central concern is the production of data. These assemblages are made of two main activities: a technical process, (operational definitions, selection, data curation) which shape the data as it is, and a socio-cultural process, which shapes the background knowledge (beliefs, instruments and other things that are shared in a scientific community). Researching big data assemblages may help to unpack digital black boxes (Pasquale 2015) and increase our knowledge about the processes of algorithms construction (Aragona and Felaco 2018), the effects of data curation on research results (Aragona et al. 2018), the values into data.

Moreover, according to evolution, neo-empiricism—the data-first method—has to be rejected. Social sciences must preserve the main tenets of the post-positivist scientific method, but at the same time promote the joint use of induction and deduction. For the advocates of evolution, scientific knowledge is pursued using "guided" computational techniques to discover hypotheses to be submitted to subsequent empirical control. The process is guided because the existing theories are used to direct the development of the discovery and not—as in quantitative computational science—to identify all the existing relationships in a database. Instead, the way in which data is constructed or re-analyzed is guided by some assumptions, supported by theoretical and practical knowledge and experience of how technologies and their configurations are able to produce valid and relevant empirical material for research. In practice, the method used is abductive (Pierce 1883), and aims to insert unexpected results in an interpretative framework.

Consequently, also the opposition between correlation and causation can be overcome. Despite the fact that quantitative computational social science has become the most widespread way of doing computational social science, it should not be forgotten that the ambitions of the authors who wrote the *Manifesto of Computational Social Science* (Conte et al. 2012) were different. Conte (2016) underlines that at the beginning there was no quantitative approach, but computational social science was mainly generative and aimed to unveil the mechanisms that produce social phenomena through simulations in informational ecosystems. This way of doing CSS has produced many theories about social phenomena such as cooperation, coordination and social conventions. Quantitative CSS, instead, it dismissed the search for the causes of social phenomena. The theoretical ambitions of the authors of the *Manifesto* have been supplanted by an emphasis on quantification and description, mainly because, as Merton first noted (1968), science goes to sectors where there is abundance of data.

Finally, evolutionary social science epistemology assumes that more or less computational analytical methods have become the standard of social research, but at the same time, it does not consider it an imperative. It may be useful to produce new visions of social phenomena through digital methods, but their methodological capacity should be constantly tested. The use of these techniques must be openly discussed, evaluating the impacts on research designs, on the formulation of questions and, when necessary, on hypothesis testing strategies. The already cited *Google Flu Trends* research is a good example of that. After early enthusiasm, the use of search queries to detect the spread of flu turned to be tricky. Over time, researchers discovered that the estimates were not so much better than that of a simple model that calculates the amount of flu based on a linear extrapolation from the two most recent measurements of flu prevalence (Goel et al. 2010). Moreover, estimates were prone to short-term failure and long-term decay. More specifically, during the Swine Flu pandemic, the trends overestimated the amount of influenza, probably because people change their search behavior during a global pandemic. It was only thank to the control of their results with those that are collected by the US Centers for Disease Control (CDC), which are regularly and systematically collected from carefully sampled doctors around US, that researchers were able to develop more precise estimates. Studies that combine big data sources with researcher-collected data will enable companies and governments to produce more accurate and timely measures.

4 Conclusions

If social sciences want to benefit from the opportunities of big data, computation and digital methods, the path to follow is adopting an epistemological perspective that not only overcomes the revolution-involution dichotomy, and the new methods-old methods one, but that also call to question some of the main dichotomy that have characterized epistemology of social sciences since recently, such as: subject-object; induction-deduction; correlation-causation.

First of all, intersubjectivity, and the attention to the context in which the representations of phenomena to be investigated are realized, constitute a fundamental starting points for an epistemology of the digital. Indeed, digital technologies have confirmed that the objects of study, and the subjects who study, both actively co-construct data. As highlighted by Lupton (2014), from the moment in which digital research techniques are used, they are theorized. Therefore, it is not possible to separate the digital analysis as an object of study, from the analysis with digital techniques itself, because both require focusing on the ways in which they are co-constituted.

A second point, linked to the first, is that a digital epistemology that wants to avoid the simplistic positions of neoempirists must pay more attention to the process. Although big data and computational techniques are able to analyze social phenomena in real time, most of the digital data represents a set of snapshots of events that update very quickly. Nothing that registers big data can capture the processes or mechanisms that determine the changes that are detected by the data (O'Sullivan 2017). Causation cannot be obtained exclusively through big data.

Moreover, technological determinism should be overcome. Digital methods may be an interesting and promising way to inspire social sciences only if we are able to inspect the theoretical premises that are embedded in the data, and the socio-technical processes that determined their final form. Recognizing the role of technology in the configurations of social research does not imply technological determinism, and that technology must guide scientific knowledge.

Epistemology of the digital needs to become concrete through the definition of a creative and critical method, that elaborates research designs with and against the digital (Marres 2017). These designs try to exploit what digital techniques can give as added value, but at the same time are going to test their reliability, alongside others techniques, including qualitative ones.

It is only in an abductive, intersubjective and critical epistemological framework and through a mixed and creative method that the current technological character of the digital social inquiry can be profitably conveyed within the different paradigmatic traditions that coexist in our disciplines.

References

Abbott, A. (2011). *Google of the past. Do keywords really matter?* Lecture of the Department of Sociology, Goldsmith, 15th March.

Aragona, B. (2017). New data science: The sociological point of view. In E. Amaturo, B. Aragona, M. Grassia, C. Lauro, & M. Marino (Eds.), *Data science and social research: Epistemology, methods, technology and applications*. Heidelberg: Springer.

Aragona, B., & Felaco, C. (2018). The socio-technical construction of algorithms. *The Lab's Quarterly*, *44*(6), 27–42.

Aragona, B., Felaco, C., & Marino, M. (2018). The politics of Big Data assemblages. *Partecipazione e conflitto*, *XI*,(2), 448–471.

Boyd, D., & Crawford, K. (2012). Critical questions for big data: Provocations for a cultural, technological, and scholarly phenomenon. *Information, Communication & Society*, *15*(5), 662–679.

Chandler, J., Rosenzweig, C., & Moss, A. J. (2019). Online panels in social science research: Expanding sampling methods beyond Mechanical Turk. *Behavior Research Methods*, *51*, 2022–2038.

Conte, R. (2016). Big Data: un'opportunità per le scienze sociali? *Sociologia e Ricerca Sociale*, *CIX*, 18–27.

Conte, R., Gilbert, N., Bonelli, G., Cioffi-Revilla, C., Deffuant, G., Kertesz, J., et al. (2012). Manifesto of computational social science. *The European Physical Journal Special Topics CCXIV*, 325–346.

Daniels, J., & Gregory, K. (Eds.). (2016). *Digital sociologies.* Bristol: Policy Press.

Ginsberg, J., Mohebbi, M. H., Patel, R. S., Brammer, L., Smolinski, M. S., & Brilliant, L. (2009). Detecting influenza epidemics using search engine query data. *Nature CDLVII, 7232,* 1012.

Goel, S., Hofman, J. M., Lahaie, S., Pennock, D. M., & Watts, D. J. (2010). Predicting consumer behavior with Web search. *Proceedings of the National Academy of Sciences, 107*(41), 17486–17490.

Halford, S., Pope, C., & Weal, M. (2013). Digital futures? Sociological Challenges and opportunities in the emergent semantic web. *Sociology XLVII, 1,* 173–189. https://doi.org/10.1177/0038038512453798

Halford, S., & Savage, M. (2017). Speaking sociologically with big data: Symphonic social science and the future for big data research. *Sociology, LI*(6), 1132–1148.

Iliadis, A., & Russo, F. (2016). Critical data studies: An introduction. *Big Data & Society, I*(2), 1–8.

Kitchin, R. (2014a). Big Data, new epistemologies and paradigm shifts. *Big Data & Society, I*(1), 1–12.

Kitchin, R. (2014). *The data revolution: Big Data Open Data Data Infrastructures and Their Consequences.* London: Sage.

Kitchin, R., & Lauriault, T. P. (2015). Small data in the era of big data. *GeoJournal, 80*(4), 463–476.

Lazer, D., Brewer, D., Christakis, N., Fowler, J., & King, G. (2009). Life in the network: The coming age of computational social science. *Science CCCXXIII, 5915,* 721–723.

Lupton, D. (2014). *Digital sociology.* London: Routledge.

Marres, N. (2017). *Digital sociology: The reinvention of social research.* New York: Wiley.

Mayer-Schönberger, V., & Cukier, K. (2012). *Big Data: A revolution that transforms how we work, live, and think.* Boston: Houghton Mifflin Harcourt.

Merton, R. K. (1968). *Social theory and social structure.* Glencoe (IL): Free Press.

Orton-Johnson, K., & Prior (a cura di), N. (2013). *Digital sociology: Critical perspectives.* Heidelberg: Springer.

O'Sullivan, D. (2017). *Big Data: why (oh why?) this computational social science?* www.escholarship.org. Accessibile all'URL https://escholarship.org/uc/item/0rn5n832. Ultimo accesso 1 febbraio 2018.

Pasquale, F. (2015). *The black box society.* Cambridge (MA): Harvard University Press.

Peirce, C. S., & (a cura di), . (1883). *Studies in logic, Boston (MA).* Brown and Company: Little.

Rogers, R. (2013), *Digital methods.* Cambridge (MA): MIT press.

Ruppert, E. (2013). Rethinking empirical social sciences. *Dialogues in Human Geography, III*(3), 268–273.

Salganik, M. J. (2018). *Bit by bit: Social research in the digital age.* London: Princeton.

Savage, M., & Burrows, R. (2007). The coming crisis of empirical sociology. *Sociology, 41*(5), 885–899.

Torabi Asr, F., & Taboada, M. (2019). Big Data and quality data for fake news and misinformation detection. *Big Data & Society, 6*(1).

Van Dijck, J., Poell, T., & De Waal, M. (2018). *The platform society: Public values in a connective world.* Oxford: Oxford University Press.

Restricted Cumulative Correspondence Analysis

Pietro Amenta, Antonello D'Ambra, and Luigi D'Ambra

Abstract In the context of the non-iterative procedures for performing a correspondence analysis with linear constraints, a new approach is proposed to impose linear constraints in analyzing a contingency table with one ordered set of categories. At the heart of the approach is the partition of the Taguchi's statistic which has been introduced in the literature as simple alternative to Pearson's index for contingency tables with an ordered categorical variable. It considers the cumulative frequency of cells in the contingency tables across the ordered variable. Linear constraints are then included directly in suitable matrices reflecting the most important components, overcoming also the problem of imposing linear constraints based on subjective decisions.

1 Introduction

Correspondence Analysis (CA) is a popular tool to obtain a graphical representation of the dependence between the rows and the columns of a contingency table (Benzecri 1980; Greenacre 1984; Lebart et al. 1984; Nishisato 1980; Beh 2004; Beh and Lombardo 2012, 2014). This representation is obtained by assigning scores in the form of coordinates to row and column categories. CA is usually performed by applying a singular value decomposition to the matrix of the Pearson ratios or the

P. Amenta (✉)
Department of Law, Economics, Management and Quantitative Methods,
University of Sannio, Benevento, Italy
e-mail: amenta@unisannio.it

A. D'Ambra
Department of Economics, University of Campania "L.Vanvitelli", Capua , Italy
e-mail: antonello.dambra@unicampania.it

L. D'Ambra
Department of Economics, Management and Institutions ,
University of Naples "Federico II", Naples, Italy
e-mail: dambra@unina.it

P. Mariani and M. Zenga (eds.), *Data Science and Social Research II*,
Studies in Classification, Data Analysis, and Knowledge Organization,
https://doi.org/10.1007/978-3-030-51222-4_2

standardized residuals of a two-way contingency table. This decomposition ensures that the maximum information regarding the association between two categorical variables are accounted for in one or two dimensions of a correspondence plot. However, little attention in literature has been paid to the case where the variables are ordinal. It is well known that the Pearson chi-squared statistic (likewise CA) can perform poorly in studying the association between ordinal categorical variables (Agresti 2007; Barlow et al. 1972). An approach dealing with this theme (Beh et al. 2011), in a CA perspective, is based on the Taguchi's statistic (Taguchi 1966, 1974) considering the cumulative frequency of cells in the contingency tables across the ordered variable. Taguchi's statistic has been introduced in the literature as a simple alternative to Pearson's index for contingency tables with an ordered categorical variable. Beh et al. (2011) developed this variant of CA in order to determine graphically how similar (or not) cumulative (ordinal) response categories are with respect to (nominal) criterion ones.

Note that the interpretation of the multidimensional representation of the row and column categories, for both approaches, may be simplified if additional information about the row and column structure of the table is available and incorporated in the analysis (Böckenholt and Böckenholt 1990; Takane et al. 1991; Böckenholt and Takane 1994; Hwang and Takane 2002). Differences between constrained and unconstrained solutions may highlight unexplained features of the data in the exploratory analyses of a contingency table. In the classical analysis, Böckenholt and Böckenholt (1990) (B&B) considered the effect of concomitant variables (given by the external information) partialling them out from the CA solution according to the null-space method. The aim of this paper is to consider an extension of the B&B's approach (Böckenholt and Böckenholt 1990) to contingency tables with one ordered set of categories by using additional information about the structure and association of the data. This extension is achieved by considering the variant of CA based on the decomposition of the Taguchi's statistic (Beh et al. 2011). A new explorative approach named *Restricted Cumulative Correspondence Analysis* is then introduced.

2 Basic Notation

Consider a two-way contingency table **N** describing the joint distribution of two categorical variables where the $(i,\ j)$-th cell entry is given by n_{ij} for $i = 1, ..., I$ and $j = 1, ..., J$ with $n = \sum_{i=1}^{I} \sum_{j=1}^{J} n_{ij}$. Let $n_{i\bullet}$ and $n_{\bullet j}$ be the ith row and jth column marginal frequencies, respectively. The $(i,\ j)$-th element of the probability matrix **P** is defined as $p_{ij} = n_{ij}/n$ so that $\sum_{i=1}^{I} \sum_{j=1}^{J} p_{ij} = 1$. Suppose that **N** has one ordered set of categories with row and column marginal probabilities given by $p_{i.} = \sum_{j=1}^{J} p_{ij}$ and $p_{.j} = \sum_{i=1}^{I} p_{ij}$, respectively. Moreover, let $\mathbf{D}_I$ and $\mathbf{D}_J$ be the diagonal matrices whose elements are the row and column masses $p_{i.}$ and $p_{.j}$, respectively. Lastly, z_{is} is the cumulative frequency of the i-th row category up to the s-th column category.

3 Visualizing the Association Between a Nominal and an Ordinal Categorical Variable

CA of cross-classifications regarding the association (using X^2 as its measure) between two categorical variables has been used by the data analysts from a variety of disciplines over the past 50 years. It is a widely used tool to obtain a graphical representation of the dependence between the rows and columns of a contingency table. CA can be usually performed by applying the Singular Value Decomposition (SVD) on the Pearson's ratios table $\tilde{\mathbf{P}} = \mathbf{D}_I^{-1/2}\mathbf{P}\mathbf{D}_J^{-1/2}$ (Goodman 1996) of generic term $\alpha_{ij} = p_{ij}/p_{i.}p_{.j}$ with $i = 1, \ldots I$ and $j = 1, \ldots, J$. That is, for the $I \times J$ correspondence matrix $\mathbf{P}$, CA amounts to the decomposition $\mathbf{D}_I^{-1/2}\mathbf{P}\mathbf{D}_J^{-1/2} = \tilde{\mathbf{A}}\tilde{\Delta}\tilde{\mathbf{B}}^T$ with $\tilde{\mathbf{A}}^T\tilde{\mathbf{A}} = \mathbf{I}$, $\tilde{\mathbf{B}}^T\tilde{\mathbf{B}} = \mathbf{I}$ and $\tilde{\Delta} = \text{diag}(1, \lambda_1, \ldots, \lambda_K)$ where the singular values λ_m are in descending order ($m = 1, \ldots, K$ with $K = \min(I, J) - 1$) and matrices $\tilde{\mathbf{A}}$ and $\tilde{\mathbf{B}}$ contain the left and the right singular vectors, respectively. If we omit the trivial solutions then CA amounts to the SVD of the matrix

$$\Pi = \mathbf{D}_I^{-1/2}(\mathbf{P} - \mathbf{D}_I\mathbf{1}\mathbf{1}^T\mathbf{D}_J)\mathbf{D}_J^{-1/2} = \mathbf{A}\Lambda\mathbf{B}^T$$

with $\mathbf{A}^T\mathbf{A} = \mathbf{I}$, $\mathbf{B}^T\mathbf{B} = \mathbf{I}$ and Λ diagonal matrix where singular values λ_m are in descending order. The theoretical developments and applications of CA have grown significantly around the world in nearly all disciplines. However, little attention in literature has been paid to the case where the variables are ordinal. Indeed, Pearson's chi-squared statistic test of independence between the variables of a contingency table does not perform well when the rows/columns of the table are ordered (Agresti 2007, Sect. 2.5; Barlow et al. 1972).

Taguchi's statistic has been introduced in the literature (Taguchi 1974, 1966; Nair 1986, 1987) as simple alternatives to Pearson's index for contingency tables with an ordered categorical variable. Taguchi's statistic takes into account the presence of an ordinal categorical variable by considering the cumulative sum of the cell frequencies across the variable. To assess the association between the nominal and ordered column variables, Taguchi (1966, 1974) proposed the following statistic

$$T = \sum_{s=1}^{J-1} \frac{1}{d_s(1 - d_s)} \sum_{i=1}^{I} n_{i\bullet} \left(\frac{z_{is}}{n_{i\bullet}} - d_s \right)^2 \tag{1}$$

with $0 \leq T \leq n(J - 1)$ and where $d_s = \sum_{i=1}^{I} z_{is}/n = z_{\bullet s}/n$ is the cumulative column proportion up to s-th column. Both Nair (1986) and Takeuchi and Hirotsu (1982) showed that the T statistic is linked to the Pearson chi-squared statistic so that $T = \sum_{s=1}^{J-1} X_s^2$ where X_s^2 is Pearson's chi-squared statistic computed on the generic contingency tables $\mathbf{N}_s$ of size $I \times 2$. This table is obtained by aggregating the columns (categories) $1, \ldots, s$ and the remaining ones $s + 1, \ldots, J$ of table $\mathbf{N}$, respectively. For this reason, (Nair 1986) referred to Taguchi's statistic T as the *cumulative chi-squared statistic* (CCS). By generalizing (1), Nair (1986) considers the class of CCS-type tests

$$T_{CCS} = \sum_{s=1}^{J-1} w_s \left[\sum_{i=1}^{I} n_{i\bullet} \left(\frac{n_{is}}{n_{i\bullet}} - d_s \right)^2 \right] \tag{2}$$

and corresponds to a given set of weights $w_s > 0$. The choice of different weighting schemes defines the members of this class. A possible choice for w_s is to assign constant weights to each term ($w_s = 1/J$), Nair (1986, 1987) shows that, for this choice, the statistic T_{CCS} becomes $T_N = \frac{1}{J} \sum_{s=1}^{J-1} \sum_{i=1}^{I} N_{i\bullet} \left(\frac{Z_{is}}{N_{i\bullet}} - d_s \right)^2$ and has good power against ordered alternatives. We can also assume that w_s is proportional to the inverse of the conditional expectation of the s-th term under the null hypothesis of independence (i.e. $w_s = [d_s(1 - d_s)]^{-1}$). T_{CCS} subsumes then T as a special case. Moreover, Nair (1987) showed that the distribution of T can be approximated using the Satterthwaite's method (1946). See D'Ambra et al. (2018) for additional T_{CCS} properties.

Beh et al. (2011) perform CA when cross-classified variables have an ordered structure by considering the Taguchi's statistic to determine graphically how similar (or not) cumulative response categories are with respect to criterion ones. This approach has been named "Cumulated Correspondence Analysis" (hereafter TCA). Let $\mathbf{W}$ be the $((J-1) \times (J-1))$ diagonal matrix of weights w_j and $\mathbf{M}$ a $((J-1) \times J)$ lower unitriangular matrix of 1's. TCA amounts to the SVD$[\mathbf{D}_I^{\frac{1}{2}}(\mathbf{D}_I^{-1}\mathbf{P} - \mathbf{1}_I\mathbf{1}_J^T\mathbf{D}_J)\mathbf{M}^T\mathbf{W}^{\frac{1}{2}}] = \mathbf{U}\boldsymbol{\Lambda}\mathbf{V}^T$ with $\mathbf{U}^T\mathbf{U} = \mathbf{I}$ and $\mathbf{V}^T\mathbf{V} = \mathbf{I}$, such that

$$\frac{T_{CCS}}{n} = \text{trace}\,(\mathbf{D}_I^{\frac{1}{2}}(\mathbf{D}_I^{-1}\mathbf{P} - \mathbf{1}_I\mathbf{1}_J^T\mathbf{D}_J)\mathbf{M}^T\mathbf{W}\mathbf{M}(\mathbf{D}_I^{-1}\mathbf{P} - \mathbf{1}_I\mathbf{1}_J^T\mathbf{D}_J)^T\mathbf{D}_I^{\frac{1}{2}}) = \sum_{i=1}^{I} \lambda_i^2$$

Moreover, we highlight that above SVD is also equivalent to perform SVD of matrix $\mathbf{D}_I^{-\frac{1}{2}}(\mathbf{P} - \mathbf{D}_I\mathbf{1}\mathbf{1}^T\mathbf{D}_J)\mathbf{M}^T\mathbf{W}^{\frac{1}{2}}$. TCA decomposes then the Taguchi's statistic T when w_s is proportional to the inverse of the conditional expectation of the s-th term under the null hypothesis of independence ($w_s = [d_s(1 - d_s)]^{-1}$).

To visually summarize the association between the row and the column categories, TCA row and column principal coordinates are defined by $\mathbf{F} = \mathbf{D}_I^{-\frac{1}{2}}\mathbf{U}\boldsymbol{\Lambda}$ and $\mathbf{G} = \mathbf{W}^{-\frac{1}{2}}\mathbf{V}\boldsymbol{\Lambda}$, respectively. Here, $\mathbf{F}$ and $\mathbf{G}$ are matrices of order $I \times M$ and $(J-1) \times M$, respectively. The s-th row of matrix $\mathbf{G}$ contains the coordinates of category $y_{(1:s)}$ in the M dimensional space (with $M = \text{rank}(\mathbf{D}_I^{\frac{1}{2}}(\mathbf{D}_I^{-1}\mathbf{P} - \mathbf{1}_I\mathbf{1}_J^T\mathbf{D}_J)\mathbf{M}^T\mathbf{W}^{\frac{1}{2}})$). Therefore, if there is approximately zero predicability of the column categories given the row categories then $\mathbf{F} \approx 0$ and $\mathbf{G} \approx 0$. To provide a more discriminating view of the difference between each cumulate rating category, authors consider also rescaling the row and column profile coordinates to obtain biplot-type coordinates (Goodman 1996): $\mathbf{F} = \mathbf{D}_I^{-\frac{1}{2}}\mathbf{U}\boldsymbol{\Lambda}^{\alpha}$ and $\mathbf{G} = \mathbf{W}^{-\frac{1}{2}}\mathbf{V}\boldsymbol{\Lambda}^{(1-\alpha)}$ with $0 \leq \alpha \leq 1$. These coordinates are related to the factorisation (for categorical data) proposed by Gabriel (Gabriel 1971) for the construction of the biplot.

Interested readers to this variant, which is linked with the partition of Taguchi's cumulative chi-squared statistic, can refer to (Beh et al. 2011; Sarnacchiaro and D'Ambra 2011; D'Ambra and Amenta 2011; D'Ambra et al. 2018) which discuss the technical and practical aspects of TCA in depth.

4 Restricted Cumulative Correspondence Analysis

Several authors (Böckenholt and Böckenholt 1990; Takane et al. 1991; Böckenholt and Takane 1994; Hwang and Takane 2002) pointed out that the interpretation of the multidimensional representation of the row and column categories may be simplified if additional information about the row and column structure of the table is available. Indeed, by incorporating this external information through linear constraints on the row and/or columns scores, a representation of the data may be obtained that is easier to understand and more parsimonious.

According to the principle of Restricted Eigenvalue Problem (Rao 1973), B&B (Böckenholt and Böckenholt 1990) proposed a canonical analysis of contingency tables which takes into account additional information about the row and column categories of the table. We name this approach "Restricted CA" (RCA). Additional information are provided in the forms of linear constraints on the row and column scores. Let $\mathbf{H}$ and $\mathbf{G}$ be the matrices of linear constraints of order $I \times E$ and $J \times L$ of ranks E and L, respectively, such that $\mathbf{H}^T\mathbf{X} = \mathbf{0}$ and $\mathbf{G}^T\mathbf{Y} = \mathbf{0}$ where $\mathbf{X}$ and $\mathbf{Y}$ are the standardized row and column scores. RCA scores are obtained by a SVD of the matrix

$$\tilde{\boldsymbol{\Pi}} = \{\mathbf{I} - \mathbf{D}_I^{-\frac{1}{2}}\mathbf{H}(\mathbf{H}^T\mathbf{D}_I^{-1}\mathbf{H})^{-1}\mathbf{H}^T\mathbf{D}_I^{-\frac{1}{2}}\}\boldsymbol{\Pi}\{\mathbf{I} - \mathbf{D}_J^{-\frac{1}{2}}\mathbf{G}(\mathbf{G}^T\mathbf{D}_J^{-1}\mathbf{G})^{-1}\mathbf{G}^T\mathbf{D}_J^{-\frac{1}{2}}\}$$

that is $\tilde{\boldsymbol{\Pi}} = \mathbf{U}\boldsymbol{\Lambda}\mathbf{V}^T$ where $\mathbf{U}^T\mathbf{U} = \mathbf{I}$, $\mathbf{V}^T\mathbf{V} = \mathbf{I}$ and $\boldsymbol{\Lambda}$ is a diagonal matrix with eigenvalues λ in descending order. Standardized row and column scores are given by $\mathbf{X} = \mathbf{D}_I^{-1/2}\mathbf{U}$ and $\mathbf{Y} = \mathbf{D}_J^{-1/2}\mathbf{V}$, respectively, such that $\mathbf{X}^T\mathbf{D}_I\mathbf{X} = \mathbf{I}$, and $\mathbf{Y}^T\mathbf{D}_J\mathbf{Y} = \mathbf{I}$, with $\mathbf{1}^T\mathbf{D}_I\mathbf{X} = \mathbf{0}$ and $\mathbf{1}^T\mathbf{D_J}\mathbf{Y} = \mathbf{0}$. The classical approach to CA is obtained when $\mathbf{H} = (\mathbf{D}_I\mathbf{1})$ and $\mathbf{G} = (\mathbf{D}_J\mathbf{1})$ which represents the case of absence of linear constraints.

It is evident that, following the B&B's approach, we can also obtain an easier to understand and more parsimonious TCA graphical representation of the association between a nominal and an ordinal categorical variable. In this case we consider only additional information about the row (nominal) categories of the table. The additional information about the ordinal nature of the column variable is used by considering the cumulative sum of the cell frequencies across it. We use the same matrices of linear constraints $\mathbf{H}$ which ensures that the weighted average of the row TCA scores equal zero. Restricted CA of cumulative frequencies (RTCA) amounts then to the SVD

$$[\mathbf{I} - \mathbf{D}_I^{-\frac{1}{2}}\mathbf{H}(\mathbf{H}^T\mathbf{D}_I^{-1}\mathbf{H})^{-1}\mathbf{H}^T\mathbf{D}_I^{-\frac{1}{2}}]\boldsymbol{\Pi}_{(T)} = \mathbf{U}\boldsymbol{\Lambda}\mathbf{V}^T \tag{3}$$

where $\boldsymbol{\Pi}_{(T)} = \mathbf{D}_I^{\frac{1}{2}}(\mathbf{D}_I^{-1}\mathbf{P} - \mathbf{1}_I\mathbf{1}_J^T\mathbf{D}_J)\mathbf{M}^T\mathbf{W}^{\frac{1}{2}}$ and such that $\mathbf{U}^T\mathbf{U} = \mathbf{I}$ and $\mathbf{V}^T\mathbf{V} = \mathbf{I}$. Standardized row and column RTCA scores are now given by $\mathbf{X} = \mathbf{D}_I^{-1/2}\mathbf{U}$ and $\mathbf{Y} = \mathbf{W}^{-\frac{1}{2}}\mathbf{V}$, respectively. TCA is then obtained when $\mathbf{H} = (\mathbf{D}_I\mathbf{1})$ which represents the case of absence of linear constraints. Note that single column categories are additionally plotted as supplementary points. Their column coordinates will be given by $\mathbf{Y}^+ = \boldsymbol{\Pi}_{(T)}^T\mathbf{U}$.

We point out that matrix $\mathbf{H}$ imposes the same constraints for all singular vectors $\mathbf{u}$ of SVD of identity (3), but it could be interesting to define different constraints on each singular vector. This aspect can be obtained by using a successive approach based on a rank-one reduction of the initial matrix $\boldsymbol{\Pi}_{(T)}$. Let $\boldsymbol{\Pi}_{(T)}^{(1)} = (\mathbf{I} - \mathbf{u}_1\mathbf{u}_1^T)\boldsymbol{\Pi}_{(T)}$ be the rank-one reduced matrix with respect to the first singular vector $\mathbf{u}_1$ corresponding to λ_1. This matrix is then substitute to $\boldsymbol{\Pi}_{(T)}$ in the SVD (3) and a new solution for $\mathbf{u}_2$, $\mathbf{v}_2$ and λ_2 is computed according to new linear constraints that we want to impose on this solution. New standardized row and column RTCA scores on this axis are so obtained. A new rank-one reduced matrix $\boldsymbol{\Pi}_{(T)}^{(2)}$ is then computed and substitute to $\boldsymbol{\Pi}_{(T)}$ in the SVD (3) for a solution with new constraints. New scores on $\mathbf{u}_2$ are consequently obtained and the approach is reiterated for the next axis $\mathbf{u}_3$, and so on. This iterative rank-one reduction approach is such that $\mathbf{X}^T\mathbf{D}_I\mathbf{X} = \mathbf{I}$ and $\mathbf{Y}^T\mathbf{W}\mathbf{Y} = \mathbf{I}$.

5 Example

In this section we illustrate the proposal method considering a data set (Table 1) mentioned in Agresti (2007).

The study is aimed at testing the effect of the factors urbanization and location on the ordered response preference for black olives of Armed Forces personnel. In particular, we have considered the case in which there is an asymmetric relationship between two categorical variables used as predictor variables (Urbanization and Region) and an ordinal response variable which is categorized into six ordered classes. The predictor variable Urbanization is characterized by two levels: Urban and Rural areas, whereas the predictor variable Region is characterized by three levels: North West, North East, and South West. The ordinal response variable is characterized by six growing ordered categories: A = dislike extremely, B = dislike moderately, C = dislike slightly, D = like slightly, E = like moderately, F = like extremely.

Since the ratings are ordered, a partition of Tagughi's inertia (of 0.1749) is applied by an unrestricted TCA which yields the singular values $\lambda_1 = 0.163$ and $\lambda_2 = 0.010$. TCA coordinates are displayed in Fig. 1. For the column categories, the label "(1)" reflects the "cumulative total" of rating 1 with those "(2:6)" of ratings 2, 3, 4, 5,

Table 1 Data table

		Dislike extremity (1)	Dislike moderately (2)	Neither like nor dislike (3)	Like slightly (4)	Like moderately (5)	Like extremilly (6)	Total
Urban	North West	20	15	12	17	16	28	108
	North East	18	17	18	18	6	25	102
	South West	12	9	23	21	19	30	114
Rural	North West	30	22	21	17	8	12	110
	North East	23	18	20	18	10	15	104
	South West	11	9	26	19	17	24	106
Total		114	90	120	110	76	134	644

Table 2 Eigenvalues

TCA and RTCA Eigenvalues						
Axis	Unconstrained			Constrained		
	Eigenvalue	%	Cum. %	Eigenvalue	%	Cum. %
(1)	0.163	93.206	93.206	0.070	95.480	95.480
(2)	0.010	5.587	98.793	0.003	3.822	99.302
(3)	0.002	1.169	99.962	0.001	0.698	100.000
(4)	0.000	0.027	99.989	0.000	0.000	100.000
(5)	0.000	0.010	100.000	0.000	0.000	100.000

and 6 given the Urban/Rural levels. Labels "(1:4)" and "(5:6)" reflect instead the comparison made of the cumulative total of ordered rating from 1 to 4 with 5 and 6, respectively, given the Urban/Rural levels. Similarly, labels "(1:3)" and "(4:6)" reflect the comparison made of the cumulative total of the ordered ratings from 1 to 3 with those of the remaining predictor categories, respectively, given the Urban/Rural levels. The remaining labels can be interpreted in a similar manner.

Note that Taguchi's analysis allows to identify how similar (or different) cumulate ordered column response categories are for each row category. Consider then Fig. 1 which graphically depicts about 98.79% of the association that exists between the two variables (see Table 2). It shows clearly that all the pairs of cumulated ratings are quite distinct, indicating that there is a perceived difference between these cumulate categories. The source of the variation between these ratings is dominated by all the Urban/Rural levels except for Urban North East (U.NE). The apparent difference between the most positive ratings and the others can be attributed mainly to U.NW and

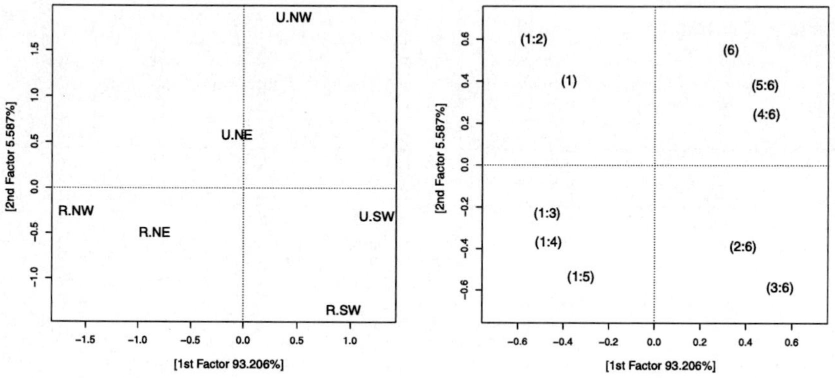

Fig. 1 TCA plots

U.SW whereas the lowest values are characterized by R.NW and R.NE. First TCA axis depicts about 93.20% of global association and clearly contrasts the medium-low ratings with the high ones but there is not an evident separate domination by the Urban and Rural categories as well as by their three levels: North West, North East, and South West. Additional drawback of this plot is that it does not highlight the contribution of the high ratings "6" of U.NE (see Table 1). Indeed, it is a negligible category because the position of this point is closest to the origin.

In order to better highlight a contrast between the Urban and Rural categories on the first axis with respect to North West and South West levels, a RTCA solution is then computed by setting $\mathbf{H} = (\mathbf{D}_I \mathbf{1} | \mathbf{H}_R)$ with

$$\mathbf{H}_R^T = \begin{pmatrix} -1 & 0 & 1 & 0 & 0 & 0 \\ 0 & 0 & 0 & -1 & 0 & 1 \end{pmatrix}$$

RTCA plot (Fig. 2) now graphically depicts 99.30% of the global association and well highlights a contrast between the Urban and Rural categories as source of variation between the cumulated ratings. Rural categories are now source of the low ratings (left hands of Fig. 2), according to their frequencies of Table 1, as well as the Urban categories for the most positive ratings (right hands of Fig. 2). Moreover, this figure now points out a not negligible position of U.NE label showing the contribution of the high ratings "6" taken by this category.

We point out that introducing linear constraints in the TCA solution has then brought several advantages:

- a more parsimonious analysis is obtained where all the global association is depicted by only 3 axes (5 with TCA);
- an easier interpretation of the results is obtained highlighting a clear contrast between the Urban and Rural categories on the first axis as source of variation of the cumulated categories;

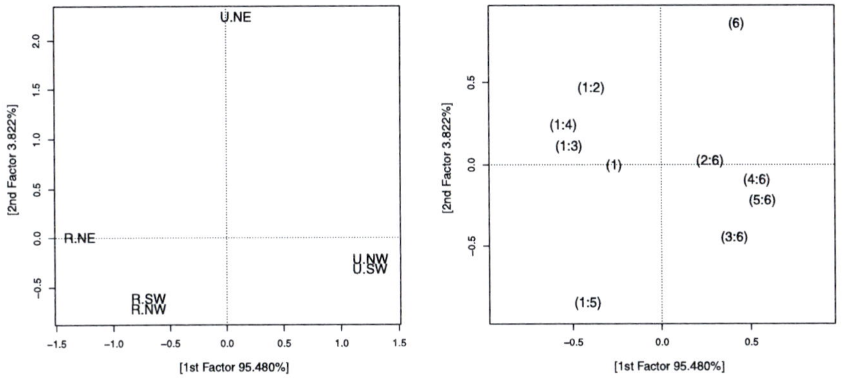

Fig. 2 RTCA plots

- we observe an increase in the explained variability of the first axis: 95.48% of RTCA versus 93.21% of TCA;
- no predictive row category is now poorly represented.

6 Conclusion

Several authors highlighted that, introducing linear constraints on the row and column coordinates of a correspondence analysis representation, may greatly simplify the interpretation of the data matrix. Imposing also different constraints for each singular value may be useful in developing a parsimonious representation of a contingency table. B&B (Böckenholt and Böckenholt 1990) presented a generalized least squares approach for incorporating linear constraints on the standardized row and column scores obtained from a canonical analysis of a contingency table. This approach is based on the decomposition of a restricted version of the matrix of the Pearson ratios. Unfortunately the Pearson chi-squared statistic (likewise correspondence analysis) can perform poorly in studying the association between ordinal categorical variables (Agresti 2007). Beh et al. (2011) deal with this theme (Beh et al. 2011) by developing a CA extension (TCA) based on the Taguchi's statistic (Taguchi 1966, 1974). This statistic considers the cumulative frequency of cells in the contingency tables across the ordered variable and it has been introduced in the literature as simple alternative to Pearson's index for contingency tables with an ordered categorical variable.

A restricted extension of the Beh's approach (RTCA) has been here suggested to obtain a more parsimonious representation of the association and easier to explain. Natural forms of constraints may often appear from specific empirical questions asked by the researchers regarding the problem of their fields. In the exemplary application, introducing linear constraints in the TCA solution has brought several advantages in terms of interpretability and axes inertia rate. RTCA extends the

Cumulative Correspondence Analysis by taking into account external information (as linear constraints) and supplies a complementary interpretative enrichment of this technique as well as of the original CA approach.

References

Agresti, A. (2007). *An introduction to categorical data analysis*. Wiley.

Barlow, R. E., Bartholomew, D. J., Bremner, J. M., & Brunk, H. D. (1972). *Statistical inference under order restrictions*. New York: Wiley.

Beh, E. (2004). Simple correspondence analysis: A bibliographic review. *International Statistical Review*, *72*(2), 257–284.

Beh, E. J., & Lombardo, R. (2012). A genealogy of correspondence analysis. *Australian & New Zealand Journal of Statistics*, *54*(2), 137–168.

Beh, E. J., & Lombardo, R. (2014). *Correspondence analysis: Theory*. Practice and New Strategies: Wiley.

Beh, E. J., D'Ambra, L., & Simonetti, B. (2011). Correspondence analysis of cumulative frequencies using a decomposition of Taguchi's statistic. *Communications in Statistics. Theory and Methods*, *40*, 1620–1632.

Benzecri, J. P. (1980). *Practique de l'analyse des donnees*. Paris: Dunod.

Böckenholt, U., & Böckenholt, I. (1990). Canonical analysis of contingency tables with linear constraints. *Psychometrika*, *55*, 633–639.

Böckenholt, U., & Takane, Y. (1994). Linear constraints in correspondence analysis. In: M. Greenacre & J. Blasius (Eds.), *Correspondence analysis in the social sciences: Recent developments and applications* (pp. 70–111). New York: Academic Press.

D'Ambra, A., & Amenta, P. (2011). Correspondence Analysis with linear constraints of ordinal cross-classifications. *Journal of Classification*, *28*, 1–23.

D'Ambra, L., Amenta, P., & D'Ambra, A. (2018). Decomposition of cumulative chi-squared statistics, with some new tools for their interpretation. *Statistical Methods and Applications*, *27*(2), 297–318.

Gabriel, K. R. (1971). The biplot graphic display of matrices with application to principal component analysis. *Biometrika*, *58*, 453–467.

Greenacre, M. J. (1984). *Theory and applications of correspondence analysis*. New York: Academic Press.

Goodman, L. A. (1996). A single general method for the analysis of cross-classified data: Reconciliation and synthesis of some methods of Pearson, Yule, and Fisher, and also some methods of correspondence analysis and association analysis. *Journal of the American Statistical Association*, *91*, 408–428.

Hwang, H., & Takane, Y. (2002). Generalized constrained multiple correspondence analysis. *Psychometrika*, *67*, 215–228.

Lebart, L., Morineau, A., & Warwick, K. M. (1984). *Multivariate descriptive statistical analysis: Correspondence analysis and related techniques for large matrices*. New York: Wiley.

Nair, V. N. (1986). Testing in industrial experiments with ordered categorical data. *Technometrics*, *28*(4), 283–291.

Nair, V. N. (1987). Chi-squared type tests for ordered alternatives in contingency tables. *Journal of American Statistical Association*, *82*, 283–291.

Nishisato, S. (1980). *Analysis of categorical data: Dual scaling and its applications*. Toronto: University of Toronto Press.

Rao, C. R. (1973). *Linear statistical inference and its applications*. Wiley.

Sarnacchiaro, P., & D'Ambra, A. (2011). Cumulative correspondence analysis to improve the public train transport. *Electronic Journal of Applied Statistical Analysis: Decision Support System and Services, 2*, 15–24.

Satterthwaite, F. (1946). An approximate distribution of estimates of variance components. *Biometrical Bullettin, 2*, 110–114.

Taguchi, G. (1966). *Statistical analysis*. Tokyo: Maruzen.

Taguchi, G. (1974). A new statistical analysis for clinical data, the accumulating analysis, in contrast with the chi-square test. *Saishin Igaku, 29*, 806–813.

Takane, Y., Yanai, H., & Mayekawa, S. (1991). Relationships among several methods of linearly constrained correspondence analysis. *Psychometrika, 56*, 667–684.

Takeuchi, K., & Hirotsu, C. (1982). The cumulative chi square method against ordered alternative in two-way contingency tables. Technical Report 29, Reports of Statistical Application Research. Japanese Union of Scientists and Engineers.

Determining the Importance of Hotel Services by Using Transitivity Thresholds

Pietro Amenta, Antonio Lucadamo, and Gabriella Marcarelli

Abstract Customers' preferences related to the quality, the change, and the progress of their expectations have turned the quality in an indispensable competitive factor for hotel enterprises. The hotels have to evaluate the customer satisfaction and to assign to each factor a weight, expressing its importance for their customers. The aim of this paper is to evaluate the importance of hotel services. Our analysis involves more than 300 customers that answered to a survey and it takes into account five criteria: Food, Cleanliness, Staff, Price/benefit, and Comfort. To derive the ranking of preferences we used pairwise comparisons. The main issue linked to pairwise comparisons is the consistency of judgements. Transitivity thresholds recently proposed in literature give meaningful information about the reliability of the preferences. Our study shows how the use of ordinal threshold may provide a ranking of services different from that obtained by applying traditional consistency Saaty thresholds.

1 Introduction

Pairwise Comparison Matrices (PCMs) are widely used for representing preferences in multi-criteria decision problems. Given a set of n elements, to derive the ranking of preferences by means of pairwise comparisons, a positive number a_{ij} is assigned to each pair of elements (x_i, x_j) with $i, j = 1, \ldots, n$. This number expresses how much x_i is preferred to x_j as regards a given criterion. By comparing all the elements, a positive square matrix $A = \left(a_{ij}\right)$ of order n is then obtained. The value $a_{ij} > 1$ implies that x_i is strictly preferred to x_j, whereas $a_{ij} < 1$ expresses the opposite preference, and $a_{ij} = 1$ means that x_i and x_j are indifferent (Saaty 1980, 1994). In order

P. Amenta · A. Lucadamo (✉) · G. Marcarelli
DEMM - University of Sannio, Piazza Arechi II, Benevento, Italy
e-mail: antonio.lucadamo@unisannio.it

P. Amenta
e-mail: amenta@unisannio.it

G. Marcarelli
e-mail: gabriella.marcarelli@unisannio.it

P. Mariani and M. Zenga (eds.), *Data Science and Social Research II*,
Studies in Classification, Data Analysis, and Knowledge Organization,
https://doi.org/10.1007/978-3-030-51222-4_3

to derive the ranking of alternatives, we may apply one of the following prioritization methods: the Eigenvector Method (EVM), the Arithmetic Mean Method (AMM), the Row Geometric Mean Method (RGMM), the logarithmic Least Squares method, the Singular Value Decomposition, to cite just a few (Aguaron and Moreno-Jimenez 2003; Gass and Rapcsák 2004; Peláez and Lamata 2003; Saaty 1980). Regardless of the method chosen for the prioritization procedure, before applying any methods, it is then necessary to check the consistency of these judgements. Consistency may be ordinal or cardinal. The cardinal consistency implies that the judgements are transitive and proportional: a decision-maker is perfectly consistent in making estimates if his or her judgements satisfy the following consistency condition $a_{ij} * a_{jk} = a_{ik}$ for each $i, j, k = 1, 2, \ldots, n$ (Saaty 1980). For example, if $a_{12} = 2$ and $a_{23} = 3$ then a_{13} must be equal to 6 to ensure that a 3×3 pairwise comparison matrix is perfectly consistent.

The ordinal consistency implies instead only the transitive property; meaning that, if $a_{ij} > 1$ and $a_{jk} > 1$, then $a_{ik} > 1$. Transitivity is a condition weaker than consistency. Perfect consistency is unattainable in practice, but a degree of inconsistency can be considered acceptable. The consistency of judgements is strictly connected with the reliability of the preferences. If the judgements are not consistent, then the methods used to derive the ranking (prioritization methods) could provide different results. If the judgements are instead only ordinally consistent (that is, only transitive), then most methods provide vectors (priority vectors) representing the same ranking, expressing in this way the same preferences: only the intensity of the preferences can vary (Siraj et al. 2015). Due to its relationship with the reliability of the preferences, the consistency of judgements has been widely analyzed by many authors. Several indices have been proposed to measure the degree of consistency of the judgements expressed by the decision-maker. Each index is a function that associates pairwise comparisons with a real number that represents the degree of inconsistency in the judgements (Aguaron and Moreno-Jimenez 2003; Crawford and Williams 1985; Koczkodaj 1993; Salo and Hamalainen 1997). Here we consider the Consistency Index (CI), given by

$$CI = \frac{\lambda_{max} - n}{n - 1}, \tag{1}$$

for $i, j = 1, \ldots, n$, where λ_{max} represents the maximum eigenvalue of the pairwise comparison matrix. If the matrix is perfectly consistent, then $CI = 0$. Saaty suggested also the Consistency Ratio

$$CR = \frac{CI}{RI}, \tag{2}$$

where RI is the Random Index, which is obtained as the mean value of the CI derived from randomly generated matrices of order n.

Consistency ratio and its thresholds may be useful to face cardinal consistency but they do not take into account the ordinal consistency (transitivity). Transitivity thresholds, proposed by Amenta et al. (2020), could be useful because it may provide meaningful information about the reliability of the preferences and it may also allow us to avoid the revision of judgements. If the decision-maker is interested in the ordinal ranking of elements and not in the intensity of preferences, then a transitivity threshold represents an important tool for this task: an index value less than the transitivity threshold ensures (with a high probability) that the ranking of preferences is unique on varying the prioritization methods, only the intensity of preferences may be different.

In this paper we focus on the ordinal consistency. We analyze a case study based on a group decision problem concerning the evaluation of appropriate hotel services.

The paper is organized as follow: Sect. 2 illustrates the ordinal consistency thresholds; in Sect. 3, through a case study, we show that by using ordinal thresholds we obtain a ranking of alternatives different from that provided by using traditional consistency thresholds; in Sect. 4 some concluding remarks are provided.

2 Ordinal Consistency Thresholds

Saaty's consistency threshold has been criticized because it may allow us to accept many intransitive matrices or reject many transitive ones. For this reason, we use the method proposed by Amenta et al. (2020) to verify if a PCM is transitive or not.

To calculate the transitivity thresholds we generate 500,000 random comparison matrices of size n and, using an algorithm based on the approach introduced by Gass (1998), we check how many transitive and intransitive matrices are generated (the proportion of two categories varies as n varies). We then consider CR, for all matrices and, in order to define the thresholds associated with these indices, we introduce the following notation:

- let λ and $1 - \lambda$ be the proportion of random generated intransitive and transitive matrices, respectively;
- let α be the percentage of random intransitive matrices that are accepted according to the threshold value to be set;
- let β be the percentage of random transitive matrices that are rejected according to the same value.

According to Amenta et al. (2020) the threshold is the value that minimizes the quantity $\lambda\alpha + (1 - \lambda)\beta$. Table 1 shows the transitivity–intransitivity thresholds for CR index and the corresponding percentage of misclassified matrices for sizes 3–8.

If the decision-maker is interested in the ordinal ranking of elements and not in the intensity of preferences, then a transitivity threshold represents an important tool for this task: an index value less than the transitivity threshold ensures (with a high probability) that the ranking of preferences is unique while on varying the prioritization methods, only the intensity of preferences may be different. In this case, even though the index is higher than the consistency threshold, the decision-maker may avoid to revise his/her judgments.

Table 1 Transitivity–intransitivity thresholds and percentage of intransitive and transitive matrices, respectively, for CR index and different matrix size orders (n)

n	Threshold	Misclassification rate	λ
3	1.405	2.155	0.2494
4	0.647	6.118	0.6248
5	0.440	4.959	0.8815
6	0.327	1.645	0.9787
7	0.256	0.221	0.9977
8	0.254	0.015	0.9998

3 The Selection of the Ranking of Hotel Services

3.1 The Problem

Tourism industry is one of the most rapidly growing sectors in the world. Hotels constitute the main units of tourism sectors. In these enterprises, the production and consumption of touristic goods and services occur simultaneously. One of the aim for hotels is to be superior to their competitors. Customers' preferences related to the quality, the change, and the progress of their expectations have turned the quality in an indispensable competitive factor for hotel enterprises. Hotel enterprises have to adopt a quality approach that will satisfy the expectations of customers. The hotels that have adopted this approach are more advantageous compared to their competitors in respect of business profitability and continuity by gaining customer pleasure (Sezgin and Tokay 2016). Because of the intangibility of the products produced by these enterprises, hotels struggle much on this subject. The hotels have to evaluate the customer satisfaction and to assign to each factor a weight, expressing its importance for their customers.

Our analysis concerns some hotels in the city of Benevento (south of Italy). It involves 368 customers that answered to a survey in April 2019 and takes into account the following five criteria: Food, Cleanliness, Staff, Price/benefit, and Comfort. Customers have to assign a judgment to each pair of criteria, by using Saaty scale.

3.2 Results and Discussion

By comparing the pairs of criteria, each customer has to express a judgment (a_{ij}), representing the importance he/she assigns to the i-th criterion with respect to the j-th criterion as regards the main objective of the analysis.

Before deriving the global weights vector, we analyze the consistency of all 368 pairwise comparison matrices (n = 5). Several matrices have a Consistency Ratio greater than the Saaty threshold (CR > 0.1). Among these, there are some matrices that can be considered transitive. For example, in Tables 2, 3, and 4 we show three pairwise comparison matrices associated with three customers.

The matrix A1 has CR value equal to 0.06937917, so it is considered strictly consistent. According to the classical procedure it can be used in the analysis.

For matrix A2 CR index is equal to 0.34941. This value leads to consider the matrix as inconsistent according to Saaty threshold and so it has to be removed (because the questionnaire is anonymous and the customers cannot revise their judgments).

If we consider the transitivity threshold (0.440 for n = 5), the matrix is classified as ordinally consistent (that is transitive). We may be sure that, with a high probability, the methods proposed to derive the priority vector will provide the same ranking of preferences; only the intensity of preferences could vary. Matrix A3 instead has a CR value equal to 0.7554. This indicates that the matrix is neither consistent nor

Table 2 Pairwise comparison matrix with CR < 0.1

A1 =		*Food*	*Cleanliness*	*Staff*	*Price/benefit*	*Comfort*
	Food	1	1/6	2	6	1/2
	Cleanliness	6	1	8	9	7
	Staff	1/2	1/8	1	2	1/3
	Price/benefit	1/6	1/9	1/2	1	1/3
	Comfort	2	1/7	3	3	1

Table 3 Pairwise comparison matrix with 0.1 < CR < 0.440

A2 =		*Food*	*Cleanliness*	*Staff*	*Price/benefit*	*Comfort*
	Food	1	5	2	8	4
	Cleanliness	1/5	1	1/8	8	8
	Staff	1/2	8	1	8	2
	Price/benefit	1/8	1/8	1/8	1	1/8
	Comfort	1/4	1/8	1/2	8	1

Table 4 Pairwise comparison matrix with CR > 0.440

A3 =		*Food*	*Cleanliness*	*Staff*	*Price/benefit*	*Comfort*
	Food	1	1/5	1/2	2	4
	Cleanliness	5	1	1/6	7	1/9
	Staff	2	6	1	7	1/2
	Price/benefit	1/2	1/7	1/7	1	1/4
	Comfort	1/4	9	2	4	1

transitive. It is not possibile to consider these judgments in our analysis, because the ranking of preferences depends on the prioritization method applied.

In Table 5 we show the vectors obtained using EVM, AMM, and RGMM. It is evident that the three methods give three different rankings of preferences. For this reason, we removed from the analysis the matrices that overcome also the transitivity threshold.

The matrices that have a CR < 0.1 are only 54. By aggregating these matrices, via Weighted Geometric Mean Method (WGMM) we obtain the following synthesis Matrix (SyntMat1) (Table 6):

Applying the classical eigenvector method we derive the priority vector providing the ranking of the services shown in Table 7.

We propose instead to consider in the analysis all matrices that have a CR lower than the transitivity threshold (0.440). In this case we consider 286 matrices and apply WGMM obtaining the matrix in Table 8 (SyntMat2).

The priority vector associated with SyntMat2 is in Table 9. By comparing the results, obtained in Tables 7 and 9, we can remark that the food and the staff do not change their ranks: in both analyses they are in 5th and 3th position, respectively. The cleanliness ranks first when we consider only the consistent matrices and second

Table 5 Prioritization values and ranking (in parentheses) according to different prioritization methods for the matrix A3

Methods/Alternatives	Food	Cleanliness	Staff	Price/benefit	Comfort
EVM	0.2057 (3)	0.1826 (4)	0.2588 (2)	0.0335 (5)	0.3195 (1)
AMM	0.1381 (4)	0.2381 (3)	0.2959 (1)	0.0365 (5)	0.2914 (2)
RGMM	0.1575 (3)	0.1510 (4)	0.3479 (1)	0.0499 (5)	0.2936 (2)

Table 6 Synthesis of the individual matrices with CR < 0.1

$SyntMat1 =$

	Food	*Cleanliness*	*Staff*	*Price/benefit*	*Comfort*
Food	1.0000000	0.7423385	0.799638	0.9857524	0.7303906
Cleanliness	1.347094	1.0000000	1.138722	1.1731605	1.1234617
Staff	1.250566	0.8781772	1.000000	0.9629850	0.8607162
Price/benefit	1.014454	0.8523983	1.038438	1.0000000	0.9792385
Comfort	1.369130	0.8901060	1.161823	1.0212016	1.0000000

Table 7 Priority vector for the synthesis matrix calculated considering only matrices with CR < 0.1

Alternatives	Food	Cleanliness	Staff	Price/benefit	Comfort
Priority vector	0.1680123	0.2286986	0.1951168	0.1940183	0.2141540
Ranking	5	1	3	4	2

Table 8 Synthesis of the individual matrices with CR < 0.440

*SyntMat*1 =

	Food	*Cleanliness*	*Staff*	*Price/benefit*	*Comfort*
Food	1.0000000	0.7423385	0.799638	0.9857524	0.7303906
Cleanliness	1.347094	1.0000000	1.138722	1.1731605	1.1234617
Staff	1.250566	0.8781772	1.000000	0.9629850	0.8607162
Price/benefit	1.014454	0.8523983	1.038438	1.0000000	0.9792385
Comfort	1.369130	0.8901060	1.161823	1.0212016	1.0000000

Table 9 Priority vector for the synthesis matrix calculated considering individual matrices with CR < 0.440

Alternatives	Food	Cleanliness	Staff	Price/benefit	Comfort
Priority vector	0.1856310	0.2010218	0.1963203	0.2216098	0.1954171
Ranking	5	2	3	1	4

if we take into account all the transitive matrices. Furthermore price/benefit ratio, that ranks 4th in the first analysis, represents the most important service according to customers' judgments aggregated in SyntMat2. Finally the comfort, that is the second most preferred service according to SyntMat1, ranks 4th in the second matrix.

It is important to evaluate the real ranking of preferences, because it allows the managers of the hotels to have information about the importance that customers give to the services. In this way the managers know which services they need to focus their efforts on, in order to increase the customer satisfaction. We think that, in this kind of survey, it is important to consider the customers whose judgments that are at least rational from the point of view of transitivity.

4 Concluding Remarks

Some criticisms on the consistency threshold for CR, particularly regarding its inability to capture the ordinal consistency, have been highlighted in literature. Transitivity thresholds may allow to avoid the revision of the judgements if we are only interested in the qualitative ranking of decision-makers' preferences. If the value assumed by the consistency ratio is ranged between the consistency and the transitivity threshold values, then we are confident about the reliability of the priority vector. In this case, the decision-maker avoids the need to revise his or her judgements.

In order to show the usefulness of transitivity thresholds, in this paper we analyze a case study concerning the customer satisfaction related to the services offered by some hotels.

References

Aguaron, J., & Moreno-Jimenez, J. (2003). The geometric consistency index: Approximated threshold. *European Journal of Operational Research, 147*, 137–145.

Amenta, P., Lucadamo, A., & Marcarelli, G. (2020). On the transitivity and consistency approximated thresholds of some consistency indices for pairwise comparison matrices. *Information Sciences, 507*, 274–287.

Crawford, G., & Williams, C. (1985). A note on the analysis of subjective judgment matrices. *Journal of Mathematical Psychology, 29*, 387–405.

Gass, S. I. (1998). Tournaments, transitivity and pairwise comparison matrices. *The Journal of the Operational Research Society, 49*(6), 616–624.

Gass, S., & Rapcsák, T. (2004). Singular value decomposition in AHP. *European Journal of Operational Research, 154*, 573–584.

Koczkodaj, W. (1993). A new definition of consistency of pairwise comparisons. *Mathematical and Computer Modelling, 18*, 79–84.

Peláez, J., & Lamata, M. (2003). A new measure of consistency for positive reciprocal matrices. *Computer and Mathematics with Applications, 46*, 1839–1845.

Saaty, T. (1980). *Multicriteria decision making: The analytic hierarchy process*. New York: McGraw-Hill.

Saaty, T. (1994). *Fundamental of decision making and priority theory with the AHP*. Pittsburgh: RWS Publications.

Salo, A., & Hamalainen, R. (1997). On the measurement of preference in the analytic hierarchy process. *Journal of Multi-Criteria Decision Analysis, 6*, 309–319.

Sezgin, M., & Tokay, S. (2016). Determining the most appropriate hotel with multi-criteria decision making techniques: The example of Mersin Province. *International Journal of Scientific and Research Publications, 6*(6).

Siraj, S., Mikhailov, L., & Keane, J. (2015). Contribution of individual judgments toward inconsistency in pairwise comparisons. *European Journal of Operational Research, 242*, 557–567.

Staging Cancer Through Text Mining of Pathology Records

Pietro Belloni, Giovanna Boccuzzo, Stefano Guzzinati, Irene Italiano, Carlo R. Rossi, Massimo Rugge, and Manuel Zorzi

Abstract Valuable information is stored in a healthcare record system and over 40% of it is estimated to be unstructured in the form of free clinical text. A collection of pathology records is provided by the Veneto Cancer Registry: these medical records refer to cases of melanoma and contain free text, in particular, the diagnosis. The aim of this research is to extract from the free text the size of the primary tumour, the involvement of lymph nodes, the presence of metastasis, and the cancer stage of the tumour. This goal is achieved with text mining techniques based on a supervised statistical approach. Since the procedure of information extraction from a free text can be traced back to a statistical classification problem, we apply several machine learning models in order to extract the variables mentioned above from the text. A gold standard for these variables is available: the clinical records have already been assessed case-by-case by an expert. The most efficient of the estimated models is the gradient boosting. Despite the good performance of gradient boosting, the classification error is not low enough to allow this kind of text mining procedures to be used in a Cancer Registry as it is proposed.

P. Belloni (✉) · G. Boccuzzo
Department of Statistical Sciences, University of Padua,
Via Cesare Battisti 241, Padova , Italy
e-mail: pietro.belloni.1@phd.unipd.it

S. Guzzinati
Veneto Tumour Registry, Azienda Zero, Via Jacopo Avanzo 35, Padova , Italy

I. Italiano · M. Rugge · M. Zorzi
Azienda ULSS6 Euganea, Via Enrico degli Scrovegni 14, Padova , Italy

C. R. Rossi
Department of Surgery, Oncology and Gastroenterology, University of Padua,
Via Nicolò Giustiniani 2, Padova , Italy

Veneto Oncologic Institute, Via Gattamelata 64, Padova , Italy

P. Mariani and M. Zenga (eds.), *Data Science and Social Research II*,
Studies in Classification, Data Analysis, and Knowledge Organization,
https://doi.org/10.1007/978-3-030-51222-4_4

1 Introduction

A significant amount of useful information is stored in a healthcare record system. For the most part, information is structured, thus easily accessible for mere consultation or even statistical modelling. However, approximately 40% of it is estimated to be unstructured but pertaining to clinical text (Dalianis 2018): clinical text mining (the adaptation of text mining to clinical text) is employed to access it. The aim of clinical text mining is to extract information from clinical text, hence bridging the gap between structured and unstructured information and allowing access to more data (Spasic et al. 2014). In the healthcare setting, free text can often be found in medical records, diagnoses, reports, and forms: this text, if properly analysed, can provide useful information. For instance, it contains information on patients' health conditions, symptoms, or recommended treatments.

The aim of this research is to extract information from a collection of free text employing a supervised statistical approach, in order to structure the information contained in text. In particular, text from cancer diagnoses is employed to extract cancer stage, the size of the primary tumour, the involvement of lymph nodes, and the presence of metastasis. The text used is drawn from pathology records of incident cases in 2013, collected by the Veneto Tumour Registry (RTV) on a specific type of cancer: skin melanoma. Despite being widely discussed by the pertaining literature, text mining studies mostly deal with social media, and clinical setting is not often concerned. The first attempt to apply automatic information extraction to clinical text can be found in Pratt and Pacak (1969) but without the employment of statistical procedures. The studies applying a statistical approach to clinical text (similarly to this article) are more recent (McCowan et al. 2006, 2007; Nguyen et al. 2007; Martinez et al. 2013). Text is in English for the most part but, since language is a fundamental part of text mining, studies on text in other languages can lead to widely varied results (Ehrentraut et al. 2012; Angelova et al. 2017). As far as Italian is concerned, only one clinical text mining study was conducted, yet unsupervised (Alicante et al. 2016) so, as far as we know, this is the first research employing supervised clinical text mining in Italian.

All the operations (data manipulation, preprocessing, models estimation) have been carried out through the software R (ver. 3.4.4), except for the neural networks estimation that has been carried out with TensorFlow (ver. 1.8.0).

2 Clinical Text Mining

Text mining is a broad class of linguistic, statistical, and machine learning procedures aimed at analysing large sets of text in order to extract information (Ceron et al. 2014). Text mining is often referred to as a branch of data mining, which is the extraction of information from large sets of data, in this instance in the form of text. The added difficulty of text mining consists in having to turn text into statistical features.

2.1 The Distinctive Traits of Clinical Text

The first field of application of text mining was not the healthcare setting, but social media and sentiment analysis, hence the kind of text which text mining instruments were developed for is completely different from the type of text covered hereafter. The adaptation of text mining instruments to the clinical setting is a minor area of research and the pertaining literature is reduced. Clinical text is different from other types of text traditionally analysed using text mining for many factors. First, clinical text is written by professionals (doctors, nurses, radiologists, health workers, researchers…), so it features highly specific jargon, different from common vocabulary: this leads to greater text complexity. Another difficulty is due to clinical text often being hurriedly written: for instance, sometimes diagnoses are made by doctors able to devote just a few minutes to each patient, and involve abbreviations, misspellings, broken sentences, implied verbs, and acronyms (Allvin 2011; Patrick and Nguyen 2011). In the clinical context, abbreviations, acronyms, or misspellings are estimated to account for up to 30% of text, a variable highly depending on what language the text is written in Pakhomov et al. (2005). Overall, the extraction of information from text can be based on two approaches (Aggarwal and Zhai 2012): a rule-based approach or a statistical approach (also known as machine learning approach). Even if text mining in social media setting is now almost entirely based on statistical approach, both approaches are still applied to clinical text.

2.2 Statistical Approach

The statistical approach to text mining is entirely drawn from data mining techniques. This approach requires a preprocessing phase where text is modified and turned into statistical features, followed by a text classification phase using data mining techniques. Some examples of research employing statistical approach to clinical text mining can be found in Ehrentraut et al. (2012) and Martinez et al. (2013). The advantage of statistical approach is undoubtedly its greater adaptability to the previously discussed problems: classification is more suitable for dealing with synonyms, abbreviations, and misspellings. Furthermore, while a set of rules can only classify text pertaining to a specific setting (rules have to change according to the setting), data mining procedures based on statistical approach can be applied to every type of text with slight modifications. The main disadvantage of this approach is that to achieve competitive results, classification must be supervised: there must be a set of texts where the information to extract is already known. This implies that for every text mining attempt based on statistical approach, information is extracted manually, which requires time and qualified personnel.

Table 1 Advantages and disadvantages of the rule-based and the statistical approach to clinical text mining

Rule-based	Advantages	Easily interpreted Electronic devices assistance such as dictionaries (especially in English) Very accurate if rules are correctly devised
	Disadvantages	Susceptible to low quality of text Affect by context of development Large time spans to devise rules
Statistical	Advantages	Adaptable to low quality of text Slightly affected by context of development Short estimate time spans
	Disadvantages	Difficult to interpret (often impossible) Need for an estimate set with previously analyse text

2.3 Rule-Based Approach

The rule-based approach to text mining is the easiest and most intuitive approach. To classify text in two or more classes, a set of rules is devised to match some words of the text (pattern) with a given class where the text is to be classified. Drawing up rules is a fundamental process that can be time-consuming and expensive, but, if properly carried out, can provide an extremely accurate text classification. Some examples of rule-based clinical text mining can be found in Napolitano et al. (2010); Hanauer et al. (2007); Zhou et al. (2006).

Rule-based approach is affected by text complexity: misspellings, synonyms, or abbreviations can heavily affect rules performance, thus compromising the subsequent classification. It is difficult to write manually a set of rules comprehending all the synonyms for terms in a text; this is why specific software can be used, with medical dictionaries including lists of synonyms, abbreviations, and common misspellings. MetaMap (https://metamap.nlm.nih.gov) is probably the most used software application, among the most effective. MetaMap is a software application developed by the National Library of Medicine, able to match medical text with UMLS (Unified Medical Language System) standard medical terms so as to find all synonyms and abbreviations. But MetaMap, like all similar software, was developed for English text and at the moment there is no satisfying Italian adaptation (Chiaramello et al. 2016). The impossibility to effectively map synonyms and abbreviations is a significant disadvantage of the rule-based approach in non-English text. In Table 1 advantages and disadvantages of the two approaches to clinical text mining can be found for a comparison.

3 Data Description

The clinical text analysed using text mining is drawn from pathology records and is related to cases of skin melanoma collected by the RTV.[1] The RTV registers incident cancer cases on the basis of three sources of information: pathology records (meaning the results of histological and cytological exams or autopsies), hospitals discharge forms (including diagnoses and other information on hospitalization), and death certificates (including the cause of death). These documents include both structured and unstructured information in the form of free text, such as diagnosis. All Pathology Laboratories of both public hospitals and of private facilities in the Veneto Region must send to the RTV their electronic archives including the aforementioned sources. Information drawn from these sources are cross-checked with those at the regional registry office to remove cases referring to citizens not residing in the region. Subsequently, the collected data are automatically processed through a record-linkage procedure reporting doubtful cases that are to be assessed manually. Only at the end of this automatic (or potentially manual) process, can a case be registered in the RTV database. Text that ought to be analysed using text mining is stored in pathology records referring to skin melanoma cases occurring in 2013 and collected by the RTV up to December 31, 2017.

3.1 Available Data

Three databases provided by the RTV have been merged, including the pathology records referring to the aforementioned melanoma cases. The first database includes 3065 medical records collected in a period of time between 2013 and 2015, the second database includes 81 medical records collected in 2016 and the third one includes 104 medical records collected in 2017. These medical records are associated with 547 patients diagnosed with a cutaneous melanoma in 2013 and undergoing treatment in a facility in the Veneto region. Hence, more than one record is often associated with each patient.

Each record includes both structured information (for example, patients' and records' ID code, date of drafting...) and unstructured information in the form of free text. In particular, the diagnosis, the result of macroscopic examinations of cancer tissue and of microscopic examinations of the same tissue are included in the form of text. Only the diagnosis text was chosen to extract information on cancer stage, it is thought to be the most comprehensive type of text describing the relevant information to extract. Not every record includes a text: in 20.8% of diagnoses is missing, hence the number of texts available for text mining procedure decreases to 2574.

[1] Data analysis is based on anonymized data that have been analysed at the Cancer Registry of the Veneto Health Care System (Azienda Zero) after a formal agreement with the University of Padua.

Table 2 Distribution of the number of characters in the diagnoses of pathology records

Minimum	1° quart.	Median	3° quart.	Maximum	Mean	Std. dev.
7	129	216	432	3314	407.20	485.68

3.2 Text Features

The difficulties arising from this text are the ones related to clinical text, listed in Sect. 2.1: abbreviations, misspellings, and technical and varied vocabulary. Other potential difficulties are discussed in the texts taken into consideration:

1. **Diverse text**: since the RTV collects records from all healthcare facilities in the region, data derive from different sources. This leads to great diversity in texts, especially regarding their form and length: some diagnoses consist in a few words while others are lengthy (Table 2), with many paragraphs or bulleted lists. Length increases computational complexity during the estimate phase of text mining models, while formatting types are handled during the data preprocessing phase.
2. **Previous formatting**: many texts present not applicable symbols. The presence of this *noise text* is probably due to some texts being formatted before entering databases, hence not being plain text. After acquiring texts from sources and saving them in databases, text formatting was lost, but not the characters defining it. The presence of this kind of noise text from previous formatting can be found in 13.3% of texts, but this problem can be solved during the data preprocessing phase.

The statistical approach to text mining corresponds to a supervised text classification, i.e., the class associated with each text corresponds to the information to extract. A classification model must be estimated using a group of texts whose classification is already known (a gold standard); without the gold standard, an unsupervised classification should have been employed: a procedure known as text clustering whose performance is usually worse than supervised procedures (Chaovalit and Zhou 2005). Our supervised text mining procedure aims at extracting: the cancer stage (according to the TNM classification as described in Balch (2001)), the size of the primary tumour (the T parameter of the TNM staging), the involvement of lymph nodes (the N parameter of the TNM staging) and the presence of metastasis (the M parameter of the TNM staging).

From a statistical point of view, the entire procedure is equivalent to estimating four classification models whose outcomes are the four variables of the gold standard identifying the tumour characteristics. For the pathology records taken into consideration, manual classification was carried out within the "Project for the high-resolution registration of skin melanoma": this project allowed the registration of the staging of the melanoma and its T, N, and M components (Guzzinati 2018). This manual classification is the gold standard for our supervised text mining models. The 2574 texts are associated with four gold standards (each to every outcome):

1. **Cancer staging**. The classes related to the cancer staging are five. Stage I corresponds to a primary tumour thickness (T parameter) less or equal 1.00 mm or between 1.01 and 2.00 mm without ulceration. Stage II corresponds to a primary tumour thickness between 1.01 and 2.00 mm with ulceration or greater or equal 2.01 mm. Stage III corresponds to a primary tumour of any thickness and the presence of at least one metastatic lymph node (N parameter). Stage IV corresponds to a primary tumour of any thickness, the presence of at least one metastatic lymph node and the presence of metastasis (M parameter). Stage X collects not definable or missing cases.
2. **Size of the primary tumour (T)**. Three classes have been considered: *low T* indicates a primary tumour thickness less or equal 1.00 mm or between 1.01 and 2.00 mm without ulceration, *high T* indicates a primary tumour thickness between 1.01 and 2.00 mm with ulceration or greater or equal 2.01 mm and *T X or missing* indicates a not definable or missing T.
3. **Involvement of lymph nodes (N)**. Similarly, three classes have been considered: *No lymph nodes* indicates that no lymph nodes are infected (N is equal to 0), *At least one lymph node* indicates a lymph nodes infection (N is greater than 0) and *N X or missing* indicates a not definable or missing N.
4. **Presence of metastasis (M)**. Also for the M parameter three classes have been considered: *Absent* (M is equal to 0), *At least one metastasis* (M is equal to 1) and *M X or missing* indicates a not definable or missing M.

A similar choice of merging classes of T, N, and M parameters can be found in McCowan et al. (2006). The distribution of all outcomes shows a strong imbalance among classes (Fig. 1). The strong imbalance of outcomes among classes causes problems to statistical models of classification that will result in having a good predictive value for more frequent classes but a worse one for less frequent classes. This problem has been thoroughly addressed by pertaining literature (Chawla et al. 2002; Chawla 2003; Chawla et al. 2004; Cieslak and Chawla 2008).

4 Methods

First of all, features were extracted by the text through a preprocessing phase. Subsequently, a series of machine learning models able to use features to classify text was employed.

4.1 Text Preprocessing

Preprocessing is a set of operations aimed at extracting variables conveying information from raw text (Kwartler 2017). Preprocessing consists in three phases: text normalization, stemming, and document-term matrix creation (Fig. 2). The result of

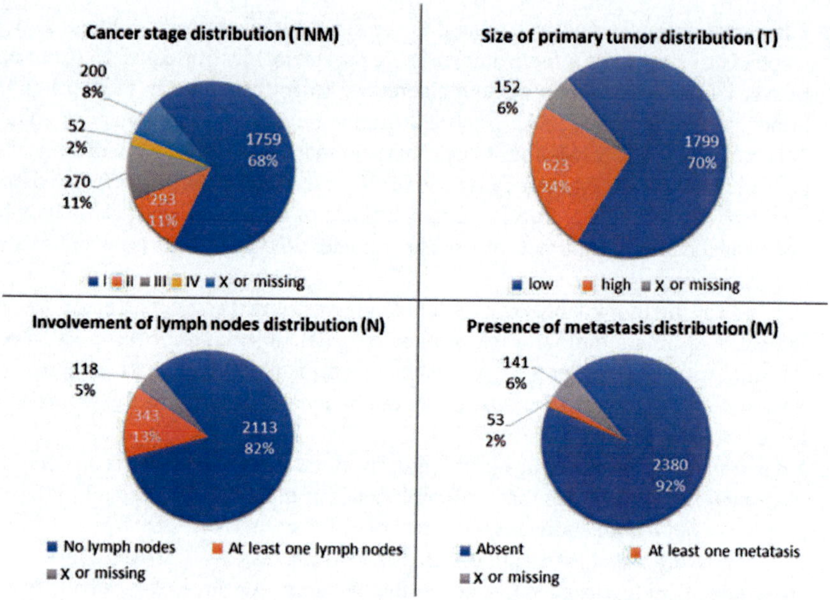

Fig. 1 Distribution of classes among gold standard

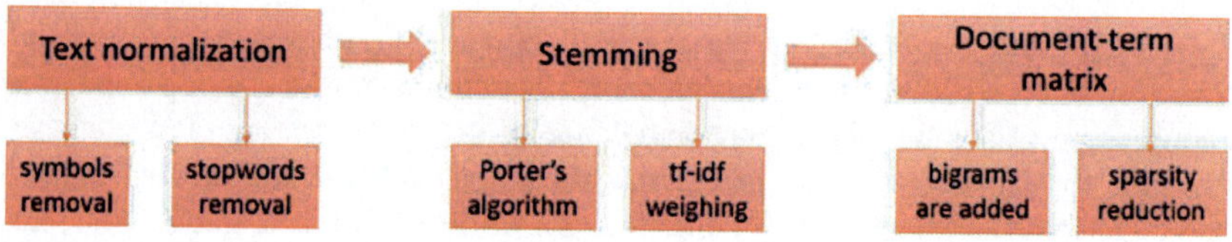

Fig. 2 Workflow of text preprocessing

the preprocessing procedure is a matrix suitable for being used as a regression matrix for machine learning classification models.

4.1.1 Normalization

At the beginning, all texts are normalized: punctuation, symbols, and capital letters are removed. Subsequently, stopwords (general words conveying no information whatsoever) are removed. The standard stopwords list for Italian, contained in the R library **tm** was used (Feinerer 2018), purposely expanded adding a list of stopwords pertaining to the topic at stake, in this case cutaneous melanoma.

4.1.2 Stemming

Stemming is the process of cutting a word to its stem. Stem is not the semantic origin, but the cut form of the word: for instance, the words "tumore", "tumori", "tumorale", etc. can be associated to the same stem, "tumour". This way, the features extracted from text are fewer and, at the same time, more texts will have the same common features. Stemming is completely automized and based on algorithms called stemmers. The first stemmer was suggested in Lovins (1968), but this study will employ the one suggested in Porter (1980), which seems to be more effective in general (Jivani 2011).

The algorithm is also implemented for Italian and contained in the R library **tm**. Each stem stands for a piece of information extracted from the text and will be dealt with as a feature: this process is the so-called bag-of-words approach, which is the most widespread approach to text mining nowadays (Jurafsky and Martin 2008; Zhang et al. 2010).

4.1.3 Weighted Document-Term Matrix

The next step is building a matrix with lines being text, columns being stems, and *tf-idf* weights of stems in text being elements (Miner et al. 2012). This matrix aims at serving as regression matrix for text classification models. *tf-idf* weights represent the importance of a stem in a text belonging to a corpus (Ramos 2003). The acronym *tf-idf* abbreviates the phrase *term frequency - inverse document frequency* and with x as word in y document, it is calculated as follows:

$$tf\text{-}idf_{x,y} = (N_{x,y}/N_{\cdot,y}) \cdot log(D/D_x) \tag{1}$$

where: $N_{x,y}$ is the number of times stem x appears in text D_y; $N_{\cdot,y}$ is the number of stems in D_y; $N_{x,y}/N_{\cdot,y}$ is term frequency; D is the overall number of texts; D_x is the number of texts where stem x appears at least once and D/D_x is the inverse document frequency. This matrix consists in 2574 lines (number of texts) and 2631 columns (number of stems). The matrix has 99.18% of zeros in it. High dimensionality and wide sparsity make this matrix less suitable for being used as regression matrix. This problem is widespread in text mining, and was solved by getting rid of all stems appearing just a few times: this proceeding reduces matrix's dimension without losing a significant amount of information (Feinerer et al. 2008; Feinerer 2018). Stems with 99% sparsity were chosen to be removed from the document-term matrix, meaning they only appeared in 1% of texts. After this proceeding, matrix's columns decreased to 316 (with 94.02% of elements equal to 0), hence the number of stems has been drastically reduced and sparsity slightly decreased. The new matrix has more suitable dimensions to be used as regression matrix for a classification model.

4.1.4 Adding Bigrams

A limit of the bag-of-words approach is that it doesn't take into consideration the order of words in texts, while bigrams allow to consider phrases made up by words pairs. Bigrams are used similarly to stems: a matrix is built with lines being texts and columns being bigrams. The matrix's elements will be equal to 0 if bigram isn't in the text and to 1 if bigram is in the text (to simplify, *tf-idf* weights were chosen not to be used for). Bigrams are different from stems because the former are created without resorting to stemming so that the ordered pair of whole words (not cut words) is taken into consideration. To decrease sparsity in document-term matrix, only bigrams appearing in at least 1% of texts were taken into consideration. The matrix originated by bigrams is juxtaposed to the one originated by stems, building a matrix made up of 2574 lines (texts) and 695 columns (variables extracted from text in the form of 316 stems and 379 bigrams). This will serve as regression matrix for text classification models.

Trigrams, quadrigrams, etc. could be also dealt with as bigrams, so that word order in sentences is taken more and more into consideration, but, in general, considering more than three words doesn't add new information and doesn't increase classification quality (Ceron et al. 2014). This is why it was chosen to apply this procedure to bigrams only.

4.2 Classification Models

Using document-term matrix as regression matrix allows to estimate statistical models of supervised classification whose outcomes are variables made available by the gold standard. Usually, the classification problems connected to text mining are dealt with using machine learning models because they are able to better face high dimensionality and sparsity in regression matrices (Hastie et al. 2013).

Four classifications, one for each outcome, were estimated using a variety of models. Usually, for instances of supervised clinical text mining the model of choice is support vector machines (McCowan et al. 2007; Nguyen et al. 2007; Martinez et al. 2013), but their performances were compared to the other models such as classification tree, random forest, gradient boosting in its variety called XGBoost, and a neural network with two fully connected hidden layers. Data were randomly divided into two sets: train set (1500 samples) and test set (1074 samples), all models have been trained in the train set with the 20-folds cross-validation, subsequently accuracy error has been calculated on the test set (Stehman 1997). Table 3 shows the comparison between models' accuracy errors for each outcome. Gradient boosting with XGBoost algorithm outperforms any other classification model. More in general, XGBoost algorithm makes it an efficient model when it comes to noisy and unbalanced data (Nielsen 2016).

Table 3 Classification error of machine learning models

Outcome (%)	Baseline error (%)	Support vector machines (%)	Single tree (%)
TNM	31.7	26.3	28.5
T	30.1	18.7	26.5
N	17.9	14.4	16.2
M	7.5	6.5	–

Outcome (%)	Random forest (%)	XGBoost (%)	Deep neural network (%)
TNM	25.2	20.6	24.5
T	21.9	13.7	18.0
N	13.9	10.1	12.4
M	6.1	5.3	5.9

4.2.1 How Gradient Boosting Works

Boosting is a method allowing to combine a number of different classifiers to increase their effectiveness, especially using classification tree. Freund and Schapire introduced in Freund and Schapire (1996) the first boosting algorithm *AdaBoost*. Then, this model was modified by other authors and its performance was improved until it became the gradient boosting (Breiman 1997; Friedman 2001). This was in turn the subject of research focused on machine learning: many articles have been written on this theme recently, included (Chen et al. 2016) which originated the XGBoost algorithm.

Boosting with the *AdaBoost* algorithm

For the sake of simplicity, consider the problem of classifying a dichotomic variable Y able to assume -1 and 1 as values using a set of explanatory variables X. Given a set of observations $(X_i; Y_i)$ belonging to a train set, where $i = 1 \dots n$, $G(x)$ is defined as the classification originated by a tree classifier $G(X)$. At the beginning, data are assumed to have the same weight in originating classification: $w_i = 1/n$, with $i = 1 \dots n$. Now, suppose to estimate a sequence of classifiers $G_m(x)$ where $m = 1 \dots M$ and to combine it in a weighted average leading to the final classification:

$$G(x) = \text{sign}\left(\sum_{m=1}^{M} \alpha_m G_m(x)\right). \quad (2)$$

The idea leading to boosting is: modify with an iterative approach weights $w_1 \dots w_n$ associated to single observations so as to attach more importance to incorrectly classified observations, and determine weights $\alpha_1 \dots \alpha_M$ associated to classificator so as to attach more importance to accurate classifier in the final weighted average. In particular, formulas to calculate α_m and update w_i are described in the Algorithm 10.1 p. 399 in Hastie et al. (2013). For a generic step m of the algorithm, classifier $G_m(x)$ is estimated according to the weights obtained by the previous step. Hence, it will attach more importance to observations that had a greater classification error. This way, algorithm "learns" from data, focusing each time on elements that present classification problems.

This characteristic makes boosting a very effective model to solve problems with varying responses whose classes are unbalanced. The elements of less frequent classes will have a greater classification error due to their underrepresentation. After a number of steps, *AdaBoost* algorithm will attach a greater weight to these observations, with a tendency to counterbalance their underrepresentation. Lastly, in the final classification the same algorithm will attach less importance to the classifications of the first steps (the ones where underrepresented observations haven't gained greater weight yet).

From Boosting to Gradient Boosting

Suppose to generalize (2) in the following form:

$$f(x) = \sum_{m=1}^{M} \beta_m b(x; \gamma_m). \tag{3}$$

This model is attributable to an additional model where $f(x)$ replaces the final classifier, functions $b(x; \gamma_m)$ are single classifiers and β_m is the weight attached to each of them. γ_m contains all parameters estimated from data by the classification model. In this model, parameters β and γ can be estimated jointly minimizing a function $L(y_i; f(x_i))$ measuring the difference between predicted values and estimated values, known as loss function:

$$\max_{\beta,\gamma} \sum_{i=1}^{n} L\left(y_i, \sum_{m=1}^{M} \beta_m b(x; \gamma_m)\right). \tag{4}$$

This joint minimization is often hard from a computational point of view and is preferably approximated as follows: the $m - th$ classifier is estimated minimizing its loss function $L(y_i, \beta_m b(x_i; \gamma_m))$ and is added to the sum of previously estimated classifiers $f_{m-1}(x)$. $f_m(x)$ is estimated, then this process is repeated in an iterative way. This procedure is reflected by Algorithm 10.2 p. 342 in Hastie et al. (2013), also known as *stagewise modelling*, which generalizes *AdaBoost* algorithm allowing to estimate the boosting model through the loss function of single classifiers. This minimization problem is dealt with through a gradient boosting method, whence the

Table 4 Confusion matrix for TNM stage classification

Predicted class	I	II	III	IV	X	TOT
I	1694	109	98	27	105	2033
II	23	156	22	6	18	225
III	25	17	134	12	15	204
IV	3	1	2	2	4	12
X	14	10	14	5	58	101
TOT	1759	293	270	52	200	2574

Columns: **Gold standard**

Classification error: 20.6%
Errors in each class:
I: 3.7%
II: 46.8%
III: 50.4%
IV: 96.2%
X: 71.0%

Table 5 Confusion matrix for T stage classification

Pred. class	Low	High	X	TOT
Low	1726	176	82	1984
High	68	444	20	532
X	5	3	50	58
TOT	1799	623	152	2574

Columns: **Gold standard**

Classification error: 13.8%
Errors in each class:
Low: 4.1%
High: 28.7%
X: 67.1%

Table 6 Confusion matrix for N stage classification

Pred. class	N0	N>0	X	TOT
N0	2071	157	46	2274
N>0	39	182	11	232
X	3	4	61	68
TOT	2113	343	118	2574

Columns: **Gold standard**

Classification error: 10.1%
Errors in each class:
N0: 2.0%
N>0: 46.9%
X: 48.3%

name gradient boosting. Gradient boosting allows to minimize complex functions in a very effective way, hence it is frequently used in a machine learning context.

5 Discussion of Results and Future Prospects

Further details on model performance can be found in the confusion matrices obtained for the four outcomes (Tables 4, 5, 6, and 7) which are useful to comprehend in which class the classification error is more present.

Classes with a higher error are the ones with a lower frequency. Despite gradient boosting model's ability to handle well-unbalanced classes, the imbalance level is so

Table 7 Confusion matrix for M stage classification

		Gold standard			
		M0	M1	X	TOT
Pred. class	M0	2355	29	80	2464
	M1	2	22	0	24
	X	23	2	61	86
	TOT	2380	53	141	2574

Classification error:	5.3%
Errors in each class:	
M0:	1.1%
M1:	58.5%
X:	56.7%

high that it leads to a high classification error in less frequent classes. Another problem is the high error originated by X classes of outcomes: X class defines the stage as *non measurable*, meaning that whoever determined the cancer staging couldn't establish one or more T, N, or M parameters with certainty. Presumably, an X stage cancer belongs to any other stage, so the text of the pertaining diagnosis is likely to include characteristics of the cancer attributable to other classes. This leads to the difficulty to classify X stage. The obtained outcomes can be compared with similar studies of the pertaining literature. It was found that the majority of research studies on oncological text mining adopt a rule-based approach, studies with a similar approach to the one adopted in the present paper are fewer and obtain similar results when it comes to classification error (McCowan et al. 2007; Nguyen et al. 2007; Martinez et al. 2013).

5.1 *Possible Applications of the Clinical Text Mining Procedure*

Despite the good performance of gradient boosting, the classification error is not low enough to allow this kind of text mining procedures to be used in a Cancer Registry as it is proposed. A partial use of the procedure is however possible: there's a significant number of correctly classified texts and they tend to be more present in the most frequent class. If a Cancer Registry adopted a clinical text mining process like the aforementioned one, it could extract staging information from some of the texts with a certain degree of certainty and should manually analyse only the remaining ones. As a matter of fact, the classification model estimates how likely it is for a diagnosis to have a certain staging: each staging has an estimated probability and the class of choice is the one with the highest probability. If the probability of the estimated class is significantly high (for instance, higher than a threshold established by the Cancer Registry) it is possible to conclude that staging has a good degree of certainty. This would allow not to manually analyse that diagnosis and focus only on the ones whose staging had a lower probability.

5.2 *Possible Developments*

Two strategies are suggested to increase the overall effectiveness of the clinical text mining procedure:

1. **A mixed statistical and rule-based approach**: Some authors adopt a combination of the two approaches to clinical text mining described in Sects. 2.2 and 2.3 so as to harness their respective advantages (Nassif et al. 2009; Liu et al. 2012; Kovacevic et al. 2013; Aalabdulsalam et al. 2018). Usually, two software applications are employed during the preprocess phase of texts: cTAKES (http://ctakes.apache.org) and MetaMap (or equivalent ones). cTAKES is capable of part-of-speech tagging, with a particular focus on negation detection. MetaMap, on the other hand, associates every term with a medical concept: this way, synonyms, misspellings, and acronyms are identified. As a consequence, there are less variables extracted from the text through bag-of-words and they are more meaningful. The term-document matrix will have a lower dimensionality and a lower sparsity. Subsequently, a supervised classification through a machine learning model can be operated.
2. **Word embedding approach**: To turn text into statistical variables, bag-of-words can be relinquished in favour of an approach based on words vectoring, known as *word embedding*. The aim of word embedding is to obtain a firmer, more effective representation than the document-term matrix obtained through bag-of-words, by representing the words in the text as real, dense (not sparse) vectors with a reduced dimension (usually with a length of tens or hundreds), known as *word vectors*. Word vectors are vectors associating words among themselves: more similar words will have closer vectors. Every dimension of word vectors is the value assumed by words with respect to a latent variable representing a concept shared with the words taken into consideration (Mikolov et al. 2013; Pennington et al. 2014). In other words, words are vectors where every part grasps a dimension of that word's meaning. Clinical text mining studies conducted using this approach are just a few but their outcomes are promising (Wu et al. 2015).

6 Conclusions

Two thousand five hundred seventy-four diagnoses texts have been provided by Veneto Tumour Registry, pertaining to as many pathology reports. The objective was to extract TNM staging (both in its aggregate form and separating T, N, and M stages) using a clinical text mining procedure. The text mining procedure can be related to a supervised classification of these texts based on an outcome and a series of variables extracted from texts themselves. In the analysed instance, outcomes were provided by a gold standard obtained by the "Project for high-resolution registration of cutaneous melanoma". After an initial preprocessing phase, different machine learning models have been compared according to the accuracy of their classification and the best model resulted to be gradient boosting with XGBoost algorithm. The classifications obtained through a gradient boosting model still have too high a classification error to allow the whole procedure to be used in a health facility such as a tumour registry. Nevertheless, a partial application of this text mining procedure could be employed in a health facility to decrease the number of texts where staging information is manually extracted.

Acknowledgements The authors thank Professor Bruno Scarpa (Department of Statistical Sciences, University of Padua) for the valuable help, in particular concerning the statistical modelling section, and Beatrice Bezzon for the linguistic support.

References

Aalabdulsalam, A. K., et al. (2018). Automated extraction and classification of cancer stage mentions from unstructured text fields in a central cancer registry. In *AMIA Summits on Translational Science Proceedings* (pp. 16–25).

Aggarwal, C. C. & Zhai, C. (Eds.). (2012). *Mining text data*. Springer Science and Business Media.

Alicante, A., Corazza, A., Isgrò, F., & Silvestri, S. (2016). Unsupervised entity and relation extraction from clinical records in Italian. *Computers in Biology and Medicine*, *72*, 263–275.

Allvin, H., et al. (2011). Characteristics of Finnish and Swedish intensive care nursing narratives: a comparative analysis to support the development of clinical language technologies. *Journal of Biomedical Semantics*, *2*, 1–11.

Angelova, G., Boytcheva, S., & Nikolova, I. (2017). Mining association rules from clinical narratives. In *Proceedings of Recent Advances in Natural Language Processing* (pp. 130–138).

Balch, C. M., et al. (2001). Final version of the American Joint Committee on Cancer staging system for cutaneous melanoma. *Journal of Clinical Oncology*, *19*, 3635–3648.

Breiman, L. (1997). *Arcing the edge*. Technical Report. Statistics Department, University of California.

Ceron, A., Curini, L. & Iacus, S. M. (2014). *Social Media e Sentiment Analysis: L'evoluzione dei fenomeni sociali attraverso la Rete*. Springer Science & Business Media.

Chaovalit, P. & Zhou, L. (2005). Movie review mining: A comparison between supervised and unsupervised classification approaches. In *Proceedings of the 38th Annual Hawaii International Conference on System Sciences* (Vol. 112).

Chawla, N. V. (2003). C4. 5 and imbalanced data sets: Investigating the effect of sampling method, probabilistic estimate, and decision tree structure. In *Proceedings of the ICML 3* (Vol. 66).

Chawla, N. V., Bowyer, K. W., Hall, L. O., & Kegelmeyer, W. P. (2002). SMOTE: Synthetic minority over-sampling technique. *Journal of Artificial Intelligence Research*, *16*, 321–357.

Chawla, N. V., Japkowicz, N., & Kotcz, A. (2004). Special issue on learning from imbalanced data sets. *ACM Sigkdd Explorations Newsletter*, *6*, 1–6.

Chen, T. & Guestrin, C. (2016). Xgboost: A scalable tree boosting system. In *Proceedings of the 22nd ACM Sigkdd International Conference on Knowledge Discovery and Data Mining* (pp. 785–794). ACM.

Chiaramello, E., Paglialonga, A., Pinciroli, F., & Tognola, G. (2016). Attempting to use metamap in clinical practice: A feasibility study on the identification of medical concepts from Italian clinical notes. *Studies in Health Technology and Informatics*, *228*, 28–32.

Cieslak, D. A. & Chawla, N. V. (2008). Learning decision trees for unbalanced data. In *Joint European Conference on Machine Learning and Knowledge Discovery in Databases* (pp. 241–256). Springer.

Dalianis, H. (2018). *Clinical text mining: Secondary use of electronic patient records* (Vol. 192). Springer.
Ehrentraut, C., Dalianis, H., Tanushi, H., & Tiedemann, J. (2012). Detection of Hospital Acquired Infections in sparse and noisy Swedish patient records. In *Sixth Workshop on Analytics for Noisy Unstructured Text Data* (pp. 1–8).
Feinerer, I. (2018). *Introduction to the tm Package 2018.*
Feinerer, I., Hornik, K., & Meyer, D. (2008). Text mining infrastructure in R. *Journal of Statistical Software, 25.*
Freund, Y., & Schapire, R. E. (1996). Experiments with a new boosting algorithm. *ICML, 96,* 148–156.
Friedman, J. H. (2001). Greedy function approximation: A gradient boosting machine. *Annals of Statistics,* 1189–1232.
Guzzinati, S. et al. (2018). High resolution registry of melanoma and care pathways monitoring in the Veneto Region, Italy in ENCR scientific meeting.
Hanauer, D. A., Miela, G., Chinnaiyan, A. M., Chang, A. E., & Blayney, D. W. (2007). The registry case finding engine: An automated tool to identify cancer cases from unstructured, free-text pathology reports and clinical notes. *Journal of the American College of Surgeons, 205,* 690–697.
Hastie, T., Tibshirani, R., & Friedman, J. H. (2013). *The elements of statistical learning.* New York: Springer.
Jivani, A. G. (2011). A comparative study of stemming algorithms. *International Journal of Computer Technology and Applications, 2,* 1930–1938.
Jurafsky, D. & Martin, J. H. (2008). *Speech and language processing.* Pearson London.
Kovacevic, A., Dehghan, A., Filannino, M., Keane, J. A., & Nenadic, G. (2013). Combining rules and machine learning for extraction of temporal expressions and events from clinical narratives. *Journal of the American Medical Informatics Association, 20,* 859–866.
Kwartler, T. (2017). *Text mining in practice with R.* Wiley.
Liu, H., et al. (2012). Clinical decision support with automated text processing for cervical cancer screening. *Journal of the American Medical Informatics Association, 19,* 833–839.
Lovins, J. B. (1968). Development of a stemming algorithm. *Mechanical Translation and Computational Linguistics, 11,* 22–31.
Martinez, D., Cavedon, L. & Pitson, G. (2013). Stability of text mining techniques for identifying cancer staging in Louhi. In *The 4th International Workshop on Health Document Text Mining and Information Analysis.*
McCowan, I., Moore, D., & Fry, M.-J. (2006). Classification of cancer stage from free-text histology reports. *Engineering in Medicine and Biology Society,* 5153–5156.
McCowan, I., et al. (2007). Collection of cancer stage data by classifying freetext medical reports. *Journal of the American Medical Informatics Association, 14,* 736–745.
Mikolov, T., Chen, K., Corrado, G., & Dean, J. (2013). Efficient estimation of word representations in vector space (pp. 1–12).
Miner, G., Elder, J., & Hill, T. (2012). *Practical text mining and statistical analysis for non-structured text data applications.* Academic Press.
Napolitano, G., Fox, C., Middleton, R., & Connolly, D. (2010). Pattern-based information extraction from pathology reports for cancer registration. *Cancer Causes & Control, 21,* 1887–1894.
Nassif, H., et al. (2009). Information extraction for clinical data mining: A mammography case study. In *International Conference on Data Mining* (pp. 370-42).
Nguyen, A. N., Moore, D. C., McCowan, I. & Courage, M. (2007). Multi-class classification of cancer stages from free-text histology reports using support vector machines. In *29th Annual International Conference of the IEEE Engineering in Medicine and Biology Society* (pp. 5140–5143).
Nielsen, D. (2016). *Tree BoostingWith XGBoost-why does XGBoostWin "Every" machine learning competition?*

Pakhomov, S., Pedersen, T., & Chute, C. G. (2005). Abbreviation and acronym disambiguation in clinical discourse eng. *AMIA Annual Symposium Proceedings, 2005*, 589–593.
Patrick, J. & Nguyen, D. (2011). Automated proof reading of clinical notes. In *Proceedings of the 25th Pacific Asia Conference on Language, Information and Computation.*
Pennington, J., Socher, R. & Manning, C. (2014). Glove: Global vectors for word representation. In *Proceedings of the 2014 Conference on Empirical Methods in Natural Language Processing (EMNLP)* (pp. 1532–1543).
Porter, M. F. (1980). An algorithm for suffix stripping. *Program, 14*, 130–137.
Pratt, A. W. & Pacak, M. G. (1969). Automated processing of medical English. In *Proceedings of the 1969 Conference on Computational Linguistics* (Association for Computational Linguistics) (pp. 1–23).
Ramos, J. (2003). Using tf-idf to determine word relevance in document queries. In *Proceedings of the First Instructional Conference on Machine Learning* (Vol. 242, pp. 133–142).
Spasic, I., Livsey, J., Keane, J. A., & Nenadic, G. (2014). Text mining of cancerrelated information: Review of current status and future directions. *International Journal of Medical Informatics, 83*, 605–623.
Stehman, S. V. (1997). Selecting and interpreting measures of thematic classification accuracy. *Remote Sensing of Environment, 62*, 77–89.
Wu, Y., Xu, J., Jiang, M., Zhang, Y. & Xu, H. (2015). A study of neural word embeddings for named entity recognition in clinical text. In *AMIA Symposium 2015* (pp. 1326–1333). American Medical Informatics Association.
Zhang, Y., Jin, R., & Zhou, Z.-H. (2010). Understanding bag-of-words model: A statistical framework. *International Journal of Machine Learning and Cybernetics, 1*, 43–52.
Zhou, X., Han, H., Chankai, I., Prestrud, A. & Brooks, A. (2006). Approaches to text mining for clinical medical records. In *Proceedings of the 2006 ACM symposium on Applied computing* (Vol. 235).

Predicting the Risk of Gambling Activities in Adolescence: A Case Study

Laura Benedan and Gianna Serafina Monti

Abstract Adolescent gambling is internationally considered a serious public health concern, although this phenomenon is less explored than adult gambling. It is also well known that the early onset age of gambling is a risk factor for developing gambling problems in adulthood. This study examined 7818 adolescents enrolled in 16 public high schools in Lombardy, a Northwest Italy region, between March 2017 and April 2018 and it is part of a larger study aimed at investigating dysfunctional behaviours of adolescents, with the purpose of identifying the factors that increase the risk of vulnerability and the protection factors able to reduce the incidence of pathological phenomena. The objective of the present study was to investigate the susceptibility of adolescents in high school to develop gambling problems, explained by some individual and social factors, and by the association with the use of substances, such as alcohol and tobacco and other risk-taking behaviours. Various modelling methods were considered from an imbalanced learning perspective, as the prevalence rate of high school students in the sample with problem gambling is equal to 6.1%.

1 Introduction

Gambling behaviour in Lombardy is spreading at a higher rate than before (Cavalera et al. 2018). Besides, over the last few years, an increasing number of adolescents have been engaged in gambling activities. Adolescents may be more sensitive to the problem of gambling than adults (Keen et al. 2017), and a part of the occasional gambling teenagers are at risk of developing a real psychiatric disorder, known as the

L. Benedan (✉)
Bicocca Applied Statistics Center, University of Milano Bicocca, Milan, Italy
e-mail: laura.benedan@unimib.it

G. S. Monti
Department of Economics, Management and Statistics, University of Milano Bicocca, Milan, Italy
e-mail: gianna.monti@unimib.it

P. Mariani and M. Zenga (eds.), *Data Science and Social Research II*,
Studies in Classification, Data Analysis, and Knowledge Organization,
https://doi.org/10.1007/978-3-030-51222-4_5

Gambling Disorder provided for by the DSM-5 (American Psychiatric Association 2013). This phenomenon has been recognised as an important public health issue and an emerging field of research (Blinn-Pike et al. 2010), which is associated with delinquent behaviour, depression and suicide attempts (Cook et al. 2015). Health statistics in recent years have shown an increase in the frequency of gambling-related diseases among the younger age groups of the population. Several studies have been carried out in order to understand what factors determine this disorder, and how they affect its development. However, further studies are needed to better understand the role of different variables in influencing the values, beliefs and behaviours associated with this problem. The present study aimed at predicting the gambling problems of adolescents through a Balanced Random Forest (BRF) algorithm to deal with the imbalanced data classification problem, typical of such a phenomenon while considering a broad set of risk and protective factors.

2 Materials and Methods

This study examined 7818 adolescents enrolled in 16 public high schools in Lombardy, a Northwest Italy region, between March 2017 and April 2018. It is part of a more extensive research project aimed at investigating the prevalence of gambling in adolescence, the motivations and perceptions of adolescents towards this phenomenon with the purpose of identifying the factors that increase the risk of vulnerability and the protection factors able to reduce the incidence of pathological phenomena.

The goal is to predict the minority class accurately and also to perform feature selection. Imbalanced data typically refers to a classification problem where the number of observations per class is not equally distributed. Learning algorithms will be biased towards the majority group, as imbalanced data causes suboptimal classification performance. Within our setting, the class imbalance is intrinsic to the problem, i.e. there are typically very few cases of youth gambling problem as compared to the large number of adolescents without this serious condition. According to the systematic review carried out by Calado et al. (2017), the prevalence of problem gambling among adolescents varies from 0.2 to 12.3% depending on the country considered. In Italy, 2.3–2.6% of teenagers would be classified as problem gamblers (Calado et al. 2017). Such a minority class is the most important from the data mining perspective, in spite of its rareness, it may carry important and useful knowledge. For these reasons, we require methods to improve its recognition rates.

2.1 How to Remedy to the Imbalance Problem?

Several machine learning approaches could be applied to copy with imbalance data classification. We can distinguish methods based on data level resampling, and meth-

Table 1 Confusion matrix

	Predicted minority class	Predicted majority class
Actual minority class	TP (true positive)	FN (false negative)
Actual majority class	FP (false positive)	TN (true negative)

ods related to the algorithmic level. On the one hand, among the data level resampling techniques, we can identify three different methods:

- Downsampling: randomly subset all the classes in the training set so that their class frequencies match the least prevalent class.
- Oversampling: randomly sample (with replacement) the minority class to be the same size as the majority class.
- Hybrid methods: techniques such as SMOTE (Synthetic Minority Oversampling Technique) and ROSE (Random OverSampling Examples) synthesise new data points in the minority class.

On the other hand, from the algorithmic level, some purposes are related to cost sensitive learning, while others consider ensemble methods.

To evaluate the performance of learning algorithms on imbalanced data, different metrics (see Eq. 1) could be used as a function of the confusion matrix (Table 1):

$$
\begin{aligned}
&\text{True Negative Rate } (\text{Acc}^-) = \frac{\text{TN}}{\text{TN} + \text{FP}} \\
&\text{True Positive Rate } (\text{Acc}^+) = \frac{\text{TP}}{\text{TP} + \text{FN}} \\
&\text{G-mean} = (\text{Acc}^- \times \text{Acc}^+)^{1/2} \\
&\text{Precision} = \frac{\text{TP}}{\text{TP} + \text{FP}} \\
&\text{Recall} = \frac{\text{TP}}{\text{TP} + \text{FN}} = \text{Acc}^+ \\
&\text{F-measure} = \frac{2 \times \text{Precision} \times \text{Recall}}{\text{Precision} + \text{Recall}}
\end{aligned}
\tag{1}
$$

3 The Case Study

3.1 Description of the Data

The data were collected by "Casa del Giovane di Pavia" and "Fondazione Exodus" within the project "Semi di Melo", which aimed at getting to know today's teenagers in order to understand better them, their problems and difficulties.

Several schools were recruited, and students from all classes took part in the research. All participants filled out an online questionnaire during school hours, under the supervision of teachers. This questionnaire was designed to investigate, among others, their attitude and involvement in gambling defined as involvement in activities such as scratch cards, slots, bets and online gambling.

The final data set consisted of $n = 7818$ adolescents (55% males) and $p = 45$ features. After applying a coherent approach to the data cleaning process, i.e. removing messy data from the records and variable transformations, we found that 6% of the sample belonged to the minority class. We considered as problem gamblers ($Y = 1$) those adolescents who play at least twice a week or usually spend more than 10.00 euros in at least one of the gambling activities under investigation (i.e. scratch cards, slot machines, online gambling and bets).

The data collected during this phase were combined and subjected to some exploratory analyses to highlight the critical points. The distribution by age groups is shown in Table 2. These analyses also showed that 48% of the young people interviewed gambled at least once in their life. However, it should be noted that the definition of gambling includes different types of games, including scratch cards, bets, slot machines and online games. Table 3 depicts each gambling activity separately. Considering the different types of gambling, 3.3% of the sample gambles two or more times a week and that 1.6% of adolescents spend more than 20.00 euros weekly for scratch cards, bets and slot machines, reaching over 50.00 euros per week. Furthermore, 20% of those who gamble at least twice a week regularly participate in at least two different types of games. In line with the literature on Italian adults and international studies conducted with adolescent participants, results showed gender differences in gambling participation: 56% of males gambled at least once in their life, while only 39% of females did the same. Similarly, 81% of problem gamblers in our sample were males. When considering the differences between males and

Table 2 Distribution of the sample by age

Age group	Participants per age group		Problem gamblers within each age group	
	Absolute frequencies	Relative frequencies (over total) (%)	Absolute frequencies	Relative frequencies (within the age) (%)
14	1196	15.30	63	5.27
15	1735	22.19	109	6.28
16	1684	21.54	89	5.29
17	1525	19.51	101	6.62
18	1183	15.13	78	6.59
19	495	6.33	40	8.08

Table 3 Absolute (n_j) and relative (%) frequency of gambling activities

		Slots		Online games		Bets		Scratch cards	
		n_j	(%)	n_j	(%)	n_j	(%)	n_j	(%)
	Have you ever gambled?	613	(7.8)	392	(5.0)	1661	(21.2)	3062	(39.2)
How often do you gamble?	Never	7459	(95.40)	7538	(96.40)	6320	(80.80)	5442	(69.60)
	Less than once a month	280	(3.60)	161	(2.10)	751	(9.60)	2055	(26.30)
	From 1 to 4 times a month	46	(0.60)	57	(0.70)	554	(7.10)	259	(3.30)
	From 2 to 4 times a week	12	(0.20)	22	(0.30)	148	(1.90)	34	(0.40)
	Every day	21	(0.30)	40	(0.50)	45	(0.60)	28	(0.40)
How much do you spend a week on gambling activities?	Never gambled	7393	(94.60)	7529	(96.30)	6308	(80.70)	5380	(68.80)
	Less than 10 euros	360	(4.60)	212	(2.70)	1316	(16.80)	2282	(29.20)
	Between 11 and 20 euros	23	(0.30)	31	(0.40)	115	(1.50)	101	(1.30)
	Between 21 and 50 euros	12	(0.20)	9	(0.10)	27	(0.30)	23	(0.30)
	More than 50 euros	30	(0.40)	37	(0.50)	52	(0.70)	32	(0.40)

females, it was found that 9% of male teenagers were considered as problem gamblers, compared to 3% of girls (the approximate p-value for the difference between proportions in the two genders is <0.001).

3.2 *Random Forest (RF)*

Random forest (Breiman et al. 1984) is an ensemble of unpruned classification, induced from bootstrap samples of the training data, using random feature selection in the tree induction process. Prediction is made by aggregating (majority vote for classification) the predictions of the ensemble. A Balanced Random Forest (BRF) algorithm by SMOTE-based oversampling (Chawla et al. 2011; Blagus and Lusa 2013) the minority class was implemented to weaken the effect of the skewed class distribution in the learning process. All analyses were performed using the software R (R Core Team 2019).

3.3 *Bagging Method Using Decision Trees (TB)*

Bagging (Bootstrap Aggregating) algorithm is used to improve model accuracy in regression and classification problems. Bagging algorithm builds multiple models from separated subsets of train data and constructs a final aggregated and more accurate model. A Balanced bagging method using decision trees (BTB) by SMOTE-based oversampling the minority class was implemented. Bagging improves prediction accuracy at the expense of interpretability.

3.4 *Performance Comparison*

Performance comparison derived from the test set for the default (0.5) and alternate cutoffs, derived using the ROC curve, i.e. a graphical representation of the tradeoff between (1-Acc^-) and Acc^+ for every possible cutoff.

Table 4 shows that procedures based on data treated for imbalance improve the prediction accuracy of the minority class, and have favourable performance. All the classifiers tend to be biased towards the majority class, but the used approach SMOTE-based oversampling is efficient in reducing the extremely imbalanced data problem.

Table 4 Performance comparison derived from the test set for the default (0.5) and alternate cutoffs, derived using the ROC curve

Methods	Acc$^+$	Acc$^-$	Precision	F-measure	G-mean	Weighted accuracy
RF cutoff = 0.056	0.753	0.779	0.176	0.286	0.766	0.766
RF cutoff = 0.3	0.208	0.996	0.774	0.328	0.455	0.602
RF cutoff = 0.5	0.082	1.000	1.000	0.152	0.287	0.541
TB cutoff = 0.07	0.706	0.708	0.132	0.222	0.707	0.707
TB cutoff = 0.3	0.364	0.970	0.431	0.394	0.594	0.667
TB cutoff = 0.5	0.242	0.996	0.800	0.372	0.491	0.619
BRF cutoff = 0.218	0.797	0.737	0.160	0.266	0.766	0.767
BRF cutoff = 0.3	0.645	0.866	0.232	0.342	0.747	0.756
BRF cutoff = 0.5	0.307	0.980	0.490	0.378	0.549	0.644
BTB cutoff = 0.25	0.740	0.734	0.149	0.248	0.737	0.737
BTB cutoff = 0.3	0.671	0.815	0.185	0.290	0.739	0.743
BTB cutoff = 0.5	0.424	0.944	0.321	0.366	0.633	0.684

3.5 Results

Balanced Random Forest (BRF) variable importance plot is reported in Fig. 1, while Fig. 2 shows the variable importance plot based on the univariate ROC curves. Variables are ranked in terms of importance, with variables of highest importance at the top, in classifying between problem gamblers and non-problem gamblers. The ranked list of variables displays the importance of each variable in classifying data. The figures show the top 20 variables in importance of classification from a total of 44.

In line with the literature (Dowling et al. 2017), the male gender is a crucial factor. In other words, male teenagers have a higher risk of becoming problem gamblers than their female counterpart. Besides, the social context plays a fundamental role: the presence of friends or relatives who are frequent players is a significant risk factor for gambling in adolescence, as well as the perceived influence of friends towards the engagement in risky behaviours in general, and in particular towards

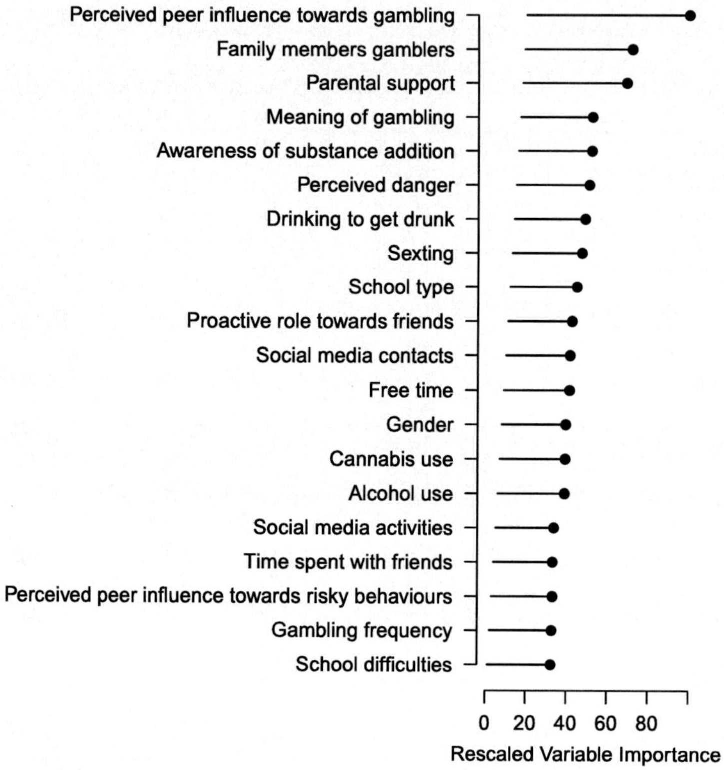

Fig. 1 Variable importance plots based on BRF model fit results (Top 20). The importance values are in percentage

gambling. Furthermore, the places where teenagers spend their free time, what they do on social media, the amount of time spent with friends and the number of social media contacts affect adolescents' tendency to gamble. On the cognitive side, some noteworthy factors emerged from the analysis: the meaning attributed to the concept of gambling, be it a positive idea of amusement and profit, rather than a negative one of illness and risk; the awareness that some behaviours and substances can cause addiction, and the consequent perceived danger and a proactive disposition to help friends exposed to these same dangers (e.g. when a friend gambles excessively). In addition, some risky behaviours are important risk factors, including drinking alcohol regularly, getting drunk on a regular basis or sending sexual images (sexting). Finally, it seems that the school type is an important variable to consider when examining gambling behaviours.

The variable importance measures based on model information are more informative than those based on ROC curve, as a model-based approach is able to incorporate the correlation structure between the features into the importance calculation. The BRF results highlighted the role of parental support, defined as the willingness to

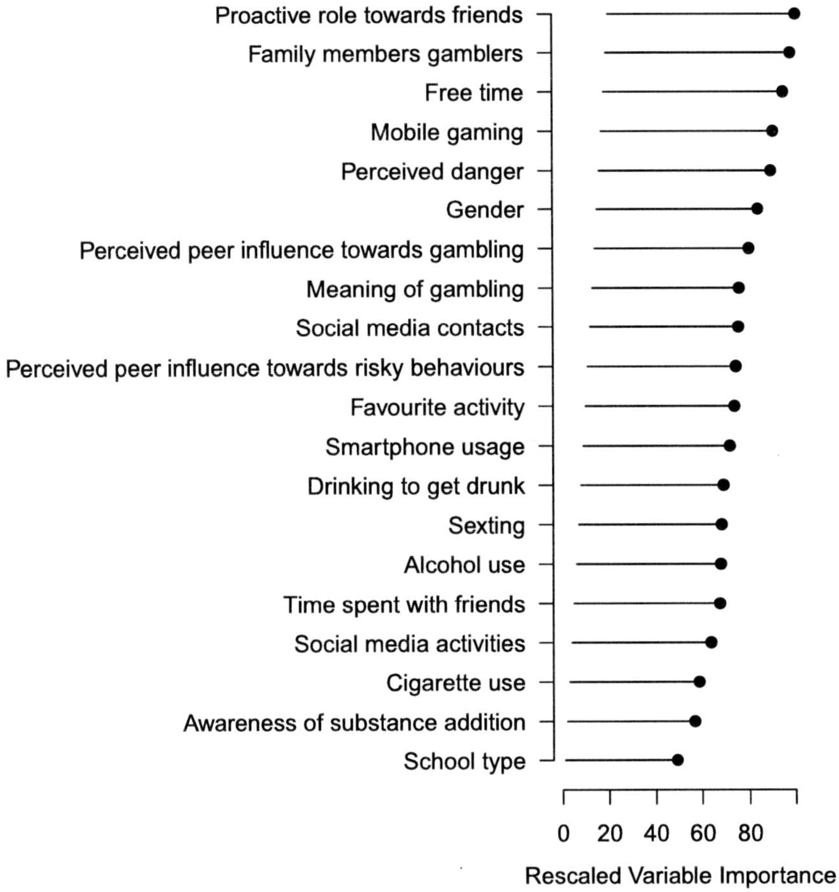

Fig. 2 Variable importance plots based on ROC Method (Top 20). The importance values are in percentage

tell parents about their problems because of the expectation of being understood and getting advice, as a protective factor against gambling problems. Furthermore, cannabis use and school difficulties may be considered as risk factors.

3.6 Conclusions

In conclusion, this study predicted the gambling problems of adolescents through a BRF algorithm for addressing class imbalance learning problem. Several risk and protective factors were entered in the model, and their relative importance was detected.

Although this study may help shed light on the understanding of gambling activities in adolescence, some cautionary remarks are needed. Firstly, the sample is not representative of the entire population of Lombard adolescents, even though the high number of respondents allows us to grasp a noteworthy signal. Secondly, we exclusively considered the behaviour as reported by adolescents, in terms of frequency of play and money spent, as a reference point to discriminate between problem gamblers and non-problem gamblers. Future studies should replicate these findings while considering a more sophisticated indicator of problem gambling, such as the South Oaks Gambling Screen Revised for Adolescents (Colasante et al. 2014). Finally, this study represents the first step of a more elaborate research project designed to deepen the knowledge of the complex gambling phenomenon in adolescence, highlighting the perceptions and motivations of adolescents as well as the various risk factors associated with the frequency of play. Future developments will be oriented towards a theoretical conceptualisation necessary for the design of intervention programs (Keen et al. 2017), to reduce or prevent gambling problems among adolescents.

Acknowledgments The authors warmly thank "Casa del Giovane di Pavia" and "Fondazione Exodus" for providing the survey data and for their useful comments on data collection.

Author Contributions L. Benedan discussed the results and wrote the paper; G. S. Monti performed the statistical analyses and contributed to paper writing and editing. The authors equally contributed to the present work.

References

American Psychiatric Association (2013). *Diagnostic and statistical manual of mental disorders (DSM-5®)*. American Psychiatric Pub.

Blagus, R., & Lusa, L. (2013). SMOTE for high-dimensional class-imbalanced data. *BMC Bioinformatics, 14*(106).

Blinn-Pike, L., Worthy, S. L., & Jonkman, J. N. (2010). Adolescent gambling: A review of an emerging field of research. *Journal of Adolescent Health, 47*(3), 223–236.

Breiman, L., Friedman, J., Olshen, R., & Stone, C. (1984). *Classification and regression trees*. Boca Raton, FL: CRC Press.

Calado, F., Alexandre, J., & Griffiths, M. D. (2017). Prevalence of adolescent problem gambling: A systematic review of recent research. *Journal of Gambling Studies, 33*(2), 397–424.

Cavalera, C., Bastiani, L., Gusmeroli, P., Fiocchi, A., Pagnini, F., Molinari, E., et al. (2018). Italian adult gambling behavior: At risk and problem gambler profiles. *Journal of Gambling Studies, 34*, 647–657.

Chawla, N. V., Bowyer, K. W., Hall, L. O., & Kegelmeyer, W. P. (2011). SMOTE: Synthetic minority oversampling technique. *JAIR, 16*, 321–357. arXiv:1106.1813.

Colasante, E., Gori, M., Bastiani, L., Scalese, M., Siciliano, V., & Molinaro, S. (2014). Italian adolescent gambling behaviour: Psychometric evaluation of the South Oaks Gambling Screen: Revised for adolescents (SOGS-RA) among a sample of Italian students. *Journal of Gambling Studies, 30*(4), 789–801.

Cook, S., Turner, N. E., & Ballon, B. (2015). Problem gambling among Ontario students: Associations with substance abuse, mental health problems, suicide attempts, and delinquent behaviours. *Journal of Gambling Studies, 31*, 1121–1134.

Dowling, N. A., Merkouris, S. S., Greenwood, C. J., Oldenhof, E., Toumbourou, J. W., & Youssef, G. J. (2017). Early risk and protective factors for problem gambling: A systematic review and meta-analysis of longitudinal studies. *Clinical Psychology Review*, *51*, 109–124.

Keen, B., Blaszczynski, A., & Anjoul, F. (2017). Systematic review of empirically evaluated school-based gambling education programs. *Journal of Gambling Studies*, *33*(1), 301–325.

R Core Team. (2019). *R: A language and environment for statistical computing*. Vienna, Austria: R Foundation for Statistical Computing.

Municipal Managers in Italy: Skills, Training Requirements and Related Critical Aspects

Mario Bolzan, Giovanna Boccuzzo, and Marco Marozzi

Abstract Public Administration in Italy has been experiencing noteworthy transformations necessary to meet citizens' requirements. Public managers must apply such transformations and so their skills should be redefined. The aim of this article is to identify the skills that public managers should have for a good administration of the local community and to rank them in order of importance. A survey was administered to a sample of public managers operating in the Veneto region of Italy. Then skills and training requirements underlined by the same managers have been analysed. The findings indicate that a rich set of 26 skills is required. Good teamwork, proactive behaviour and authoritativeness are the most important skills for municipal managers, whereas specific and technical knowledge does not characterize the role of managers. This is particularly true in large municipalities, in which internal structures are complex and external relationships even more so.

1 Introduction

The reform of the Public Administration (PA) in Italy is an important and urgent topic, especially in light of Italy's lack of competitiveness. The World Economic Forum in 2018 ranked Italy 31st for competitiveness, whereas other European countries like Germany, Switzerland, Netherlands, United Kingdom, Sweden and Denmark are in the top ten (www.weforum.org/issues/global-competitiveness); in some rankings the

M. Bolzan · G. Boccuzzo
Department of Statistical Sciences, University of Padua, Via Cesare Battisti, 241/243, 35121 Padova, Italy
e-mail: mario.bolzan@unipd.it

G. Boccuzzo
e-mail: boccuzzo@stat.unipd.it

M. Marozzi (✉)
Department of Environmental Sciences, Informatics and Statistics, Ca' Foscari University of Venice, Via Torino 155, 30170 Venezia Mestre, Italy
e-mail: marco.marozzi@unive.it

P. Mariani and M. Zenga (eds.), *Data Science and Social Research II*,
Studies in Classification, Data Analysis, and Knowledge Organization,
https://doi.org/10.1007/978-3-030-51222-4_6

Italian PA is even ranked below number 100. These rankings do not account for the significant differences between the North and South of Italy: the North being at the top of Europe, while the South trails at the bottom.

The Italian PA needs a deep process of reform, as requested a long time ago by the so-called 'Bassanini' law (law number 59/1997 as amended and 265/1999). The aim of this law is to simplify the administrative and bureaucratic rules and procedures (Oecd 1995).

The latest reform initiatives of the Italian PA (including Legislative Decree no. 150 of 27/10/2009, also called the Brunetta Decree, after the then PA Ministry) are in line with the privatization models, which have inspired PA reforms in other countries over the last twenty years. They follow the approach known as the New Public Management (Pollitt and Bouckaert 2004), which aims to integrate several principles regarding how the public sector should be managed and reformed using the business sector as a model (Granrusten 2015). A key-point is the modification of the public manager's role: from municipal administrator to area manager. Indeed, 'Bassanini' law maintains the central coordination, but decentralizes many tasks at the local level, where managers can pander to citizens' needs (Bell and Hindmoor 2009; Hess and Adams 2002; Maesschalck 2004; Virtanen 2000; Walker 2004, 2006; Weiss and Piderit 1999; Wright and Pandey 2011; Wright 2011).

In applying the above-mentioned Decree no. 150 of 2009, the PA is obliged to focus on enhancement of the professional skills of all employees using a widely shared system of assessment. Professional skills include knowledge, abilities, and personal characteristics in relation to a set of tasks or functions. This set of competences forms the link between the individual dimension and the expectations of the institution. The development of focus on competencies in public offices is essential for effectively serving a culturally diverse population in the modern era. Providing culturally appropriate and accessible services increases the significance and positive perception of a government office (Laud et al. 2016; Weimer and Zemrani 2017). The new legislative context requires public managers to operate with a new perspective, from a PA based on authority to a PA that must supply services to citizens, meaning that public managers also become area promoters, that is, they should retrieve resources to invest in the community. This new perspective requires new skills (Bouckaert and Halligan 2008; Boyne and Chen 2007; Brodkin 2011; Hood and Lodge 2004; Walker et al. 2011).

The competences of municipal managers aim at improving the performance of both the PA and municipalities and should consequently be developed. This point is not sufficiently addressed in Italy, although the debate is ongoing (Boyne 2010; CEDEFOP 2008; Denhardt and Denhardt 2009; De Graaf 2011; Wright 2011; Kim 2014).

In 2011 a survey was designed and carried out in collaboration with the Italian Association of Municipalities (Veneto section), to assess the perception and expectations of the role of municipalities with regard to their local area and community, and to redefine the profile of municipal managers (Bolzan 2010). A representative (by dimension of the cities of Veneto Region) sample of 30 directors and office managers were interviewed through a Delphi survey as 'privileged witnesses', in order to ascertain managers' profiles in relation to their skills and the role of municipality managers, according to their required competences and the mission of the municipality itself.

In this article, we address the following questions:

(a) What is the main mission of the local authority today?
(b) What are the more and less important skills that allow public managers to rule as the best local authority?
(c) What are the training requirements expressed by the public managers?

To address question (b) the Non-Parametric Combination method is used. It is a three-step method that can be also used for ranking combination and preference pooling. It is a very general method that comprises as a particular case the common method of averaging subject scores.

The remainder of the paper is structured as follows. Section 2 reviews the literature and gives the list of skills. Section 3 presents the survey and Sect. 4 gives the ranking of skills required for the public managers and training requirements. Section 5 discusses the most important results. Section 6 makes some concluding remarks and points for further research.

2 Literature Review

The topic of skills was introduced in the literature by David McClelland in the seventies (Draganidis and Mentzas 2006). At least 12 different definitions of skill are listed including behaviour, effectiveness, capabilities, successful performance, and knowledge. All twelve definitions describe a dynamic and multi-dimensional concept (Bouckaert 2007; Bouckaert and Halligan 2008); they include both the outcome and the way to reach it. In the UK, the term 'competence' and the plural 'competences' were adopted to indicate the set of conditions and means to give a good professional performance (Horton et al. 2002). Such a set includes knowledge (what he/she knows), abilities (how to do), and personal traits (to be). An important definition comes from Boyatzis (1982), who defined competences as '… an underlying characteristic of an individual, which is causally related to effective or superior performance in a job or a situation …'. This definition replaces the concept of job description as the carrying out of designated tasks. The new concept puts emphasis on the relationship between competences and performance.

Noordegraaf (2000) stated that competent public managers are 'professional sense-makers' who know how to perceive political cues, stimuli and triggers, which they can relate to new or existing issues (interpretative competencies); they solve problems and go ahead with policies (textual competencies). Essentially, competent public managers can manage and adapt administrative structures within rules and frameworks that are inherently ambiguous and within political situations that are inherently unstable, fluid coalitions. An important point is the context, situational or universal, of a single competence: there are universal collaborative competency dimensions *and* context matters in terms of application and interpretation (Getha-Taylor et al. 2016).

Virtanen (2000) demonstrates how the challenges of new public management have changed the expected value qualifications of public managers and caused tensions in their commitments. He constructs his own framework of four competence areas for public managers, which includes task, professionalism (itself subdivided into substantive and administrative professionalism), political and ethical competencies.

The approach based on competence models may be associated with a horizontal view of the organization structured by degree of responsibility or hierarchy (executives, head of operational units, employees), or with vertical views of the professional roles associated with the sector/service. In the first case, general skills are emphasized, while in the second, the technical-specialized competences of the various services are highlighted, as well as general ones.

In Italy, Cerase (2002) has extensively analysed the topic of skills based on interviews with seven managers and two-hundred employers in local offices of South Italy. The interviews use a simplified version of the O*Net questionnaire (http://www.onetcenter.org/questionnaires.html) adapted by the Italian Ministry of Labour. In his work, Cerase highlights a new concept of governance, based on the need to follow the collective interest under several internal and external constraints.

Competence is the merging of three aspects: knowledge, ability (to do) and personal traits (to be). Knowledge moves from basic to specific and complex arguments. Ability moves from basic cognitive capabilities to advanced and specialized skills; the ability to synthesize and evaluate, critical thinking, problem-solving and the management of human resources and trade-union relations (Dunleavy and Hood 1994; Virtanen 2000). Clearly, the boundaries among the three kinds of competences are weak, and sometimes a competence is the result of personal characteristics and a learning process. The personal traits refer to a sense of responsibility, autonomy, authority, and innovation in the development of ideas and new processes. The competences of municipal managers aim at improving the performance of both the PA and municipalities, and should consequently be developed. This point is not sufficiently developed in Italy, although the debate is ongoing (Boyne 2010; CEDEFOP 2008; Denhardt and Denhardt 2009; De Graaf 2011; Wright 2011; Bowman et al. 2016).

3 The Survey

The first step was a Delphi survey where the opinions of the municipal directors of 30 municipalities were collected to construct a list of 26 skills and related training requirements. A questionnaire was then prepared and presented to a sample of municipal directors from the 581 municipalities of the seven provinces of the Veneto. The sample size was 193. This questionnaire was split into two parts: firstly, the director was asked to assess the importance of possessing the qualities indicated by each dimension; secondly the director was asked to express an opinion in relation to the usefulness of investing in each of the dimensions. The list of competencies is

1. Good knowledge of the aims of the local authority for which the manager works.
2. Having a mind suited to government.
3. Having a sense of duty.
4. Ability to build teams and integrate others' skills.
5. Ability to communicate with the local population.
6. Ability to evaluate situations case by case, not ideologically.
7. Ability to enhance acquired knowledge.
8. Ability to give reasons for choices.
9. Ability to inspire trust.
10. Ability to interpret the local area's needs and resources.
11. Decision-making ability.
12. Ability for conflict management.
13. Ability to motivate staff.
14. Ability to obtain results from work colleagues.
15. Ability to organize planning.
16. Ability to communicate with the political world.
17. Ability for management control.
18. Sense of accountability.
19. Having a mind suited to administration.
20. Authoritative, rather than authoritarian, sense of leadership.
21. Ability to work independently of political influence.
22. Basic knowledge of cross-sector themes (e.g. informatics, statistics, quality control).
23. Creative ability (open to innovation, ability to see solutions).
24. Knowledge of administrative procedures.
25. Loyalty in relationships.
26. Technical know-how linked to role specificity.

A questionnaire, presented by e-mail with telephone follow-up, was composed of the following sections:

- Personal data: gender, level of education, work role, years of work (seniority);
- Municipal aims: respondents were asked to indicate to what extent (on a scale from 0 to 100) each of the municipality's three missions (distributor, mediator, promoter) should contribute towards characterizing its ideal identity, and to what extent (on a scale from 0 to 100) each mission currently characterizes that identity;
- Competences: respondents were first asked to assess the importance of possessing the competences indicated by the 26 dimensions, on a scale of 1 (of little importance) to 10 (very important). Then, using the same scale, they were asked to express their opinions regarding the usefulness of investing in each of the 26 dimensions in order to improve their professional profiles;
- Proposals for future actions to develop the role of municipal managers.

Regarding local authorities, the three significant dimensions or missions are identified here as

- Distributor of services: The municipality is the institutional focal point of the state's presence, the 'front office' of public authority. It must interpret local demand, meet primary needs, and give clear-cut answers to residents as regards housing, employment, security and facilities for children and the elderly. Priority is given to fundamental services such as schools and public safety, and the following are identified as good practice: provision of multipurpose offices; computerization of procedures; internal reorganization and restructuring, and equal opportunity offices, with projects aimed at immigrants and citizens of the new EU countries.
- Local mediator: The municipality mediates various needs and interests; it knows and analyses the problems of the community. Reasoning that anticipates needs must be developed and be part of a network, not isolated. The following is identified as good practice: The creation of Urban Transformation Companies and of other participating organizations, and the creation of unions.
- Promoter of local development: The municipality must be aware of the potential and resources of an area, in order to develop and strengthen them. The local authority's new role is that of interpreting local phenomena, choosing areas for intervention, and strategic planning. The importance of the principle of residents' participation in authority choices must be re-stated in order to create public awareness. The following are identified as good practice: Workgroups composed of families to promote tangible proposals for social policies; United Nations 'Agenda 21' at the local level; and discussion forums to decide upon actions of public interest and direct investment of resources.

The characteristics of the sample are shown in Table 1. Most of the managers were men (68.7%, vs. 31.3% of women). Men's seniority is higher than that of women (85.8% men vs. 74.1%). Women entered top management later, and their numbers will probably increase in the next few years. Managers with higher seniority work in medium-sized and large municipalities, in which the complexity of their position requires experience. In these cases, more time is probably necessary for career advancement. Lastly, most of the managers have university degrees (84.8%). In the smaller municipalities, 77% are graduates, whereas in large ones this percentage rises to 95%.

Table 2 shows that the main missions of municipalities are, first, that of distributor for at least half of manager, and then promoter and mediator, independently of the demographic dimension of municipalities; contrary to managers' wishes, because they hope for an increased role as promoter at the expense of that of distributor. Note that medium-sized and large municipalities have more hopes of a role as promoter.

Table 1 Characteristics of sample in % (193 municipalities)

Characteristics of managers	Size of municipality (population)			
	≤5000	5000–10000	>10000	All
	(46.7%)	(29.7%)	(23.6%)	
Gender				
M	70.2	55.6	82.9	68.7
F	29.8	44.4	17.1	31.3
Role				
Director	77.4	74.1	95.2	80.6
Office Manager	22.6	25.9	4.8	19.4
Seniority				
<8 years	27.6	7.7	10.8	17.6
≥8 years	72.4	92.3	89.2	82.4
Level of education				
Diploma	22.9	11.1	4.9	15.2
Degree	77.1	88.9	95.1	84.8

Table 2 Percentage distribution of mean weights, present and desired, assigned to three roles of municipalities, by size of municipality

		Present	Desired	Desired-present (% Variation) ([a])
≤5000 inhab.	Distributor	50.2	44.7	−11.0
	Mediator	22.3	21.0	−5.8
	Promoter	27.7	33.5	20.9
5000–10000 inhab.	Distributor	52.3	41.8	−20.1
	Mediator	22.0	22.0	0.0
	Promoter	26.2	36.6	39.7
≥10000 inhab.	Distributor	50.8	40.6	−20.1
	Mediator	24.6	22.9	−6.9
	Promoter	24.6	36.5	48.4
Total	Distributor	50.6	42.7	−15.6
	Mediator	23.0	21.9	−4.8
	Promoter	26.6	35.1	32.0

[a] [(Desired-Present)/Present] ∗ 100

Note Percentage sums among three kinds of mission do not total 100, because some respondents only gave percentages referring to one or two missions, not all three

4 Method for Constructing Importance Rankings

In this section, we propose a method for constructing importance rankings according to evaluations from a sample of experts (in our case, municipal managers) and generally expressed as scores (from 1 to 10) regarding several aspects representing partial dimensions of a given phenomenon of interest. There are 26 dimensions describing the overall range of competences.

Arithmetic means (weighted or otherwise) are mainly used for pooling preference ratings. More general methods have been proposed, see Lago and Pesarin (2000). In particular, the non-parametric combination of rankings has been shown in simulation studies to perform well.

Let us consider n subjects, who are asked to rate each of M dimensions on a scale from 1 to 10. The problem is how to obtain this ranking, i.e. how to pool subject preferences. X_{mi} is the rate of dimension m given by subject i, $i = 1, \ldots, n$. We assume that, if $X_{mi} > X_{m'i}$, then subject i rates dimension m better than dimension m'. In the literature, this problem is usually solved by averaging subject ratings $\bar{X}_m = \sum_{i=1}^{n} \frac{X_{mi}}{n}$, $m = 1, \ldots, M$ and dimension $\tilde{m}$ are such that $\bar{X}_{\tilde{m}} = \max(\bar{X}_1, \ldots, \bar{X}_M)$ is the best dimension with first rank position, dimension $\hat{m}$ is such that $\bar{X}_{\hat{m}} = \max_{\{i=1,\ldots,M, i \neq \tilde{m}\}}(\bar{X}_i)$ is the dimension with the second rank position and so on. For the sake of simplicity, it is assumed that there are no ties in ranking positions.

An alternative way of pooling preferences is the non-parametric combination ranking method, consisting of three steps (Lago and Pesarin 2000). In the first step, a score for dimension m is computed as follows:

$$\eta_{mi} = \frac{\#(X_{mi} \geq X_{m'i}) + 0.5}{M + 1}, \tag{1}$$

where $\#(X_{mi} \geq X_{m'i})$ indicates the rank transformation of X_{mi}. This step is repeated for each subject i and dimension m. Values of 0.5 and 1 are added to the relative rank transformation $\#(X_{mi} \geq X_{m'i})/M$ of X_{mi}, respectively to the numerator and denominator, to obtain η_{mi}, varying in the open interval (0, 1). The reason for these corrections is merely computational, in order to avoid later numerical problems with logarithmic transformations. Note that η_{mi} scores are one-to-one increasingly related to ranks $\#(X_{mi} \geq X_{m'i})$. Using η_{mi} after the first step makes it straightforward to obtain a (partial) ranking of the M dimensions for each subject, but it is the overall dimension rank which is of interest.

In the second step, subjects' scores assigned to dimension m are combined as follows:

$$C_m = -\sum_{i=1}^{n} \log(1 - \eta_{mi}). \tag{2}$$

This step is repeated for the remaining $M - 1$ dimensions, and a non-parametric combination of subjects' scores is made.

In the last step, the overall ranking for dimension m is computed as $R_m = \#(C_m \geq X_{m'})$. Of course, dimension $\tilde{m}$ with $R_{\tilde{m}} = M$ is the first rank position dimension, $\hat{m}$ with $R_{\hat{m}} = M - 1$ is the second and so on.

5 Discussion

Table 3 presents the results, more precisely it lists the ranking of the competences, stratified by municipality size and shows some differences among the three kinds of municipalities. *Ability to motivate staff* is the most important skill for managers in small and medium-sized municipalities, whereas *Ability to build teams and integrate others' skills* is the most important in the larger ones. In small and medium-sized municipalities, personnel numbers are lower and employees usually have several duties, often routine tasks. Consequently, it is important to motivate work colleagues/peers and also offer innovative projects, in order to avoid an excess of repetitive work. Conversely, in large municipalities, there are many employees and roles are better specified. There is a risk of thinking of everything as being in watertight compartments, whereas integration of skills and communications among colleagues are crucial.

Ability to inspire trust is considered very important only in large municipalities, and *Ability to communicate with the local population* increases in importance with the demographic size of the community. In PA, communications with local residents and stakeholders are always on the increase, and this competence is more difficult in large municipalities, where direct contacts among residents and managers are rarely feasible. The 'distance' between residents and stakeholders may induce biased spread of information. Working in PA requires competence of relationships in PA, but also outside it. Communications between PA and residents and participation of interested stakeholders are necessary. Communications with residents require listening skills, established with the institution of URP, *Offices for Public Relations with Residents* (law no. 150 of 2000). Good communications should evolve into participation and collaboration, aimed at caring for the community as a whole.

Good knowledge of the aims of the local authority for which the manager works is always important: an open, critical mind is required, able to implement operations and resources. This is related to *Decision-making ability*, third in importance for medium-sized and large municipalities and sixth for small ones: managers should be able to identify the aims of the local authority and make the correct decisions for good administrative functioning. The larger the municipality, the more complex the functioning, and thus the ability to make clear-cut decisions becomes more necessary.

At the bottom of the list we find *Having a mind suited to administration*: this is in contrast with the previous public image of municipal managers, when the ability for administrative paperwork was considered one of the most important skills.

Basic knowledge of cross-sector themes (e.g. computers, statistics, quality control) also comes near the end of the ranking, probably because it is considered more suitable for office staff. The topics of quality, accessibility and the proper use of

Table 3 Ranking of importance of competences, overall and by size of municipality

	ALL	Size of municipality		
		<5000	5000–10000	≥10000
1	Ability to motivate staff	Ability to motivate staff	Ability to motivate staff	Ability to build teams and integrate others' skills
2	Ability to build teams and integrate others' skills	Loyalty in relationships	Authoritative, rather than authoritarian, sense of leadership	Ability to motivate staff
3	Decision-making ability	Good knowledge of aims of local authority for which manager works	Decision-making ability	Decision-making ability
4	Good knowledge of aims of local authority for which manager works	Ability to build teams and integrate others' skills	Good knowledge of aims of local authority for which manager works	Good knowledge of aims of local authority for which manager works
5	Loyalty in relationships	Ability to obtain results from work colleagues	Loyal in relationships	Loyalty in relationships
6	Ability to obtain results from work colleagues	Decision-making ability	Ability to build teams and integrate others' skills	Ability to inspire trust
7	Authoritative, rather than authoritarian, sense of leadership	Authoritative leader, not authoritarian, sense of leadership	A sense of duty	Ability to obtain results from work colleagues
8	Sense of duty	Sense of duty	Ability to obtain results from work colleagues	Ability for conflict management
9	Ability to organize planning	Ability to inspire trust	Ability to organize planning	Authoritative, rather than authoritarian, sense of leadership
10	Ability to inspire trust	Ability to organize planning	Knowledge of administrative procedures	A sense of duty
11	Knowledge of administrative procedures	Knowledge of administrative procedures	Creative ability (open to innovation, ability to see solutions)	Ability to organize planning
12	Ability for conflict management	Ability to give reasons for choices	Ability to inspire trust	Ability to give reasons for choices
13	Creative ability (open to innovation, ability to see solutions)	Ability for conflict management	Technical know-how linked to role specificity	Creative ability (open to innovation, ability to see solutions)
14	Ability to give reasons for choices	Creative ability (open to innovation, ability to see solutions)	Ability to interpret the local area's needs and resources	Ability to communicate with the local population

(continued)

Table 3 (continued)

	ALL	Size of municipality		
		<5000	5000–10000	≥10000
15	Technical know-how linked to role specificity	Technical know-how linked to role specificity	Ability to evaluate situations case by case, not ideologically	Ability to enhance acquired knowledge
16	Ability to evaluate situations case by case, not ideologically	Ability to evaluate situations case by case, not ideologically	Ability to give reasons for choices	Sense of accountability
17	Ability to interpret local area's needs and resources	Ability to enhance acquired knowledge	Ability to work independently of political influence	Ability to evaluate situations case by case, not ideologically
18	Ability to enhance acquired knowledge	Ability to interpret the local area's needs and resources	Ability to communicate with the local population	Technical know-how linked to role specificity
19	Ability to communicate with local population	Ability to work independently of political influence	Ability for conflict management	Ability to interpret the local area's needs and resources
20	Ability to communicate with the political world	Ability to communicate with the political world	Ability to enhance acquired knowledge	Ability to communicate with the political world
21	Sense of accountability	Ability to communicate with the local population	Ability for management control	Knowledge of administrative procedures
22	Ability to work independently of political influence	Sense of accountability	Ability to communicate with the political world	Ability for management control
23	Ability for management control	Ability for management control	Sense of accountability	Ability to work independently of political influence
24	Having a mind suited to government	Having a mind suited to government	Having a mind suited to government	Having a mind suited to government
25	Basic knowledge of cross-sector themes (e.g. computers, statistics, quality control)	Basic knowledge of cross-sector themes (e.g. computers, statistics, quality control)	Basic knowledge of cross-sector themes (e.g. computers, statistics, quality control)	Basic knowledge of cross-sector themes (e.g. computers, statistics, quality control)
26	Having a mind suited to administration	Having a mind suited to administration	Having a mind suited to administration	Having a mind suited to administration

data are critical points which have not yet been faced by many local authorities. Research by IBM (The Economist, 2010) estimates that more than 50% of managers do not consider that the datasets used for their decision-making processes are reliable.

This result contradicts a global trend where these skills are considered central to understand and manage both public and private institutions, both local and not. In fact, technological advancements in high-throughput technologies allow to produce an enormous amount of data in most fields, the so-called 'big data'. This trend highlights the importance of being able to deal adequately with such quantities: several years ago it has been estimated that, by 2018, the United States alone may require 140,000 to 190,000 more people with good analytical skills, as well as 1.5 million managers and analysts with the know-how to use 'big data' analysis to make effective decisions (Kinsey 2011). Economy 4.0 is changing the job market and the most requested job according to both Harvard Business Review and Forbes Magazine is the data scientist. Harvard Business Review (2012) 'Data Scientist: The Sexiest Job of the twenty-fir Century' and Forbes 29/01/2018 'Data Scientist Is the Best Job In America According Glassdoor's 2018 Rankings'. *Technical know-how linked to role specificity* plots below the median. This means that administrative/practical competences are clearly viewed as less important than the overall view of the mission of the municipality and personnel management. This is clearly related to the reduced role as distributor of services, desired by several managers (see Tables 1 and 2).

Table 4 shows the ranking of training requirements that reflects to a large extent that of skills. Clearly, investment in basic or technical competences (*Basic knowledge of cross-sector themes (e.g. computers, statistics, quality), Ability for management control and Technical know-how linked to role specificity*) is not required of managers. These competences are necessary at lower levels and are usually acquired during training, see Getha-Taylor et al. (2016).

Table 4 shows that relational and personnel management skills require more training, although skills necessary for increasing effectiveness, such as *Ability to obtain results from work colleagues* (especially in medium-sized and large municipalities) are also important. This awareness is an important point of the ongoing transition in the PA: outcomes are achieved by tangible results, not by administrative acts. More investment in training is required to achieve a mentality focused on results. Instead, *Having a mind suited to administration* does not appear to be important, and no training is required. The same applies to *Knowledge of administrative procedures*.

As interpreting the main dimensions of identified competences can help in understanding the profiles of skills, we analysed correlations among competences by exploratory factor analysis. Starting from a set of correlated variables (competences), the method can identify a small number of new variables (factors), calculated as a linear combination of the original variables. The aim is to capture the highest amount of variability of a complex phenomenon with a small number of factors. The importance of each original variable to the factor is called factorial weight; only the variables with the highest weights are kept in each factor. The results of factor analysis are shown in Table 5. We chose to maintain five factors, according to the amount of variability explained, and considered only weights higher than 0.4 in interpreting them.

Factor 1 was mainly identified by the following competences: *Ability to build teams and integrate skills* (factorial weight 0.956); *Good knowledge of the aims of the local authority for which the manager works* (0.701); *Ability for conflict management*

Table 4 Ranking of requirements for development of competences, by size of municipality

	ALL	Size of municipality		
		<5000	5000–10000	≥10000
1	Ability to motivate staff	Ability to motivate staff	Ability to obtain results from work colleagues	Ability to motivate staff
2	Ability to obtain results	Ability to organize planning	Ability to motivate staff	Ability to build teams and integrate others' skills
3	Ability to build teams and integrate others' skills	Ability to build teams and integrate others' skills	Ability to build teams and integrate others' skills	Ability to obtain results
4	Ability to organize planning	Decision-making ability	Decision-making ability	Ability to organize planning
5	Ability for conflict management	Able for management control	Ability for conflict management	Decision-making ability
6	Decision-making ability	Ability for conflict management	Ability to organize planning	Ability for conflict management
7	Good knowledge of aims of local authority for which manager works	Ability to obtain results from work colleagues	Ability to inspire trust	Authoritative, rather than authoritarian, sense of leadership
8	Authoritative, rather than authoritarian, sense of leadership	Authoritative, rather than authoritarian, sense of leadership	Ability to enhance acquired knowledge	Good knowledge of aims of local authority for which manager works
9	Creative ability (open to innovation, ability to see solutions)	Ability to interpret local area's needs and resources	Accountability	Creative ability (open to innovation, ability to see solutions)
10	Ability to give reasons for choices	Creative ability (open to innovation, ability to see solutions)	Authoritative, rather than authoritarian, sense of leadership	Ability to interpret local area's needs and resources
11	Ability to interpret local area's needs and resources	Good knowledge of aims of local authority for which manager works	Creative ability (open to innovation, ability to see solutions)	Ability to give reasons for choices
12	Ability to inspire trust	Ability to communicate with local population	Ability to interpret local area's needs and resources	Ability to enhance acquired knowledge
13	Ability to communicate with local population	Ability to enhance acquired knowledge	Ability for management control	Ability to communicate the local population
14	Ability to enhance acquired knowledge	Ability to give reasons for choices	Good knowledge of aims of local authority for which manager works	Ability for management control

(continued)

Table 4 (continued)

	ALL	Size of municipality		
		<5000	5000–10000	≥10000
15	Knowledge of administrative procedures	Technical know-how linked to role specificity	Having a sense of duty	Ability to inspire trust
16	Technical know-how linked to role specificity	Knowledge of administrative procedures	Ability to give reasons for choices	Technical know-how linked to role specificity
17	A sense of duty	Sense of accountability	Ability to communicate with the local population	Sense of accountability
18	Ability to evaluate situations case by case, not ideologically	Basic knowledge of cross-sector themes (e.g. computers, statistics, quality control)	Loyal in relationships	Knowledge of how administrative processes operate
19	Ability for management control	Ability to inspire trust	Ability to evaluate situations case by case, not ideologically	A sense of duty
20	Sense of accountability	Having a mind suited to government	Having a mind suited to government	Ability to evaluate situations case by case, not ideologically
21	Ability to communicate with the political world	Loyalty in relationships	Technical know-how linked to role specificity	Loyalty in relationships
22	Loyalty in relationships	Sense of duty	Knowledge of how administrative processes operate	Having a mind suited to government
23	Having a mind suited to government	Ability to communicate with the political world	Basic knowledge of cross-sector themes (e.g. computers, statistics, quality control)	Ability to communicate with the political world
24	Ability to work independently of political influence	Ability to evaluate situations case by case, not ideologically	Ability to work independently of political influence	Basic knowledge of cross-sector themes (e.g. computers, statistics, quality control)
25	Basic knowledge of cross-sector themes (e.g. computers, statistics, quality control)	Ability to work independently of political influence	Ability to communicate with the political world	Ability to work independently of political influence
26	Having a mind suited to administration	Having a mind suited to administration	Having a mind suited to administration	Having a mind suited to administration

Table 5 Factor weights associated with first five factors covering importance of competences (weights related to factor interpretation in bold type)

	Factor 1	Factor 2	Factor 3	Factor 4	Factor 5
Technical know-how linked to role specificity	0.013	0.055	−0.018	0.194	−0.006
Basic knowledge of cross-sector themes (e.g. computers, statistics, quality control)	−0.009	**0.655**	0.015	−0.253	−0.017
Knowledge of administrative procedures	−0.097	0.319	0.032	0.149	0.266
Good knowledge of aims of local authority for which manager works	**0.701**	0.324	−0.311	−0.059	0.113
Having a mind suited to administration	−0.072	0.207	**0.693**	−0.131	0.137
Ability to organize planning	**0.605**	0.155	0.097	−0.142	0.044
Ability for management control	0.285	0.126	0.308	−0.290	0.243
Sense of accountability	0.158	0.159	0.039	−0.016	**0.557**
Ability to communicate with the political world	0.041	−0.111	0.126	0.113	**0.755**
Ability to communicate with the local population	−0.055	−0.055	−0.047	0.327	**0.642**
Ability to motivate staff	**0.636**	−0.127	0.009	0.314	0.153
Ability to build teams and integrate skills	**0.956**	−0.110	−0.081	0.071	−0.003
Ability to enhance acquired knowledge	0.301	−0.011	0.292	0.043	0.423
Ability for conflict management	**0.641**	−0.058	−0.090	−0.052	0.409
Decision-making ability	0.306	−0.091	**0.482**	0.154	0.053
Having a mind suited to government	−0.041	**0.594**	0.350	0.023	0.128
Loyalty in relationships	−0.085	0.042	−0.091	**0.713**	0.396
Ability to inspire trust	0.251	0.142	0.060	**0.645**	0.070
A sense of duty	0.068	−0.091	0.131	**0.756**	0.029
Ability to work independently of political influence	−0.218	−0.038	**0.871**	0.089	0.053
Authoritative, rather than authoritarian, sense of leadership	0.180	0.123	**0.464**	0.255	0.004
Ability to obtain results from work colleagues	**0.415**	0.322	−0.049	0.289	−0.007
Ability to give reasons for choices	**0.578**	−0.097	0.386	0.143	−0.184
Ability to evaluate situations case by case, not ideologically	**0.463**	0.066	0.319	0.227	−0.224
Creative ability (open to innovation, ability to see solutions)	0.026	**0.782**	−0.054	0.259	−0.200
Ability to interpret local areas needs and resources	0.001	**0.744**	0.063	0.117	0.023

(0.641); *Ability to organize planning* (0.605); *Ability to give reasons for choices* (0.578); *Ability to evaluate situations case by case, not ideologically* (0.463) and *Ability to obtain results* (0.415). Factor 1 is very general and we call it '*Personnel management skills*'. A municipal manager's role is clearly not self-referential, but requires multidisciplinarity and teamwork, and integrates the contributions of several colleagues working together. Managers should motivate their colleagues, emphasize their competences, and organize human resources in order to achieve set goals. They should work as a team and be able to solve any conflicts which may arise. This factor covers the 'internal' activity of managers.

Factor 2 was identified by the following competences: *Creative ability (open to innovation, ability to see solutions)* (0.782), *Ability to interpret the local area's needs and resources* (0.744), *Basic knowledge of cross-sector themes (e.g. computers, statistics, quality control)* (0.655) and a *Having a mind suited to government* (0.594). Creativity is combined with operational skills. This factor is mainly related to the '*Ability to find solutions*'.

Factor 3 was identified by *Ability to work independently of political influence* (0.871); *Having a mind suited to administration* (0.693); *Decision-making ability* (0.482) and *Authoritative leader, not authoritarian* (0.464). It represents the '*authoritative leader*', able to make decisions independently of political influence. Factor 4 is given by *Loyalty in relationships* (0.713), *Sense of duty* (0.756), and *Ability to inspire trust* (0.645) and describes the '*Work ethic*'. Factors 3 and 4 highlight aspects related to loyalty, a sense of duty, credibility and empathy. There is growing awareness of the fact that both residents and politicians are important interlocutors.

Lastly, factor 5 is given by *Ability to communicate with the political world* (0.755), *Ability to communicate with the local population* (0.642) and *Sense of accountability* (0.557). This is clearly the factor covering '*Relational skills*' at various levels and is applied to various actors: politicians, residents, etc. Unlike factor 1, factor 5 highlights the manager's role with respect to the outside world, whereas the more important factor 1 highlights the 'internal' role. Both roles are necessary, but the latter is still considered prevalent, although the aim of promoter is to emphasize the 'external' role.

To sum up, the five factors describe the following dimensions: Managerial skills, Problem-solving, Leadership, the Work Ethic and Relational Skills.

Each factor was treated as a response variable in a linear regression analysis, with the characteristics of managers and municipalities as explanatory variables. Our aim was to understand which characteristics were related to the highest values of the factors. The results highlighted only one significant aspect, the small demographic size of municipalities: those with fewer than 1500 inhabitants are important predictors for factors 1 (*Personnel management skills*), 2 (*Ability to find solutions*) and 4 (*the Work ethic*), and show how, in small communities, local authority is entrusted to managers who are active and aware of what the local population expects and wants. The mainstays of good local administration are the work of personnel (often few in number and with more than one role) and the need to be visible to the community through real solutions.

6 Conclusion

The first aspect which emerged from our research is that the role of municipal managers is complex, because they are required to interact not only with other municipal personnel but also with local residents and politicians. The relational context is composite, encompassing both the municipal world and its surrounding territory. A rich set of competences is required for this twofold role: up to 26. The need for relationships on several fronts places social dimensions as the most important aspects of municipal managers' work. A good team spirit and proactive behaviour, together with authoritativeness, are crucial skills, for both internal and external relations. Decision-making skills and a results-oriented attitude are definitely more important than *Having a mind suited to government*. Our results are consistent to the introduction of post-bureaucratic structures and the establishment of a more managerialist culture. A new type of public servant is emerging—the public manager (Weiss and Piderit 1999; Galanti 2014; Farnham and Horton 1996).

Specific and technical knowledge do not characterize the manager's role. Managers must develop relationships with actors which combine proper aims, duties and interests (i.e. private and public, individuals and groups, politicians and technical experts). This is mainly true large municipalities, in which the internal structure is particularly complex and external relationships more complex. The manager is the centre of the municipal mission and plays a crucial role for the development of an administration based on principles of efficiency and effectiveness (Santos and Passos 2013; Smith 2018, 2019).

In Italy, the figure of the manager is undergoing a transition: a new culture of responsibility and transparency is spreading among local authorities, and managers are being evaluated according to the results they produce. Consequently, the selection and evaluation of colleagues will henceforth be based on quantifiable outcomes.

Our analyses bring to light several questions, which we propose as dualisms:

- *Dualism 1: Technical or managerial profile?* How can be the two dimensions in the selection and training of managers be combined? Which weights should be assigned to technical and managerial skills? Our study shows that managerial competences are the most important, but specific and technical ones are taken for granted, because they develop during the first stages of a manager's career.
- *Dualism 2: Selection process for managers: public competition or personal contract?* Public competition offers guarantees regarding normative rules; personal contracts are based on a trust agreement. In the latter case, those who select managers are responsible for investing in a specific human resource.
- *Dualism 3: Generational turnover.* Should importance be given to experience, or to the tendency to change? Experience is always a resource which must be enhanced, but at the same time new human resources are necessary to extend a new culture.
- *Dualism 4: Specific or transversal training?* Training is a strategic action and must be properly planned. Knowledge of specific work fields is indispensable, but specific skills are not sufficient, because most problems can be solved with a new approach based on an overall vision of the community and its identity and mission.

The ability to learn becomes itself an essential competence, which must be shared at the various levels of the system. It stimulates relations and interactions which are useful for change and for solving problems encountered while pursuing expected results.

There are many possible training methods, more problem-oriented, which are different from traditional ones:

- training, work experience (stage).
- training sessions with politicians and municipal teams.
- meetings with managers from other administrations, sometimes at different levels (i.e. provincial).
- periodic briefings with colleagues and politicians.
- online information support.

Training of managers may also be enhanced by:

1. creating a shared vocabulary and tools matching the complexity and various dimensions which training in PA requires;
2. asking how training contributes to the improved skills and performances of individuals and of the system, thus demonstrating their effectiveness;
3. concentrating on the results of evaluations, understood not only as satisfaction but also as the outcome of learning, planning new activities, and identifying the most suitable training options.

In this scenario, the role played by universities may be substantial, but challenging: university teaching personnel should be able to transmit skills, not only knowledge. The training process should also be re-organized, in order to produce graduates with really useful sets of competences (Walker 2006; Demircioglu and Audretsch 2019).

References

Bell, S., & Hindmoor, A. (2009). Rethinking governance: The centrality of the state in modern society. *The Australian Journal of Public Administration, 69*(1), 103–112.

Bolzan, M. (2010). *Competenze e processi formativi per i dirigenti degli enti locali*. Padova, Italy: Cleup.

Bouckaert, G. (2007). Cultural characteristics from public management reforms worldwide. In *Cultural aspects for public management reforms* (pp. 29–64). Amsterdam: Elsevier.

Bouckaert, G., & Halligan, J. (2008). *Managing performance, international comparisons*. London: Routledge.

Bowman, J., West, J., Berman, M., & Van Wart, M. (2016). *The professional edge: Competencies in public service*.

Boyatzis, R. (1982). *The competent manager: A model for effective performance*. Wiley.

Boyne, G., & Chen, A. (2007). Performance targets and public service improvement. *Journal of Public Administration Research and Theory, 17*, 455–477.

Boyne, G. A. (2010). Performance management: Does it work? In *Public management and performance: Research directions*. Cambridge: Cambridge University Press.

Brodkin, E. Z. (2011). Putting street-level organizations first: New directions for social policy and management research. *Journal of Public Administration Research, 21*, i199–i201.

CEDEFOP. (2008). Future skills needs in Europe. Medium-term forecast: Synthesis report. Lussemburgo: Ufficio delle pubblicazioni ufficiali delle Comunita' Europee.

Cerase, F. (2002). The competencies required in public management: A case study in Italy. In *Competency management in the public sector: European variations on a theme*. Amsterdam: IOS Press.

De Graaf, G. (2011). The loyalties of top public administrators. *Journal of Public Administration Research, 21*, 285–306.

Demircioglu, M., & Audretsch, D. (2019). Public sector innovation: The effect of universities. *The Journal of Technology Transfer, 44*(2), 596–614.

Denhardt, K., & Denhardt, J. (2009). *Public administration: An action orientation*. Wadsworth Publishing Co. Inc.

Draganidis, F., & Mentzas, G. (2006). Competency based management: A review of systems and approaches. *Information Management & Computer Security, 14*(1), 51–64.

Dunleavy, P., & Hood, C. (1994). From old public administration to new public management. *Public Money & Management, 14*(3), 9–16.

Farnham, D., & Horton, A. (1996). Public managers and private managers: Towards a professional synthesis? In *New public managers in Europe: Public servants in transition* (pp. 26–52). London: Palgrave Macmillan.

Galanti, M. T. (2014). Beyond mayors and great men: Effectiveness, policy leadership and accountability in Italian local government. *Contemporary Italian Politics, 6*(2), 159–177.

Getha-Taylor, H., Blackmar, J., & Borry, E. (2016). Are competencies universal or situational? A state-level investigation of collaborative competencies. *Review of Public Personnel Administration (ROPPA), 36*(3), 306–320.

Granrusten, P. T. (2015). The freedom to choose and the legitimacy to lead. In *Thinking and Learning about Leadership*. Sydney: Community Child Care Cooperative.

Hess, M., & Adams, D. (2002). Knowing and skilling in contemporary public administration. *Australian Journal of Public Administration., 61*(4), 68–79.

Hood, C., & Lodge, M. (2004). Competency, bureaucracy, and public management reform: A comparative analysis. *Governance: An International Journal of Policy, Administration, and Institutions, 17*(3), 313–333.

Horton, S., Farnham, D., & Hondeghem, A. (2002). *Competency management in the public sector: European variations on a theme*. Amsterdam: IOS Press.

Kim, S. (2014). Assessing the influence of managerial coaching on employee outcomes. *Human Resource Development Quarterly, 25*(1), 59–85.

Lago, A., & Pesarin, F. (2000). Nonparametric combination of dependent rankings with application to the quality assessment of industrial products. *Metron, 58*, 39–52.

Laud, R., Arevalo, J., & Johnson, M. (2016). The changing nature of managerial skills, mindsets and roles: Advancing theory and relevancy for contemporary managers. *Journal of Management & Organization, 22*(4), 435–456.

Maesschalck, J. (2004). The impact of new public management reform on public servant' ethics: Toward a theory. *Public Administration, 82*, 465–489.

Mc Kinsey, G. I. (2011). Big data: The next frontier for innovation, competition and productivity. http://www.mckinsey.com/Insights/MGI/Research/Technology_and_Innovation/Big_data_The_next_frontier_for_innovation.

Noordegraaf, M. (2000). Professional sense-makers: Managerial competencies amidst ambiguity. *International Journal of Public Sector Management, 13*(4), 319–332.

OECD. (1995). Governance in transition: Public management reforms in OECD countries.

Pollitt, C., & Bouckaert, G. (2004). *Public management reform: A comparative analysis* (2nd ed.). Oxford: Oxford University Press.

Santos, C., & Passos, A. (2013). Team mental models, relationship conflict and effectiveness over time. *Team Performance Management, 19*(7–8), 363–385.

Smith, H. (2018). *Manager as coach: An exploratory study into the experience of managers dealing with team challenge*. Ph.D. Thesis, University of Chester.

Smith, H. (2019). Manager as coach characteristics for dealing with team challenge. *Journal of Work-Applied Management, 11*(2), 165–173.
Virtanen, T. (2000). Changing competences of public manager: Tension in commitment. *International Journal of Public Sector Management, 13*(4), 333–341.
Walker, R. (2004). *Innovation and organizational performance: A critical review and research agenda*. London: Advanced Institute for Management Research.
Walker, R. (2006). Innovation type and diffusion: An empirical analysis of local government. *Public Administration, 84*, 311–336.
Walker, R., Damanpour, F., & Devece, C. (2011). Management innovation and organizational performance: The mediating effect of performance management. *Journal of Public Administration Research, 21*, 367–386.
Weimer, N., & Zemrani, A. (2017). Assessing the level of cultural competencies in public organizations. *Public Administration Quarterly, 41*(2), 273–296.
Weiss, J., & Piderit, S. (1999). The value of mission statements in public agencies. *Journal of Public Administration Research, 9*, 193–223.
Wright, B. (2011). Public sector work motivation: Review of current literature and a revised conceptual model. *Journal of Public Administration Research, 11*, 529–586.
Wright, B., & Pandey, S. (2011). Public organizations and mission valence: When does mission matter? *Administration & Society, 43*(1), 22–44.

Attitudes Towards Immigrant Inclusion: A Look at the Spatial Disparities Across European Countries

Riccardo Borgoni, Antonella Carcagnì, Alessandra Michelangeli, and Federica Zaccagnini

Abstract This paper aims at investigating individuals' attitudes towards immigrants across European countries. In particular, we analyze how socio-demographic characteristics, altruism, and political preferences shape inclusive behavior towards immigrants. We use data from the European Value Study providing information at the individual level for 16 European countries, observed in 1999 and 2008. The results show relevant differences in individuals' attitudes across countries and a general tendency to be less inclusive in the more recent period.

1 Introduction

The economic strategy launched by the European Commission, known as Europe 2020, aims to promote a smart, sustainable, and inclusive growth. The latter is supposed to foster a high employment economy and enhance social and territorial cohesion in the Member States, while reducing the impact on the natural environment.

This paper investigates the integration and social inclusion of immigrants in European countries. The analysis is carried out looking at the everyday life local environment, i.e., the neighborhood. As pointed out by Reardon (2008, p. 502),

> the neighborhood is the immediate setting in which individuals may visit neighbors, shop, take their children to a playground, walk a dog, or push a child in a stroller.

R. Borgoni · A. Carcagnì · A. Michelangeli (✉) · F. Zaccagnini
DEMS, University of Milan-Bicocca, Piazza Dell'Ateneo Nuovo 1, 20126 Milan, Italy
e-mail: alessandra.michelangeli@unimib.it

R. Borgoni
e-mail: riccardo.borgoni@unimib.it

A. Carcagnì
e-mail: antonella.carcagni@unimib.it

F. Zaccagnini
e-mail: f.zaccagnini@campus.unimib.it

P. Mariani and M. Zenga (eds.), *Data Science and Social Research II*,
Studies in Classification, Data Analysis, and Knowledge Organization,
https://doi.org/10.1007/978-3-030-51222-4_7

In brief, the neighborhood is the place in which face-to-face interactions are more likely.

Several studies have investigated natives' attitudes towards interacting with immigrants within metropolitan areas (Cutler et al. 1999; Saiz 2003; Saiz and Wachter 2010; Accetturo et al. 2014; Borgoni et al. 2019). In the typical model, preferences are captured through residential choices that shape housing market dynamics. The pattern of housing prices determined by the presence of immigrants in the neighborhood reveals native's attitudes towards them. The problem of this approach is that it requires a detailed information about residential choice and the environment where the individual lives. This type of information is unavailable at the supranational level, for example, for European countries considered as a unified area.

In this paper, we rely on an alternative approach that does not require such a type of information. Individual attitudes towards immigrants are assessed through a survey covering a large number of European countries aiming at investigating ideas, beliefs, preferences, attitudes, values, and opinions of citizens all over Europe. In particular, our analysis considers the following survey question[1]:

Wouldn't you like to have immigrants/foreign workers as neighbors?

A negative answer to this question indicates a positive attitude towards immigrants, while a positive answer reveals a difficult integration between natives and foreign-born.

We estimate the probability to be tolerant towards immigrants as a function of socio-demographic factors, political preferences, and value systems. Our findings show important differences in the individual's behavior across countries and a general tendency of people to be less immigrant inclusive in the more recent period.

The rest of the paper is organized as follows. Section 2 describes data and variables. Section 3 discusses the empirical model. Section 4 presents the results. The last section concludes.

2 Data and Variables

Data are taken from European Value Study (EVS), which is a large-scale, cross-national, repeated cross-sectional survey providing information about the ideas, beliefs, preferences, attitudes, values, and opinions of citizens all over Europe. EVS covers a period of 36 years, running from 1981 to 2017. Every nine years, the survey is repeated in a variable number of countries. To the purpose of our analysis, we consider the third and fourth waves referred to 1999 and 2008, respectively. Unfortunately, the more recent wave fielded in 2017 does not provide information on the

[1]https://europeanvaluesstudy.eu/ (last accessed on January 2, 2020).

variables of interest presented below. We select 16 countries[2] that participated to both waves and have complete information on those variables we have chosen.

As mentioned in the Introduction, the assessment of individuals' attitudes towards immigrants relies on the following question:

Wouldn't you like to have immigrants/foreign workers as neighbors?

We consider two groups of variables that are expected to shape attitudes towards immigrants: (i) demographic variables (gender, age; education; marital status; number of children; employment status; country of residence); (ii) variables measuring the degree of altruism (preference for individual freedom or social equality), whether it is important that people are encouraged to learn at home tolerance and respect for other people, political preferences (individual's position on a left-right scale). Table 1 provides summary statistics of variables; Table 4 in Appendix sets out the list of variables with their definition.

Table 1 Summary statistics of variables, years

	1999		2008	
	Mean	Sd	Mean	Sd
Tolerant (*yes*)	0.9	0.3	0.91	0.29
Gender (*female*)	0.53	0.49	0.54	0.49
Age	45.46	16.83	48.71	17.64
Educational level				
Low	0.65	0.48	0.33	0.47
Medium	0.13	0.34	0.41	0.49
High	0.22	0.41	0.26	0.44
Marital status (*unmarried*)	0.43	0.49	0.48	0.49
Children	1.65	1.53	1.56	1.38
Employment (*employed*)	0.45	0.49	0.56	0.49
Political view	5.29	1.8	5.26	1.91
Educ tolerance (*yes*)	0.23	0.41	0.23	0.42
Equality and/or freedom				
Freedom	0.57	0.49	0.51	0.49
Equality	0.37	0.48	0.45	0.49
Equally important	0.06	0.23	0.04	0.19
Foreign-born	0.067	0.05	0.11	0.03

Note: Sample years 1999 and 2008. Columns 2 and 4 report the relative frequency of unitary values in the case of dichotomous variables

[2]The countries considered are the following: Austria, Belgium, Denmark, Estonia, Finland, France, Germany, Greece, Ireland, Italy, Latvia, Netherlands, Portugal, Spain, Sweden, United Kingdom.

About 53% of respondents are female and the mean age is 45 in 1999 and 48 in 2008. We consider adult respondents, aged 18 and above. Respondents are quite unevenly distributed across educational levels with a majority for the lower secondary educational level or less in 1999 and the upper secondary educational level in 2008. About 45% of respondents are employed in 1999: the average employment rate rises to 56% in 2008. About 23% of respondents state that it is important to encourage the own children to be tolerant and respect other people. For 57% of respondents in 1999 is more important freedom than equality. This percentage decreases to 51% in 2008. Finally the average percentage of foreign-born population is 6.7% in 1999 and rise to 11% in 2008.

Overall, we consider 16,023 respondents in 1999 and 14,649 respondents in 2008, distributed across 16 countries.

3 Empirical Model

The outcome variable is a binary variable denoted by y, and defined in the following way:

$$y = \begin{cases} 1 \text{ individual is tolerant} \\ 0 \text{ otherwise} \end{cases}$$

The individual is thought to have inclusive attitudes towards immigrants if he answers negatively to the question "I don't like people of different ethnicity as neighbors"; otherwise he does not show positive attitudes towards immigrants.

As it is well known, the standard linear regression model is not appropriate for such a type of dependent variable for several reasons (Greene 2018). First, the predicted probability can be below 0 or above 1. Second, the linear regression model assumes that errors are normally distributed. However, in the case of a binary dependent variable, the distribution of errors for a given independent variable has two mass points around 0 and 1 instead of a normal distribution. Third, the linear regression model assumes that errors have a constant variance. In the case of a binary dependent variable, it is possible to show that there is heteroscedasticity in the model. This leads to biased standard errors and the results of the hypothesis tests are possibly wrong.

In order to obtain reliable estimates, we move towards the Generalized Nonlinear models (Agresti 2002).

More specifically, we use a probit model that can be specified as follows:

$$Pr(y = 1|x) = \Phi(\boldsymbol{x}'\boldsymbol{\beta}) \tag{1}$$

where Φ is the standard cumulative normal distribution with mean 0 and variance 1; $\mathbf{x}$ is a vector of individual characteristics illustrated in Sect. 2 and that are expected to affect the probability to be tolerant; $\boldsymbol{\beta}$ represents a vector of estimated coefficients.

4 Results

Estimations results of the probit model introduced in Sect. 3 are shown in Table 2. Parameter estimates are obtained via Iteratively Reweighted Least Squares (IRLS) algorithm with weights obtained from the variance function of the assumed distribution (Amemiya 1985; Hoffmann 2004).[3] The covariates entered in the model by block, as shown in Table 1. Model I and model III include the group of country dummies and the demographic variables listed in Sect. 2 for the year 1999 and 2008, respectively; Model II and model IV add other variables at the individual-level providing information on altruism and political preferences for the year 1999 and 2008, respectively. Adding variables to the baseline model I and model III decreases the residual variability of around 11%, meaning that the second group of covariates is able to explain a relevant portion of variability in the dependent variable. As mentioned in Sect. 2, several variables are dummy variables and some of them have not been included in the model specification to avoid the dummy variable trap. Germany is the reference country (DE); as regards to education, the reference is lower secondary education or less (educ1); as regards to the value system, the reference is that freedom is more important than equality (Equality 1).

Age enters in the specification as a quadratic term to verify the existence of non-linearities. The results show that this is not the case, hence age a constant effect on the probability to be inclusive towards immigrants.

As it is well known, the estimated coefficients from a probit model are not indicative of the relationship between the covariates and the response variable. Table 3 provides marginal effects that can be used to infer the effect of covariates on the response variable. Marginal effects provide an absolute change in outcome probability due to a one-unit increase in explanatory variable.

As regards to statistically significant variables, women are more likely to have inclusive attitudes towards foreign-born neighbors. Education is positively related to the probability to have positive attitudes. Such a positive effect is stronger for individuals with college education than those with upper secondary education (the reference being, as mentioned above, lower secondary education or less). People, for which it is important that children are encouraged to learn tolerance and respect for others at home, are more likely to have positive attitudes towards foreign-born neighbors. The same positive relationship is observed for people supporting political parties on the left, and for people stating that freedom and equality are important, but if they have to choose one or other, they consider equality more important.

Finally, we find evidence of geographic differences in individuals' behavior that change over time. Figure 1 shows the country marginal effect on the probability to possess positive attitudes towards immigrants for a same profile of individuals across countries, observed in 1999 (panel *a*) and 2008 (panel *b*). The profile refers to a 50 years man, married with two children, with upper secondary education, employed, supporting parties on the left, thinking that it is important that children must be encouraged to learn at home tolerance and respect for others, and stating that

[3]We used the `glm` function available in R.

Table 2 Estimated coefficient of probit model

	Model I	Model II	Model III	Model IV
Intercept	1.3265***	1.5262***	1.7069***	1.7751***
	(0.1308)	(0.1412)	(0.1500)	(0.1603)
AT	0.3390***	0.3505***	−0.7400***	−0.7154***
	(0.0648)	(0.0653)	(0.0825)	(0.0836)
BE	−0.0360	−0.0751	−0.1728	−0.2132*
	(0.0575)	(0.0584)	(0.0891)	(0.0906)
DK	0.3143***	0.3122***	−0.0876	−0.0972
	(0.0807)	(0.0816)	(0.0987)	(0.1003)
EE	−0.0352	−0.0204	−1.0566***	−1.0559***
	(0.0711)	(0.0716)	(0.0830)	(0.0841)
ES	0.1934**	0.1492*	0.1418	0.0938
	(0.0713)	(0.0720)	(0.1078)	(0.1093)
FI	0.0472	0.0518	−0.5000***	−0.5032***
	(0.0703)	(0.0712)	(0.0971)	(0.0991)
FR	0.2080***	0.1672**	0.0649	0.0090
	(0.0620)	(0.0628)	(0.0961)	(0.0979)
GR	−0.2208**	−0.1495*	−0.4629***	−0.4072***
	(0.0691)	(0.0705)	(0.0855)	(0.0867)
IE	0.1952*	0.1925*	−0.4123**	−0.3897**
	(0.0759)	(0.0765)	(0.1460)	(0.1468)
IT	−0.1021	−0.1207*	−0.6162***	−0.6244***
	(0.0562)	(0.0569)	(0.0911)	(0.0924)
LV	0.5930***	0.6434***	−0.7207***	−0.6984***
	(0.0937)	(0.0943)	(0.0891)	(0.0904)
NL	0.5437***	0.5057***	−0.4859***	−0.5114***
	(0.0824)	(0.0836)	(0.0865)	(0.0878)
PT	0.4955***	0.5017***	−0.4607***	−0.4664***
	(0.0819)	(0.0828)	(0.0902)	(0.0912)
SE	0.7739***	0.7538***	−0.1723	−0.1945
	(0.1195)	(0.1208)	(0.1376)	(0.1403)
UK	0.2097**	0.1927**	−0.1480	−0.1603
	(0.0680)	(0.0685)	(0.0935)	(0.0946)
Gender	0.1322***	0.1104***	0.0962**	0.0696*
	(0.0282)	(0.0285)	(0.0309)	(0.0313)
Age	−0.0077	−0.0104	0.0018	−0.0015
	(0.0053)	(0.0053)	(0.0052)	(0.0053)
Age^2	−0.0000	0.0000	−0.0001	−0.0000
	(0.0001)	(0.0001)	(0.0001)	(0.0001)

(continued)

Table 2 (continued)

	Model I	Model II	Model III	Model IV
Educ2	0.1562***	0.1504***	0.1111**	0.1076**
	(0.0448)	(0.0452)	(0.0391)	(0.0395)
Educ3	0.3263***	0.3148***	0.2520***	0.2484***
	(0.0401)	(0.0405)	(0.0450)	(0.0457)
Marital status	−0.0461	−0.0525	−0.0071	−0.0091
	(0.0322)	(0.0324)	(0.0342)	(0.0345)
Children	0.0140	0.0153	0.0331*	0.0330*
	(0.0107)	(0.0108)	(0.0132)	(0.0133)
Employment	0.0696*	0.0719*	−0.0326	−0.0291
	(0.0330)	(0.0333)	(0.0376)	(0.0381)
Political view		−0.0592***		−0.0437***
		(0.0077)		(0.0082)
Tolerant education		0.2025***		0.2907***
		(0.0321)		(0.0342)
Equality 2		0.0921**		0.1302***
		(0.0305)		(0.0322)
Equality 3		0.1119		−0.1563*
		(0.0628)		(0.0747)

Table 3 Marginal effects of covariates

	Model I	Model II	Model III	Model IV
AT	0.0461***	0.0464***	−0.1604***	−0.1497***
	(0.0071)	(0.0069)	(0.0237)	(0.0233)
BE	−0.0061	−0.0127	−0.0276	−0.0339*
	(0.0099)	(0.0102)	(0.0156)	(0.0161)
DK	0.0427***	0.0415***	−0.0134	−0.0145
	(0.0088)	(0.0087)	(0.0159)	(0.0158)
EE	−0.0059	−0.0033	−0.2639***	−0.2585***
	(0.0122)	(0.0119)	(0.0285)	(0.0285)
ES	0.0284**	0.0221*	0.0188	0.0124
	(0.0092)	(0.0097)	(0.0130)	(0.0136)
FI	0.0076	0.0081	−0.0979***	−0.0961***
	(0.0109)	(0.0108)	(0.0241)	(0.0241)
FR	0.0305***	0.0246**	0.0090	0.0013
	(0.0080)	(0.0084)	(0.0129)	(0.0136)

(continued)

Table 3 (continued)

	Model I	Model II	Model III	Model IV
GR	−0.0413**	−0.0264	−0.0871***	−0.0724***
	(0.0145)	(0.0135)	(0.0199)	(0.0188)
IE	0.0286**	0.0277**	−0.0781*	−0.0709*
	(0.0098)	(0.0097)	(0.0345)	(0.0331)
IT	−0.0178	−0.0208*	−0.1281***	−0.1272***
	(0.0103)	(0.0105)	(0.0248)	(0.0248)
LV	0.0672***	0.0690***	−0.1575***	−0.1475***
	(0.0066)	(0.0060)	(0.0260)	(0.0255)
NL	0.0645***	0.0600***	−0.0929***	−0.0965***
	(0.0065)	(0.0068)	(0.0206)	(0.0209)
PT	0.0602***	0.0594***	−0.0874***	−0.0864***
	(0.0069)	(0.0067)	(0.0213)	(0.0211)
SE	0.0774***	0.0746***	−0.0280	−0.0311
	(0.0060)	(0.0061)	(0.0248)	(0.0253)
UK	0.0306***	0.0278**	−0.0234	−0.0248
	(0.0087)	(0.0087)	(0.0161)	(0.0160)
Gender	0.0220***	0.0180***	0.0140**	0.0098*
	(0.0047)	(0.0047)	(0.0045)	(0.0044)
Age	−0.0013	−0.0017	0.0003	−0.0002
	(0.0009)	(0.0009)	(0.0008)	(0.0007)
Age^2	−0.0000	0.0000	−0.0000	−0.0000
	(0.0000)	(0.0000)	(0.0000)	(0.0000)
Educ2	0.0239***	0.0226***	0.0158**	0.0149**
	(0.0063)	(0.0063)	(0.0055)	(0.0054)
Educ3	0.0476***	0.0452***	0.0336***	0.0320***
	(0.0051)	(0.0051)	(0.0055)	(0.0054)
Marital status	−0.0076	−0.0085	−0.0010	−0.0013
	(0.0054)	(0.0053)	(0.0049)	(0.0048)
Children	0.0023	0.0025	0.0048*	0.0046*
	(0.0018)	(0.0017)	(0.0019)	(0.0019)
Employment	0.0115*	0.0117*	−0.0047	−0.0041
	(0.0055)	(0.0054)	(0.0054)	(0.0053)
Political view		−0.0096***		−0.0061***
		(0.0012)		(0.0011)
Tolerant education		0.0353***		0.0457***
		(0.0060)		(0.0060)
Equality 2		0.0147**		0.0181***
		(0.0048)		(0.0044)
Equality 3		0.0169		−0.0243
		(0.0088)		(0.0128)

Fig. 1 Country marginal effect on the probability to have positive attitudes with foreign-born neighbors, years 1999 (**a**) and 2008 (**b**)

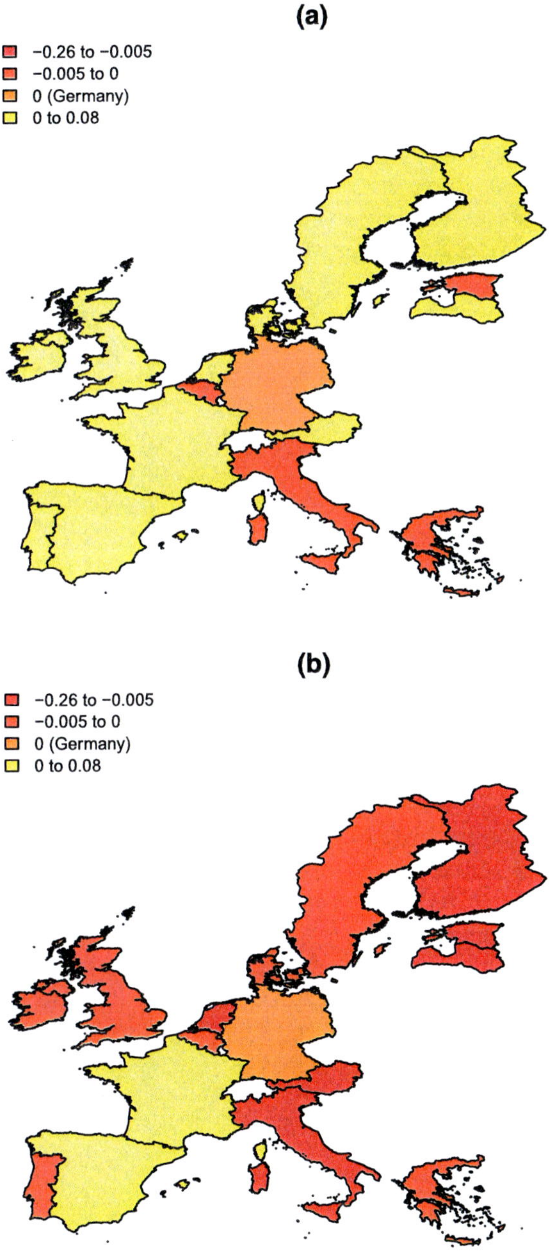

freedom is more important than equality. The values of the marginal effect have been normalized with respect to Germany, which is the reference country and its marginal effect has been set equal to zero. Substantial variations are observed over time and across countries. Looking at the two maps, the first thing that catches the eye is that people in general show a greater openness in 1999 than in 2008. In the first wave, residents in Austria, Denmark, France, Ireland, Latvia, Netherlands, Portugal, Spain, Sweden, and the United Kingdom are more likely to have positive attitudes towards immigrants, while Estonia, Belgium, Greece, and Italy show the worst attitudes towards immigrant inclusion. In the more recent wave, several countries, as Austria, Latvia, Netherlands, Portugal, Sweden, and the United Kingdom become less tolerant and open to foreign-born residents.

We could wonder whether this widespread change of attitudes could be due to the significant increase in immigrant population, rising from just over 6.7% in 1999 to more than 11% in 2008. Among the 16 European countries considered in our analysis, Germany is the largest immigrant host country (22% of total population in 2008), followed by the United Kingdom (16%), France (14%), Italy and Spain (around 10% each). The correlation coefficient between the percentage of foreign-born population by country and the probability to have positive attitudes towards foreign-born neighbors is $\rho_{1999} = 0.226$ indicating that the presence of foreign-born population is positively associated with natives' immigrant inclusive behavior. A possible interpretation is that the low percentages of foreign-born population in 1999 were not perceived as a threat. The correlation coefficient for 2008 is $\rho_{1999} = -0.348$ implying that the presence of foreign-born population is negatively correlated with positive attitudes towards immigrants. It has to be noted that in 2008 there are not only higher immigration levels than 10 years earlier but also the peak of the Great Recession that has certainly contributed to modify individuals' behavior.

5 Conclusion

The aim of the paper has been to investigate how individual features shape their attitudes towards immigrants. The analysis has been carried out using data from the European Value Study, waves 1999 and 2008. A probit model has been used to estimate the probability that it is not a problem for the individual to have immigrants/foreign workers as neighbors. The response variables were expected to be a function of socio-demographic variables and other personal variables, such as altruism and political preferences.

The results have shown that women are more likely to have inclusive attitudes towards foreign-born neighbors, as well as people with at least upper secondary education. Also the value system plays a significant role in shaping individuals' behavior.

The comparison of individuals' attitudes across countries provides information that is particular relevant to inform the debate on gaps in social cohesion and integration in Europe, while also indicating specific directions for public policies. Moreover, the analysis highlights the importance for local, national, and supranational governments to establish information systems for monitoring the factors influencing individuals' behavior, such as education and the value system. This would signifi-

Table 4 List of variables with their definition

Variable	Definition
Tolerant	1: tolerant; 0: otherwise
Country	European Country
Gender	1: female; 0: male
Age	numeric values
Education	coded on three levels
Marital status	1: unmarried; 0: otherwise
Children	number of children
Employement	1: employed; 0: otherwise
Political preference	Likert scale from 1 (left) to 10 (right)
Tolerant education	1: yes; 0: otherwise
Equality versus freedom	coded on three levels

[a] *Source* Survey European Value Study, years 1999 and 2008

cantly improve one's ability to detect disparities in integration across countries and identify appropriate policy actions.

Acknowledgements We thank participants to Data Science and Social Research Conference (DSSR 2019, Milan). Financial support by the Italian Ministry of University and Research is gratefully acknowledged. The usual disclaimer applies.

Appendix

Table 4 sets out the list of variables with their definition.

References

Accetturo, A., Manaresi, F., Mocetti, S., & Olivieri, E. (2014). Don't stand so close to me: The urban impact of immigration. *Regional Science and Urban Economics*, *45*(C), 45–56.

Agresti, A. (2002). *Categorical data analysis* (2nd ed.). New York: Wiley.

Amemiya, T. (1985). *Advanced econometrics*. Harvard University Press.

Borgoni, R., Degli Antoni, G., Faillo, M., & Michelangeli, A. (2019). Natives, immigrants and social cohesion: Intra-city analysis combining the hedonic approach and a framed field experiment. *International Review of Applied Economics*, *33*(5), 697–711.

Cutler, D. M., Glaeser, E. L., & Vigdor, J. L. (1999). The rise and decline of the American Ghetto. *Journal of Political Economy*, *107*(3), 455–506.

Greene, W. H. (2018). *Econometric analysis* (8th ed.). New York: Pearson.

Hoffmann, J. P. (2004). *Generalized linear models: An applied approach*. Boston: Pearson.

Reardon, S. F., Matthews, S. A., O'Sullivan, D., Lee, B. A., Firebaugh, G., Farrell, C. R., et al. (2008). The geographic scale of metropolitan racial segregation. *Demography*, *45*(3), 489–514.

Saiz, A. (2003). Room in the kitchen for the melting pot: Immigration and rental prices. *Review of Economics and Statistics*, *85*(3), 502–521.

Saiz, A., & Wachter, S. (2011). Immigration and the neighborhood. *American Economic Journal: Economic Policy*, *3*, 169–188.

A Bibliometric Study of the Global Research Activity in Sustainability and Its Dimensions

Rosanna Cataldo, Maria Gabriella Grassia, Carlo Natale Lauro, Marina Marino, and Viktoriia Voytsekhovska

Abstract The scientific production on "sustainability" has been increasing in recent years. To better understand and characterize this trend, a bibliometric study of international papers on this subject has been developed. A total of 3,994 articles from 1985 to 2018 have been selected and analyzed in order to discover the research trends in this field and the main dimensions and words related to the term "sustainability" that are most commonly employed in the scientific literature. The research has been conducted in the Web of Science from ISI Web of Knowledge database with the aim of identifying the major themes, authors, areas, types of documents and the sources, titles, years of publication and countries of these publications, as well as the main themes related to the topic "sustainability".

1 Introduction

Today sustainability is highly interdisciplinary in nature and yet evolving, a complex multidimensional phenomenon, which has already been studied for a couple of decades. It can be observed from different perspectives and angles, but the most accu-

R. Cataldo (✉) · M. G. Grassia · M. Marino
Department of Social Sciences, University of Naples "Federico II", Vico Monte Della Pietà, 1, 80138 Napoli, Italy
e-mail: rosanna.cataldo2@unina.it

M. G. Grassia
e-mail: mgrassia@unina.it

M. Marino
e-mail: mari@unina.it

C. N. Lauro
Department of Economics and Statistics, University "Federico II", via Cintia, Monte Sant'Angelo, 80126 Naples, Italy
e-mail: clauro@unina.it

V. Voytsekhovska
Department of Economics of Enterprise, Lviv Polytechnic National University, Lviv, Ukraine
e-mail: viktoriia.v.voitsekhovska@lpnu.ua

P. Mariani and M. Zenga (eds.), *Data Science and Social Research II*, Studies in Classification, Data Analysis, and Knowledge Organization,
https://doi.org/10.1007/978-3-030-51222-4_8

rate definition would seem to be that provided by the United Nations: "Sustainable development is the development that meets the needs of the present, without compromising the ability of future generations to meet their own needs" (Sachs 2012). The concept of development and growth has been studied in general by Solow (1994) and Daly et al. (1974), who have drawn attention, besides the economic component, also to the social and environmental dimension. A country's economic, social, and environmental performance is now seen through the lens of Sustainable Development Goals, which were proclaimed in September 2015 during the UN General Assembly (Hopwood et al. 2005). The successful implementation of the Sustainable Development Agenda 2030 and its 17 goals requires the development of a scientific ideology in order to monitor sustainable development in its three key areas. Bettencourt and Kaur (2011) explain the importance of the roles of international and regional scientific organizations that need a more solid theoretical and methodological framework than that provided by the topics of individual research projects in this area. The requirement for a further development of sustainability science is described by Anand and Sen (2000) and others Kates (2011), Sachs et al. (2016). Global scientific interest in its application has increased in recent years. Nowadays increasing attention is being focused on the topic of "sustainability" and its dimensions by policy-makers and by the scientific research community (Bettencourt and Kaur 2011; Smedt et al. 2018; Skute et al. 2019; Montella et al. 2019). Starting from the global scientific interest in this topic, it is proposed in this study to apply a bibliometric analysis in order to analyze the research trends in this field during the period from 1986 to 2018, to provide a complete picture of the concept of sustainability and to understand which aspects are most frequently highlighted in scientific research.

The article is structured as follows. In the following section, we explain the methodology employed. Next, the research design and the main results of the analysis are presented. Finally, some conclusions, limitations of this study, and future research proposal are discussed.

2 Study Methodology

This study was developed from a bibliometric research project, aimed at discovering the main dimensions and words employed in scientific publications linked with the theme of "sustainability". According to Silva (2004), bibliometric mathematical and statistical methods are used to assess the productivity of scientific outputs quantitatively. Bibliographic data are processed through a workflow: study design, data collection, data analysis, data visualization, and interpretation. Aria and Cuccurullo (2017) stated that bibliometric analysis is a cumbersome activity, which involves many producers. However, there are automated software tools that are used by information scientists or practitioners (Guler et al. 2016). By extracting descriptive and network data from bibliographic literature, you can perform a citation analysis (Darvish 2018). Recently, automated workflows to assemble specialized software into a comprehensive and organized data flow have begun to emerge for bibliometrics (for example *BibExcel* Persson et al. 2009, *Pajek* (http://vlado.fmf.uni-lj.si/pub/networks/pajek/) and *Gephi* (https://gephi.org)).

For this study, the *Bibliometrix* package (Aria and Cuccurullo 2017) in the R programming language (https://www.r-project.org/)[1] was used for the analysis and visualization of the bibliographic data from the Web of Science databases. Today *Bibliometrix* is more than just a statistical tool. It is becoming a community of international developers and users who exchange questions, impressions, opinions, and examples within an open source project. The Bibliometrix R-package (http://www.bibliometrix.org) provides a set of tools for quantitative research in bibliometrics and scientometrics, supporting scholars in three key phases of analysis: (1) data importing and conversion to the R format; (2) bibliometric analysis of a publication dataset; (3) building matrices for co-citation, coupling, collaboration, and co-word analysis. Matrices are the input data for the performance of network analysis, multiple correspondence analysis, and certain data reduction techniques (Aria and Cuccurullo 2017).

3 Data Collection

The data for this research project were collected in the Web of Science's database of the Institute for Scientific Information (ISI). Web of Science (WoS) is the world's most trusted independent global citation database. It is recognized as covering a broad range of relevant journals and peer-reviewed articles of high quality (Skute et al. 2019). This multidisciplinary platform connects regional, specialty, data. and patent indexes to the WoS Core Collection, the world's only true citation index. The comprehensive platform allows you to track ideas across disciplines and time from over 1.7 billion cited references from over 159 million records. Over 9,000 leading academic, corporate and government institutions and millions of researchers trust WoS to enable them to produce high-quality research, gain insights and make better informed decisions that guide the future of their institutional research strategy.

We identified the keyword queries for each dimension (Sustainable Development Goals—SDGs) and then used these queries to pull related scientific literature from the bibliography database. The keywords associated with a given dimension are specific to the work in that domain. In Table 1, the dimensions of sustainability for each area are reported.[2]

Next, we queried WoS using the following Topic Search (TS): TS = "sustainability" and the keywords of all 17 of the dimensions previously identified as the terms for searching titles, abstracts and/or keywords. Combinations of the keywords with the main term, "sustainability", were searched to obtain the bibliographic data of relevant research articles. The search was restricted to literature published in the

[1] Several studies have revealed the importance and role of R and its packages in vast scientific fields.

[2] For a detailed description of the individual goals, refer to the site "Sustainable Development Goals" (www.un.org/sustainabledevelopment/development-agenda/).

Table 1 SDGs and the three areas

Area	Goals	
Social area	Goal 1	No poverty
	Goal 2	Zero hunger
	Goal 3	Good health and well-being
	Goal 4	Quality education
	Goal 5	Gender equality
	Goal 6	Clean water and sanitation
Economic area	Goal 7	Affordable and clean energy
	Goal 8	Decent work and economic growth
	Goal 9	Industry, innovation, and infrastructure
	Goal 10	Reduced inequalities
	Goal 11	Sustainable cities and communities
	Goal 12	Responsible production and consumption
Environment area	Goal 13	Climate action
	Goal 14	Life below water
	Goal 15	Life on land
	Goal 16	Peace, justice, and strong institution
	Goal 17	Partnerships for the goals

period from 1986 to 2018 (incl.) This process resulted in a final sample of 3,994 articles, which constitute the core material of this study, relating to 10,297 authors and 1,235 sources.

4 Analysis and Discussion

This section presents the results of the bibliometric analysis of the term "sustainability" and its SDG keywords. Table 2 shows the main information about the bibliographic data frame over the proposed time period. Since the purpose of this research review was to gain an overview of sustainability, some inclusion and exclusion criteria were set. The database included journal articles, articles as book chapters, and articles as proceedings papers. It excluded less reliable document types such as letters, short surveys, notes, and articles in the press and editorial material. The exclusion criteria were based on the quality, reliability, and validity of the document, as these aforementioned types did not undergo a peer-review process (Marsh et al. 2008). The bibliographic data frame is composed of 3,994 articles, relating to 10,297 authors,

Table 2 Main information

Article	3,994
Sources	1,235
Authors	10,297
Authors of single-authored documents	635
Authors of muti-authored documents	9,662
Single-authored documents	745
Author's keywords	8,574

published in 1,235 sources. There are 635 authors of single-authored documents and 9,662 authors of multi-authored documents, emphasizing the need for collaboration between authors, even from different countries and/or belonging to different research domains.

The articles extracted produced 8,574 keywords that refer to the main keyword "sustainability" and to the different keywords of the 17 dimensions.

Figure 1 illustrates the annual scientific production in relation to the research topic. Comparing the quantity of publications from 1986 to 2018, it is evident that in the first years of the analysis (1986–1992) the number of publications is very low, emphasizing the fact that the topic was probably not very well developed and addressed by researchers. Starting from 1992, with the *Earth Summit—the United Nations Conference on Environment and Development in Rio de Janeiro*, we notice a slight increase in publications. Researchers are starting to take an interest in this problem and address the issue in their work. The first evidence of a notable increase in popularity comes in 2005, when the number of publications doubles compared to

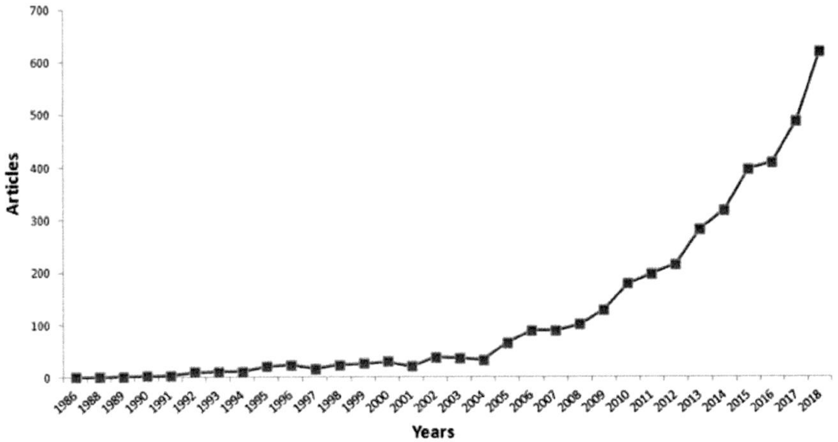

Fig. 1 Growth trajectory of the literature in sustainability, 1986–2018 (n = 3,994)

the previous year. Starting from 2005, the analysis reveals a significant exponential growth until the final year of research period, demonstrating the emergence of studies addressing this problem, especially in order to find solutions to the problems of "sustainability" with the development of new models. This growth is probably related to an important summit held in September 2000 at the UN Headquarters in New York, the *United Nations Millennium Summit* (United Nations 2015). This Summit led to the elaboration of eight Millennium Development Goals (MDGs) to reduce extreme poverty by 2015. Evidently, the summit aroused the interest of researchers in this issue, causing an almost exponential growth in production (Fig. 1). Figure 2 shows the main sources of publications related to the theme. Most studies concerning the subject were published in the journals focused on dealing with issues related to social, behavioral, technological, and management innovation, which demonstrates the relevance of this theme to concerns inherent in innovation and business models (Neumayer 2007; Stiglitz et al. 2017; Sachs 2012; Gan et al. 2017; Smedt et al. 2018). From Fig. 2 it is possible to note that the five most relevant sources, based on the number of articles, are *Sustainability*, the *Journal of Cleaner Production*, the *International Journal of Sustainability in Higher Education*, *Energy Policy* and *Environmental Education Research*, journals whose aims are to provide up-to-date information on new developments and trends in relation to this topic.

As regards provenance, we have examined the research activity of countries in terms of their publication output on this theme. Figure 3 shows the top 20 most productive countries in terms of publication output and the scientific collaboration during the period 1986–2018. The authors who have distinguished themselves by their publications related to this topic come mainly from the United States, the United Kingdom, China, and Italy. These are also the main countries that collaborate with each other, thus providing a recognition of the importance of the relationships between researchers involved in collaborative activities. It is noteworthy that authors from other countries collaborate principally with the most productive nations and farless frequently with the others.[3] While the United States is clearly leading in terms of publication output in sustainability, with over half of the total publications, so creating a large gap between itself and the rest of the selected countries, it should be noted that China takes third place in terms of publications, indicating that China has significantly increased its publication output in relation to sustainability in recent years (see also Fig. 4).

In relation to the individual author we have examined the researchers' production (in terms of number of publications, and total citations per year) over the research time period.

As can be seen in Fig. 4, these authors started to publish articles relating to this theme in 1994 (Bastianoni S.) with a significant increase in production in 2012. Over the past three years, there has been a noticeable increase in articles submitted

[3]In the country network the bigger are the node and word, the larger is the weight. The distance between two nodes reflects the strength of the relation between two nodes, a shorter distance generally indicating a stronger relation.

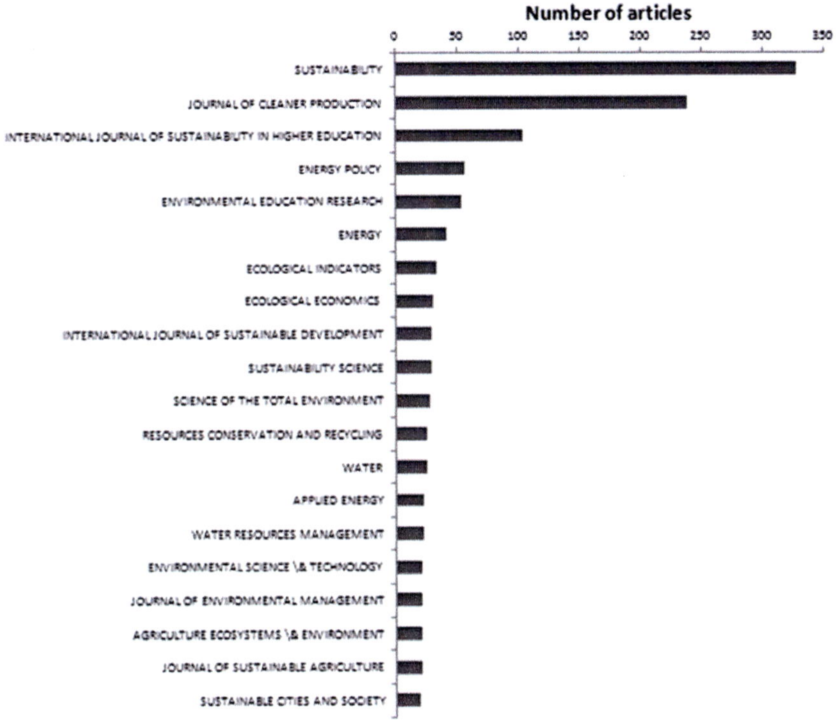

Fig. 2 Corresponding author by country

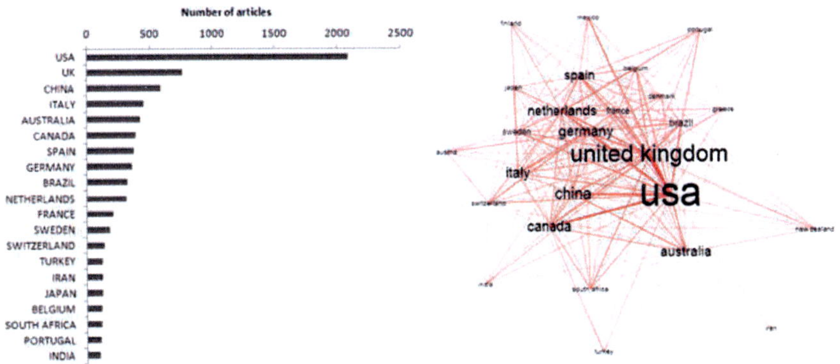

Fig. 3 Corresponding author by country and Scientific Collaboration by country

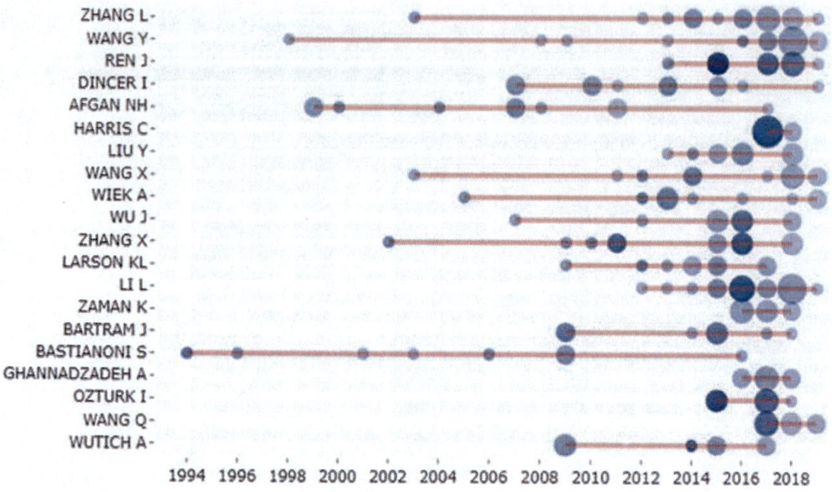

Fig. 4 Top authors' productivity during the period 1994–2018

by Chinese authors (Zhang L., Wang Y., Ren J.) highlighting the growing interest in this topic in China.[4]

A keyword co-occurrence network map is shown in Fig. 5.[5] Keyword co-occurrence can effectively reflect research hotspots in different disciplines, providing auxiliary support for scientific research (Li et al. 2016) and information about emerging topical trends. Considering the 3,994 articles, we obtained 8,574 keywords in total. Among these 5,358 keywords appeared only once, accounting for 62.5%. The node, "sustainability", has thicker lines connecting with "environment", "climate change" and "higher education", whose link strengths are greater than 10. Other links to highlight are the connections with "health", "governance", "water", "food", "energy", and "renewable energy". It is important to note that among the keywords the nodes "indicators" and "sustainability indicators" appear, emphasizing the need to construct indicators to study the various dimensions of this multidisciplinary topic and to create synthetic indicators starting from the over one hundred indicators that exist today. It seems curious that in this map also the node "China" appears, evidencing by the fact that China is now the world's leading renewable energy producer now, outperforming competitors in Denmark, Germany, Spain, and the United States, with the largest wind turbine and solar panel producer (Bradsher 2010). It is possible that this will be considered as a reference by researchers in this area of sustainability.

---

[4]In the figure the size of the dots is linked to the number of citations per year while the color intensity is related to the number of articles produced.

[5]In the network in Fig. 5 a line between two keywords represents the fact that they have appeared together. The thicker is the line, the greater is the co-occurrence that they have (Gu et al. 2017). The link strength between two nodes refers to the frequency of the co-occurrence.

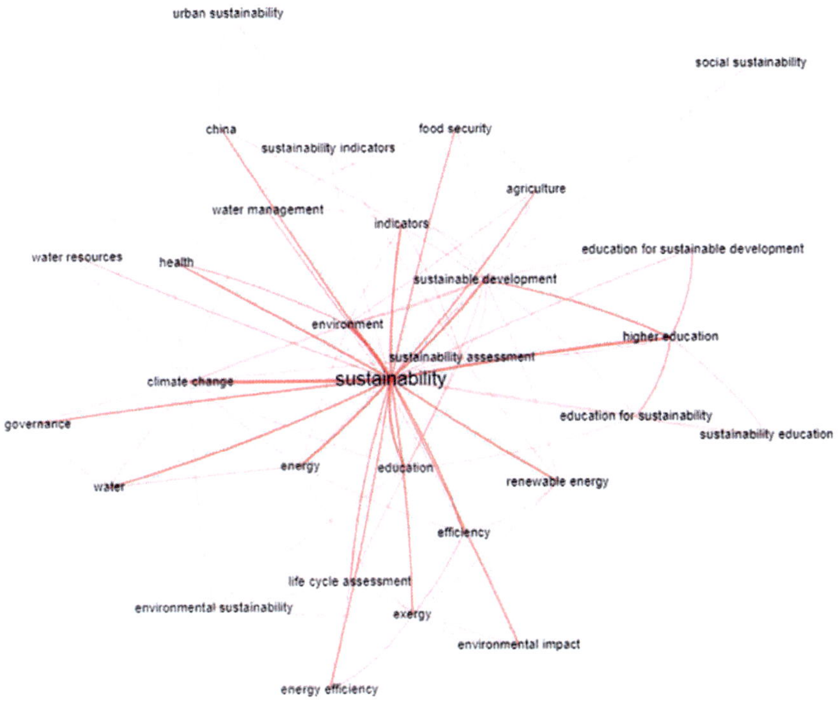

Fig. 5 Keyword co-occurrences

The last figure, Fig. 6, shows keywords as themes, classified by different levels of density and centrality in the network of scientific keywords. A thematic map is a very informative plot, enabling us to analyze themes according to the quadrant in which they are placed: (1) the upper-right quadrant: motor-themes; (2) the upper-left quadrant: very specialized/niche themes; (3) the lower-left quadrant: emerging or disappearing themes; and (4) the lower-right quadrant: basic themes.

As we can see from the thematic map, the themes with a higher centrality are "sustainable development", "climate change", "environment", and "energy". This suggests that these topics appear ubiquitously in different scientific works and can be considered a common synthesis of the content expressed in the literature. It is also possible to observe the emergence of "water quality", "water supply", "renewable energy", and "public health" as very specialized topics in the scientific works. Additionally, in this map "indicators" appear as a motor theme, emphasizing how in the last few years researchers have focused their attention on this theme. Finally, it is possible to note also here the presence of "China" as emerging theme in the articles of the analyzed bibliographic data frame. It can be deduced that this country, being now the world's leading renewable energy producer now, is considered as an example to be monitored by researchers on this topic.

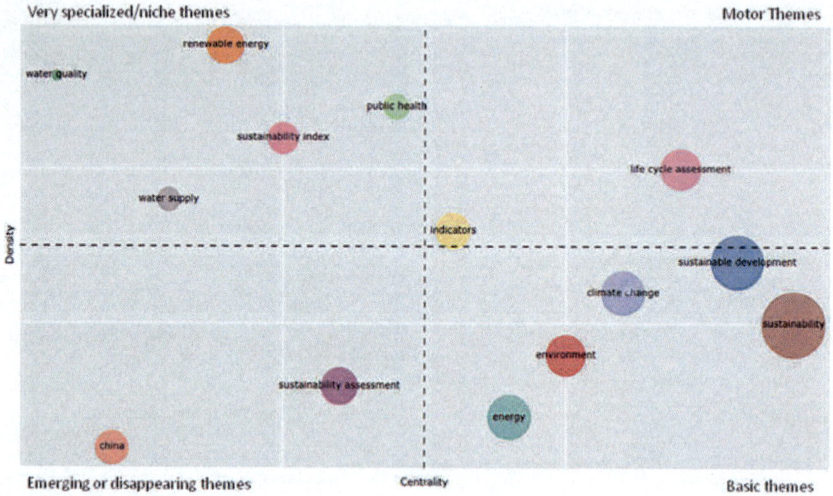

Fig. 6 Thematic map

5 Final Remarks

This research project has presented a bibliometric method with the objective of reviewing scientifically the topic of sustainability, using the Web Of Science database over the time period 1986–2018 and analyzing 3,994 publications related to this theme. In the selected time period of time the scientific production at first gradually and then almost exponentially increased over the more than thirty years analyzed. The USA leads the ranking of countries that have published articles on this issue, followed by the United Kingdom and China, there also being a great number of collaborations among all countries in studies of this theme. The term "sustainability" is strongly linked to the environment, climate change, energy, and water, which are the main issues addressed not only by researchers but, as we all know, also by institutional decision-makers. Finally, in the networks concerning the keywords, the presence of the term "China", both as a term strongly linked to the concept of sustainability and as an emerging term in scientific research, indicates that, perhaps, other states should consider this country as a reference point in order to solve a number of problems related to sustainable development. As a limitation of this study, it should be stressed that it has been undertaken using only one specific database. For this reason, it is suggested that it may be necessary to query other scientific databases in order to provide a complete picture of the theme of sustainability.

References

Anand, S., & Sen, A. (2000). Human development and economic sustainability. *World Development, 28*, 2029–2049.

Aria, M., & Cuccurullo, C. (2017). Bibliometrix: An R-tool for comprehensive science mapping analysis. *Journal of Informetrics, Elsevier, 11*(4), 959–975.

Bettencourt, L., & Kaur, J. (2011). Evolution and structure of sustainability science. *Proceedings of the National Academy of Sciences of the United States of America, 108*, 19540–19545.

Bradsher, K. (2010). China leading global race to make clean energy. *The New York Times, 30*(01).

Daly, H., Cobb, J., John, B., Cobb, J. B., & Cobb, C. W. (1974). *For the common good: redirecting the economy toward community, the environment, and a sustainable future* (p. 73). Boston: Beacon Press.

Darvish, H. (2018). Bibliometric analysis using Bibliometrix an R Package.

De Smedt, M., Giovannini, E., & Radermacher, V. (2018). *Measuring sustainability, for good measure advancing research on well-being metrics beyond GDP: Advancing research on well-being metrics beyond GDP* (p. 241). OECD Publishing.

Gan, X., Fernandez, I. C., Guo, J., Wilson, M., Zhao, Y., Zhou, B., et al. (2017). When to use what: Methods for weighting and aggregating sustainability indicators. *Ecological Indicators, Elsevier, 81*, 491–502.

Gu, D., Li, J., Li, X., & Liang, C. (2017). Visualizing the knowledge structure and evolution of big data research in healthcare informatics. *International Journal of Medical Informatics, 98*, 22–32.

Guler, A. T., Waaijer, C. J., & Palmblad, M. (2016). Scientific workflows for bibliometrics. *Scientometrics, 107*(2), 385–398.

Hopwood, B., Mellor, M., & OB́rien, G. (2005). Sustainable development: Mapping different approaches. *Sustainable Development, 13*(1), 38–52.

Kates, R. W. (2011). What kind of a science is sustainability science? *Proceedings of the National Academy of Sciences, 108*(49), 19449–19450.

Li, H., An, H., Wang, Y., Huang, J., & Gao, X. (2016). Evolutionary features of academic articles co-keyword network and keywords co-occurrence network: Based on two-mode affiliation network. *Physica A: Statistical Mechanics and its Applications, 450*, 657–669.

Marsh, H. W., Jayasinghe, U. W., & Bond, N. W. (2008). Improving the peer-review process for grant applications: Reliability, validity, bias, and generalizability. *American Psychologist, 63*(3), 160.

Montella, A., Marzano, V., Mauriello, F., Vitillo, R., Fasanelli, R., Pernetti, M., et al. (2019). Development of macro-level safety performance functions in the city of Naples. *Sustainability, 11*(7), 1871.

Neumayer, E. (2007). Sustainability and well-being indicators. In *Human well-being* (pp. 193–213). Springer.

Persson, O., Danell, R., & Schneider, J. W. (2009). How to use Bibexcel for various types of bibliometric analysis. In *Celebrating scholarly communication studies: A Festschrift for Olle Persson at his 60th Birthday* (Vol. 5, pp. 9–24).

Sachs, J. D. (2012). From millennium development goals to sustainable development goals. *Lancet, 379*(9832), 2206–2211.

Sachs, J., Schmidt-Traub, G., Kroll, C., Durand-Delacre, D., & Teksoz, K. (2016). *SDG index and dashboards: A global report*. Bertelsmann Stiftung.

Silva, M. R. D. (2004). Análise biliométrica da produção científica docente do Programa de Pós-Graduação em Educação Especial da UFSCar: 1998–2003.

Skute, I., Zalewska-Kurek, K., Hatak, I., & de Weerd-Nederhof, P. (2019). Mapping the field: A bibliometric analysis of the literature on university-industry collaborations. *The Journal of Technology Transfer, 44*(3), 916–947.

Solow, R. M. (1994). Perspectives on growth theory. *Journal of Economic Perspectives, 8*(1), 45–54.

Stiglitz, J. E., Sen, A., & Fitoussi, J.-P. (2017). Report by the commission on the measurement of economic performance and social progress.
United Nations. (2015). United Nations Conference on Sustainable Development, Rio+20.

Big Data Marketing: A Strategic Alliance

Federica Codignola

Abstract The progress of technology and science goes along with the vast richness of the cultural and material existence of individuals, so that each person's behavior at any moment generates large quantities of traceable information and data. Through the rise of the Internet, the quantity of data is geometrically increasing. It is then quite complex to manage it with traditional data systems. Today, Big Data touch every side of any economic activity, from public transportation, communications, bank securities, insurance, to government, health, education and other public utilities. Concomitantly with the rise of cloud computing, cloud applications, the multiplicity of mobile devices, and the maturity of e-commerce giants' data marketing systems such as Amazon and Google, Big Data marketing is becoming increasingly focused and is being used by most companies. Following the development of digital growth, the aim of the marketing industry is to collect large amounts of assorted customer-related behavior sales data on which to develop their marketing strategies. Facing the era of Big Data, most firms have abandoned their traditional marketing strategies and have decided to opt for a powerful system of Big Data analysis so as better to identify their own customer target and consequently increase sales and profit. This chapter emphasizes the urge to observe the practical implications of an in-depth investigation on the strategic relationship between Big Data and marketing.

1 Introduction

This is an era of data. With the introduction of the notion of 'Big Data' during the first decade of the 2000s, the institutional, economic, commercial, and academic consideration for Big Data has progressively improved, and associated research has grown. Following such a trend, Big Data marketing (from now on BDM) has been increasingly studied and employed by firms. Nevertheless, the existing literature has generally been relegated to business media content, and the interpretation of BDM

F. Codignola (✉)
DEMS, Università degli Studi di Milano – Bicocca, Milan, Italy
e-mail: Federica.codignola@unimib.it

P. Mariani and M. Zenga (eds.), *Data Science and Social Research II*,
Studies in Classification, Data Analysis, and Knowledge Organization,
https://doi.org/10.1007/978-3-030-51222-4_9

controlled by firm internals rather than academics. Academic research on this topic is usually devoted to the application of data-mining technology in marketing (Chen and Zhang 2014). Nonetheless today data touch on new complex necessities. Consequently, this chapter investigates BDM through an observation of the contemporary marketing scenario.

2 The Big Data Marketing Scenario

As the strategic basis for all consumers' expectations and needs, marketing deals with experience in order to gain relevant information on customers. Firms manage several business activities through marketing processes and planning. At the same time, marketing delivers coordinated product, price, place and promotion strategies to the market. Moreover, while pursuing corporate and business objectives, marketing delivers demand-adequate products and services (Dumbill 2018). Through marketing strategies, rational ideas which are capable of directing the firm's development are produced (Ducange et al. 2018).

As far as marketing literature is concerned, the primary Porterian 4P model relies on the tenet that firms should conduct every single investigation basing it on the external environment. This will allow them to develop an efficient marketing strategy mixture that will help manage the firm's development (Constantinides 2006). Such a strategy must be built around the four features of product, price, place and promotion in order to meet every consumer need and to accomplish the firm's objectives (Athanasiadis and Ioannides 2015). Subsequent research has formulated the 4C model built on the following assets: customer, cost, convenience and communication. Such a model is based on the assumption that, in order to compete, a firm must strategically manage communication channels with customers while considering costs. By relying on previous literature, a number of studies have suggested the 4R model, which balances the features of customer recognition, relevance, relationship and reward (Rygielski et al. 2002). These last four marketing features have been identified by taking into account the firm's competitive environment and by mixing different marketing strategies. These features, which aim to create benefits for both firms and consumers, are based on the relationship marketing theory.

BDM represents an application of the essential technology of Big Data to the marketing discipline (Shaw et al. 2001). It is based on the valuable information acquired throughout the gathering, extraction, organization and investigation of Big Data. However, at the very beginning of these procedures, customers must be allowed to engage and interrelate with firms through the use of their products and services. Big Data derive from the significant information that emerges from market surveys or/and various network channels. They contribute better to comprehend any firm's or industry's market feature while supplying critical directions for the creation of efficient marketing strategies. By contributing to the firm's marketing performance, Big Data indirectly convey a general advantage to the firm. The implementation of Big Data has in fact broadly changed any marketing process (Utkarsh and Santosh

2015). For instance, through the tracking of a customer Internet path, it is possible to recognize precise consumer behaviors, which are useful in order to construct specific marketing strategies. As an example, the Chinese firm Taobao proposes a number of desired products to the consumer after tracking his or her browsing history. Another Chinese company, Qunar, presents special offers by calculating air tickets' price through the analysis of Big Data. These examples represent some typical BDM applications (Kshetri 2016).

In the next paragraph we will examine the context of BDM. However, here are four elements that provide the necessary background.

(1) *Achieving an effective tool for market analysis.* Within the limits imposed by economic and time costs, marketing has normally tended to perform market analysis by following some standardized theoretical pathways and by conducting surveys through general research design paths (Huffaker and Whittlesey 2000). Yet, as sample data generally suffer from significant shortcomings, a comprehensive and realistic picture of the market often proves difficult to obtain. Market analysis through Big Data helps overcome such a limitation. In fact, once widespread data are accessible and well-organized analysis methods are employed, the traditional use of market-data sampling proves to be less critical. Consumers regularly supply a comprehensive variety of Big Data analytics built on mobile applications, terminal sensors, website clicks, etc. (Rappa 2004). In order to conduct accurate data analysis, firms must combine network data with diverse consumer-related data. Thus, if on the one hand BDM presents precious insights on consumers' attitudes, desires and preferences, on the other hand it can help produce accurate consumer-driven goods.
(2) *Expanding the marketing focus.* The focus of marketing is switching from a consumer perspective to a more comprehensive life-ecosystem. Obviously, this fact represents a new angle for BDM. In other words, as marketing is increasingly focusing on consumers' emotional, psychological, and social aspects rather than just reflecting on consuming economic features, implications for BDM are significant (Hamel 2003). For instance, marketing can now serve as a tool for interacting with customers rather than just functioning as a business model. In fact, the idea of connecting BDM with contemporary issues such as "consumer sentiment" can help create long-run relationships with customers. Thanks to a broader, accurate collection of consumer life-behavior data (obtained from Big Data), the purchase cycle can be better defined, consumer goods more clearly positioned, and consumer behavior more easily and precisely predicted. Conversely, the more the marketing strategy is accurate, the more it can be efficiently pushed, and the more such a strategy is successful.
(3) *Balancing the power between firms and consumers.* Big Data have increased the power of consumers who now share almost the same power-level of the firms. This entails an increasingly active and effective participation of consumers through many marketing processes (Amit and Zott 2001). The introduction of consumers' participation in marketing activities and processes can significantly enhance the firm's reputation. At the same time, it can impact marketing

costs. For instance, as the progress of information technology has transformed marketing communication (e.g., advertising is no longer conveyed by traditional media through traditional firm-driven content), consumers directly or indirectly participate in the construction of marketing communication's contents and their dissemination. These results significantly reduce the costs linked to marketing-communication management through communication and media agencies, and so on. An example is offered by social media platforms such as Instagram, Facebook, Weibo, TikTok, WeChat, etc. that transform consumers' word-of-mouth in publicity. Consumers may in fact have a strong impact on brand management and brand performances.

(4) *Using the new hyper one-to-one marketing capacity.* The price-related convenience of the Internet allows firms to gain incredible amounts of specific consumers data. This ranges from information on the degree of a consumer's propensity to information on the consumption cycle or processes, and so on. After appropriate technical data processing, firms can shape perfect marketing strategies and appropriately target them to the market (Afuah and Tucci 2001). Thanks to Big Data, firms may in fact more accurately search for potential customers and seize the degree of the customer's intention to buy. Moreover, firms can use information in order better to develop future services or products and successfully communicate and promote them. Thanks to previous purchase records, hyper personalized products, services, and linked promotions may be offered to already existing customers. In the same way, today's firms have more opportunities to improve demand segmentation. The knowledge of specific consumers' features allows firms to deliver extra-personalized offers. In sum, BDM can improve various marketing aspects and processes that go from product creation up to customer relationship, etc.

3 Applying Big Data Marketing

Peterovic defines BDM as the progress of traditional marketing in the Big Data framework (Peterovic et al. 2001). So far we have shown how, in order to carry on proficient marketing processes, BDM focuses on the identification of consumers' intentions instead of merely comprehending consumer behavior.

A study by Müller and Jensen (2017) highlights explicit connections between value creation and the application of Big Data. It also assesses how the value derived from Big Data does not depend on information and technology only but is strongly connected to the firm's internal and external context and to its management. Consequently, the management must: (1) be aware of the business value of Big Data; (2) be professionally trained to understand and deal with them. Only then will managers be able to transform business processes into data-driven business alternatives and to increase the development of business decisions and actions in the light of data.

In order better to understand the possibilities of a profitable application of Big Data, three general marketing objectives will be here summarized.

(1) *Enhancing marketing effectiveness*. The increase of Big Data has changed the traditional marketing channels. Individuals may employ the Internet and mobile data to find information on services or goods (Dubosson Torbay et al. 2002). At the same time, in order to obtain information, consumers offer their own data. These will in turn be used by firms. Firms must enhance their marketing channels at a qualitative and quantitative level in order for them to reach multiple kinds of consumers. From the consumers' perspective, BDM offers very accurate standards of recommendation. Internet consumer behavior can be perfectly tracked. The same is true for the collection of customer data. Finally, even offline consumer behavior data can be accurately gathered thanks to various systems such as POS machines or membership cards. In sum, Big Data can track and examine consumer behavior in various ways and with different goals. While traditional marketing starts from the product in order to reach the consumer, BDM does exactly the opposite. It starts with knowledge of the consumer and only then offers the appropriate products (Mitchell and Coles 2003). Through supplementary and widespread data acquirement methods, BDM has deeply enhanced the accuracy level of market information.
(2) *Enhancing consumer experience*. The advance in technology and science and the speed in the industrialization progress have contributed to a vast intensification of commodities production and offers. The stunning assortment of available goods has significantly affected consumer behavior. Consumers face a massive supply of comparable products or services which they are compelled to select and choose (Morris et al. 2005). They have to take into account and compare various features (e.g. aesthetics, brand, performance, price, etc.), but to do so they must make an effort and spend time. BDM can help consumers who face this kind of issues to make their choices more easily. Through in-depth investigation on definite consumers, the firm is able to suggest customized products and services that fit the consumer's desires and requests in accordance with their individual features. This significantly restricts the variety of consumers' alternatives and optimizes customer experience. During his or her experience, the customer may quickly report to the firm (e.g., pull communication) information on one's use of the service or product, allowing valuable product or service enhancement. For example, the Chinese firm of mobile phones Xiaomi uses BDM to constantly improve their hardware thanks to customer evaluation. At the same time, BDM is used to boost a method that responds to any operational customers' need (Shih et al. 2014).
(3) *Implementing integrative marketing platform*. The advancement of the Net allows firms to obtain an increasing quantity of data on consumers. Individuals have in fact incorporated the Internet into their everyday life, and by communicating through social media platforms, etc., they let their actions be tracked by third-party software (Osterwalder et al. 2005). Thus, if marketing has always been used to reorder individuals' disorganized information, to obtain a representation of the consumer, and eventually to market it, BDM goes a step further. Thanks to its features, it can easily exploit the combination of the most important network platforms and reach standards of marketing integration. In

addition to network platform integration, Big Data can merge online and offline platforms, achieving an effectual marketing integration. For instance, this is true of the merging of traditional communication tools and media with the Net.

4 Strategic Relationship Between Big Data and Marketing

With the appearance of the consumer-driven model and the rise of multi-channel consuming model (e.g., e-commerce), the needs and the functions of consumers have significantly altered. Our planet is now intelligent and interconnected (Magretta 2002). Big Data have allowed the quantifying and forecasting of people's behavior. At the same time, they have permitted to qualitatively investigate consumers (their opinions, etc.). Consequently, new opportunities have arisen for consumer-driven businesses. As already stated, traditional marketing processes are aimed at gathering market information through market research analysis. This facilitates the firm's production, marketing, and promotion (Zott et al. 2011). Thanks to social media platforms and Big Data, present-day consumers drive marketing in the sense that, for instance, they dynamically seek out product and service information; purchase through multi-channel's logics; make rigorous selections and spread out their evaluations on the consuming experience to the public. This is the reason why the brand image of a firm not only depends on the firm's publicity or on other corporate communication strategies, but also relies on consumer-driven systems such as the reputation of virtual communities (those of e-commerce websites or of social media platforms, etc.). Consumers' shopping behavior is also affected. In a similar way, by publicly communicating their individual behavior and preferences consumers affect the production, design, sales strategies, etc., of their chosen services and products. In the Big Data era, thanks to the improvement of technologies, new challenging opportunities arise for marketing strategies that are extremely accurate (Linder and Cantrell 2002). If, on the one hand, Big Data can recognize precise opportunities of market segments, on the other they can also exactly differentiate each consumer through extremely tailored observations.

In-depth analysis and data mining can help firms collect important information that can be used to recognize consumers' thinking and behavioral paths. This is even truer for today's consumers, as they express themselves and their personality through consumption attitudes. For instance, as consumers are today more loyal to brands than they were in the past, firms must try to maximize customer value. However, as each consumer is different and has different needs, the aforesaid strategy cannot be standardized. Big Data analysis can help comprehend each consumer's preferences and behavior and offer insights for the development of perfect marketing strategies.

In conclusion, a strategic relationship between Big Data and marketing highlights the need to create value placing consumers at the center. The traditional strategic marketing perspective affirms that a large-scale production implies standardized and non-personalized production models, and that a customized production implies individual production and small-scale customization (Malone et al. 2006). In other words, customization and mass production cannot be blended. BDM, on the contrary, appears to overcome this incongruity. It seems more and more feasible for big firms to offer one-to-one customer relationship opportunities such as the ones typical of customary small convenience stores, combined with customized recommendations and real-time or Artificial Intelligence devices (Capatina et al. 2020).

5 Big Data Marketing Analysis

Big Data analysis can help firms in consumer-driven orientation. Data analysis is already consumer-oriented (Erevelles et al. 2016). Consumer's needs are investigated in order to improve production, product design, and marketing. Osterwalder employs the concept of *human-oriented ontology*, where the criteria of rationalization and confidentiality are applied to consumers' data (Osterwalder 2004). Nevertheless, only by truly protecting privacy the Internet and Big Data allow a healthy progress concerning information technology and consequently create concrete opportunities for consumer-oriented strategies. In this respect, it is important to strategically manage the opposition between general and mass data on the one hand and central data on the other. Big Data are intrinsically connected amounts of data, huge and heterogeneous, collected and managed through high-speed standards. Consequently, firms must consider merely the data that reflect market demand and consumer behavior critical (Wall et al. 2013). Superfluous and redundant data analysis cannot but negatively influence firms' strategies (in terms of costs, time, and other resources). Firms must identify the appropriate standards to extract, manage, and store central and critical data. In order to do so, a good strategy must employ specialized data analysts who will help obtain exploitable and useful results. Concomitantly, in order to realize value creation, companies should integrate the value chain by sharing data. As on their own firms' internal data no longer suffice to meet consumers' needs (Chesbrough and Rosenbloom 2002), the strategy of Big Data sharing is already a fact. For example, it is now possible to extend the data of regular upstream and downstream channels, because it is possible to set up links with social media platforms' data. Data from social media represent a central font of peripheral data. Yet, if the data are not connected to the firm's individual marketing strategy or to the data publisher through the data-gathering process, these data lose their value (Zott and Amit 2007).

6 Using SWOT Analysis for Big Data Marketing

In order to benefit from the use of Big Data, firms must be aware of data limits and strengths (Ahmadi et al. 2016). The value of any analysis depends on the accessibility of precise and pertinent data. Different applications, processes and systems produce Big Data. There are diverse sources of Big Data counting Internet, business processes, transactions, cloud, social media, etc. Referring to its features such as variety, velocity and volume Big Data may assure the SWOT analysis input needs for businesses. The main sources of Big Data are customer profiles on apps and websites, online surveys, interactions through social media, reviews or feedback, and economic transactions. Therefore it is crucial for businesses to understand how they can best harness these data for SWOT analysis. On the other side Ahmadi's study demonstrates how the SWOT model may itself be efficaciously exploited to identify the advantages, disadvantages, opportunities, risks and threats in BDM.

Strengths. Cloud computing is an adjuvant technology used to conceive and realize practical applications of Big Data. It offers great opportunities, It permitting, as it does, Big Data processing, analysis, and storage (Ahmadi et al. 2016). The cloud-computing storage function helps store massive data, but also semi-structured and unstructured data (texts, videos, audios, images). At the same time, cloud-computing data analysis helps quickly evaluate Big Data and extract the most significant information. In sum, cloud computing supplies firms with effective data processing potentialities, supporting them on a highly competitive data background. Big Data's strong provisional and analysis qualifications make the value of data noticeable (Ahmadi et al. 2016). As mentioned above, application opportunities for BDM are varied. They range from services and products customization, to customized communication; customer relationship management; product cross selling; etc. Finally, as Big Data provisional results are time-sensitive, firms discover new ways to successfully respond to the market and its publics.

Weakness. The first shortcoming is related to the investments in Big Data platforms. Having understood the value of Big Data and the need to expand Big Data applications, firms need to create Big Data platforms to stock up, analyze, and manage Big Data (Ahmadi et al. 2016). Such a fact implies a great upfront investment (hardware, software, human resources, etc.). Not to make these investments implies risks such as obtaining unreal or non-comprehensive data, data errors, the managing of incomplete data. This leads to uncertain data analysis results or incorrect Big Data forecast. Obviously, as firms' marketing decisions increasingly depend on Big Data, such a level of ambiguity represents a danger. In addition, the quality of data is something that is hard to certify; for instance, the increasing size of data leads to an amplification of noises. As noted above, BDM is based on various data collected in the different business activities of a firm, or in network data external to the firm (Yuksel and Dagdeviren 2007). Having to deal out of necessity with data deriving from multifaceted sources and selected from huge quantities, control of the level of quality is hard to reach. The need to form specialized human resources, capable of managing marketing issues (such as computers or network and data analysis and mining) is a last-not-least liability (Yuksel and Dagdeviren 2007).

Opportunities. Today's firms deal with a gradually more competitive background (Gurel and Tat 2017). At the same time, the demand for customized services or products is also growing. Offering goods that respond to the needs of the target market more effectively than competitors make firms succeed. However, this implies the necessity of perfectly identifying these needs. Big Data provide firms with real consumers' needs, and offer to consumers customized goods while increasing market opportunities. The empowerment of Internet along with artificial intelligence embodies a huge opportunity as well as sustaining the increase of applications in BDM (Gurel and Tat 2017). Internet or AI-based interactions involving individuals and firms are increasingly common and offer new and innovative ways to develop firms' marketing effectiveness. By such interactions, firms may reinforce contact with customers, enhance the levels of brand image, multiply word-of-mouth opportunities and increase all kind of marketing benefits. Thanks to the expansion of new mobile and AI technologies, many new data sources are available. In sum, the Big Data era has offered firms new opportunities to develop new marketing strategies based on precise data features connected to consumer life cycle behavior.

Threats. A primary concern touches on the risks connected to consumers' privacy. Big Data on consumers involve an enormous amount of personal traits and consumer behaviors. When firms manage their customers' data (by mistake or intentionally) ambiguously, erroneously, or illegally, they not only cause damage to customers, but also injure their brand or company perception. As a result, the corporate image suffers. Firms must manage customers' information security while preserving their data. This implies a careful BDM application management that protects customers' privacy. A further threat derives from missing data and from false data which are responsible for false positives. Huge data amounts do not necessarily correspond to authentic or exhaustive information on consumers (e.g., consumers' privacy concerns) (Gasparotti 2009). Incorrect or approximate data lead to ambiguity in the data analysis results, affecting forecasting among other things.

7 Conclusion

Marketing has always been based on the attainment, processing, and employment of information coming from the market. Big Data have transformed and reshaped such model and the whole market environment. Thus, the information coming from the market, which firms employ in order to build their marketing strategies, has been radically transformed in terms of data quantity, configuration, and processes. Consequently, the possibility of creating and efficiently managing a Big Data system allows today's firms to focus on critical information only, and to do it quickly. Such a new marketing strategy provides the firms with efficient support as it offers them new opportunities to discover new markets, customers, competitive strategies, etc. Moreover, it may facilitate the correction and the amelioration of brand management decisions while enhancing incomes. As a vital division of business operations, marketing should always be attentive and reacting to any market environment alteration. By taking into account today's market environment, firms must then integrate Big Data in every marketing process. This will help the firm better to respond to consumers and to seize new opportunities while facing competition. As for all novelties, BDM needs further analysis leading to further results. For instance, further research may help firms face a number of existing limits (e.g., lack of specialized human resources, organizational restructuring, difficulties in the effective sharing of data between departments, data-quality complexity, etc.). The insertion of marketing analysis in a Big Data scenario represents a new challenging research and managerial topic. Such a merging plays a fundamental role in the long-term firms' development and in the creation of highly competitive values.

References

Afuah, A., & Tucci, C. L. (2001). *Internet business models and strategies: Text and cases.* New York, NY: McGraw-Hill.

Ahmadi, M., Dileepan, P., & Wheatley, C. (2016). A SWOT analysis of Big Data. *Journal of Education for Business, 91*(5), 289–294. https://doi.org/10.1080/08832323.2016.1181045

Amit, R., & Zott, C. (2001). Value creation in e-business. *Strategic Management Journal, 22*(6/7), 493–520.

Athanasiadis, I., & Ioannides, D. (2015). A statistical analysis of big web market data structure using a Big Dataset of wines. *Procedia Economics and Finance, 33,* 256–368. https://doi.org/10.1016/S2212-5671(15)01710-4

Capatina, A., Kachour, M., Lichy, J., Micu, A., Micu, A. E., & Codignola, F. (2020). Matching the future capabilities of an artificial intelligence-based software for social media marketing with potential users' expectations. *Technological Forecasting and Social Change, 151.* https://doi.org/10.1016/j.techfore.2019.119794

Chen, C. P., & Zhang, C. Y. (2014). Data-intensive applications, challenges, techniques and technologies: A survey on Big Data. *Information Sciences, 275,* 314–347. https://doi.org/10.1016/j.ins.2014.01.015

Chesbrough, H. W., & Rosenbloom, R. S. (2002). The role of the business model in capturing value from innovation: Evidence Xerox Corporation's technology spin-off companies. *Industrial and Corporate Change, 11*(3), 529–555. https://doi.org/10.1093/icc/11.3.529

Constantinides, E. (2006). The Marketing mix revisited: Towards the 21st century marketing. *Journal of Marketing Management, 22*(3–4), 407–438. https://doi.org/10.1362/026725706776861190

Dubosson Torbay, M., Osterwalder, A., & Pigner, Y. (2002). E-business model design, classification and measurements. *Thunderbird International Business Review, 44*(1), 5–23. https://doi.org/10.1002/tie.1036.

Ducange, P., Pecori, R., & Mezzina, P. (2018). A glimpse on Big Data analytics in the framework of marketing strategies. *Methodologies and Application, 22*(1), 325–342. https://doi.org/10.1007/s00500-017-2536-4

Dumbill, E. (2018). *What is Big Data.* Retrieved January 24, 2018, from https://strata.oreilly.com/2012/01/what-is-big-data.

Erevelles, S., Fukawa, N., & Swayne, L. (2016). Big Data consumer analytics and the transformation of marketing. *Journal of Business Research, 69*(2), 897–904. https://doi.org/10.1016/j.jbusres.2015.07.001

Gasparotti, C. (2009). The internal and external environment analysis of Romanian naval industry with SWOT model. *Management and Marketing, 4*(3), 97–110.

Gurel, E., & Tat, M. (2017). SWOT Analysis, a theoretical review. *The Journal of International Social Research, 10*(51), 994–1006. https://doi.org/10.17719/jisr.2017.1832

Hamel, G. (2003). Innovation as a deep capability. *Leader to Leader, 27*(27), 19–24.

Huffaker, R., & Whittlesey, N. (2000). The role of Prior Appropriation in allocating water resources into the 21st century. *International Journal of Water Resources Development, 16*(2), 287–289. https://doi.org/10.1080/07900620050003161

Kshetri, N. (2016). Big Data's role in expanding access to financial services in China. *International Journal of Information Management, 36*(3), 297–308. https://doi.org/10.1016/j.ijinfomgt.2015.11.014

Linder, J., & Cantrell, S. (2002). *Changing business models: Surveying the landscape, Accenture.* Retrieved June 2018, from Semantic Scholar: https://pdfs.semanticscholar.org/3505/d56b0c632879a32715d56a131261ba7deac1.pdf?_ga=2.33715302.633741437.1574697618-1585449437.1574697618.

Magretta, J. (2002). Why business models matter. *Harvard Business Review, 80*(5), 86–92.

Malone, T., Weill, P., Lai, R., et al. (2006). *Do some business models perform better than others.* MIT, Working Paper. Retrieved April 2018, from MPRA: https://mpra.ub.uni-muenchen.de/4752/1/MPRA_paper_4752.pdf.

Mitchell, D., & Coles, C. (2003). The ultimate competitive advantage of continuing business model innovation. *Journal of Business Strategy, 24*(05), 15–21.

Morris, M. H., Schindehutte, M., & Allen, J. (2005). The Entrepreneur's business model: Toward a unified perspective. *Journal of Business Research, 58*(6), 726–735. https://doi.org/10.1016/j.jbusres.2003.11.001

Müller, S. D., & Jensen, P. (2017). Big Data in the Danish industry: Application and value creation. *Business Process Management Journal, 23*(3), 645–670. https://doi.org/10.1108/BPMJ-01-2016-0017

Osterwalder, A. (2004). *The business model ontology. A proposition in a design science approach.* Dissertation, Université de Lausanne.

Osterwalder, A., Pigneur, Y., & Tucci, C. L. (2005) Clarifying business models: Origins, present, and future of the concept. *Communications of the Information Systems, 16*(1). https://doi.org/10.17705/1CAIS.01601.

Peterovic, O., Kittl, C., & Teksten, D. D. (2001). Developing business models for eBusiness. *SSRN Electronic Journal.* https://doi.org/10.2139/ssrn.1658505

Rappa, M. A. (2004). The utility business model and future of computing services. *IBM Systems Journal, 43*(1), 32–42.

Rygielski, C., Wang, J. C., & Yen, D. C. (2002). Data mining techniques for customer relationship management. *Technology in Society, 24*(4), 122–132. https://doi.org/10.1016/S0160-791X(02)00038-6

Shaw, M. J., Subramaniam, C., Woo Tan, G., et al. (2001). Knowledge management and data mining for marketing. *Decision Support Systems, 31*(1), 335–341. https://doi.org/10.1016/S0167-9236(00)00123-8

Shih, C. C., Lin, T. M., & Luarn, P. (2014). Fan-centric social media: The Xiaomi phenomenon in China. *Business Horizons, 57*(3), 349–358. https://doi.org/10.1016/j.bushor.2013.12.006

Utkarsh, S., & Santosh, G. (2015). Impact of Big Data analytics on banking sector: Learning for Indian banks. *Procedia Computer Science, 50,* 643–652. https://doi.org/10.1016/j.procs.2015.04.098

Wall, D., Epstein, E., & Hagen, C. (2013). *Big Data and the creative destruction of today's business models.* Retrieved January 2019, from ATKEARNEY: https://www.atkearney.it/analytics/article?/a/big-data-and-the-creative-destruction-of-today-s-business-models.

Yuksel, I., & Dagdeviren, M. (2007). Using the analytic network process (ANP) in a SWOT analysis. *Information Sciences, 177*(16), 3364–3382. https://doi.org/10.1016/j.ins.2007.01.001

Zott, C., & Amit, R. (2007). Business model design and he performance of entrepreneurial firms. *Organization Science, 18*(2), 181–199. https://doi.org/10.1287/orsc.1060.0232

Zott, C., Amit, R., & Massa, L. (2011). The business model: Recent developments and future research. *Journal of Management, 37*(4), 126–137. https://doi.org/10.2139/ssrn.1674384

Data Processing in a Healthcare National System

(With the Analysis of the Italian HNS)

Manlio d'Agostino Panebianco and Anna Capoluongo

Abstract In modern society and economy, a "personal data" can be considered as an "asset" with an own intrinsic value: since the increasing speed of technological evolution (added to the borderless context given by Globalisation) led the International Regulators to consider both how to guarantee the rights of individual natural person and the impact of Big-Data processing and management on the society and on economical markets, including in this range both the public and private scope. Nowadays we are assisting to a natural evolution from Big-Data to Smart-Data, especially in medical's and pharma's fields – due to a large treatment of sensitive data – in which it is fundamental to focus on the balancing between advantages and obligations, through the correct application of the "accountability principle" of the GDPR.Artificial Intelligence, Machine Learning, IoT, and Smart Data, due to their nature and customization, give an added value to what can be called "Health 4.0", i.e. that mechanism of close collaboration between the operators of the integrated health and pharmaceutical system, through the interaction of information and data. Nevertheless, many legal and ethical aspects have not been exploited yet, still giving some uncertainties on possible future evolutions.

1 Introduction

In a global economical and social context (strongly influenced by rapid technological evolution), often data processing (and its regulation) is undervalued compared to its real extent, and actually, it cannot be *perceived* as "minor" or "secondary", since nowadays a "personal data" can be considered as an "asset" with its own intrinsic

M. d'Agostino Panebianco (✉)
BASC, University of Milano Bicocca, Milan, Italy
e-mail: mdagostino@manliodagostino.com

CESINTES, University of Rome Tor Vergata, Rome, Italy

A. Capoluongo
Milan, Italy

P. Mariani and M. Zenga (eds.), *Data Science and Social Research II*,
Studies in Classification, Data Analysis, and Knowledge Organization,
https://doi.org/10.1007/978-3-030-51222-4_10

value, and it is undeniable how in the new digital economy, data has become one of the main sources of value creation (Clemente 2019).

In order to complete the frame and to share the correct meaning and definition, according to Article 4 of Regulation (EU) 2016/679 of the European Parliament and of the Council of 27 April 2016 (so-called, General Data Protection Regulation or even GDPR), a *personal data* is any information relating to an identified or identifiable natural person ('data subject'); an identifiable natural person is one who can be identified, directly or indirectly, in particular by reference to an identifier such as a name, an identification number, location data, an online identifier or to one or more factors specific to the physical, physiological, genetic, mental, economic, cultural or social identity of that natural person.

As a matter of facts, the increasing speed of technological evolution (added to the borderless context given by Globalisation) led the scientific community and international Regulators to consider both how to guarantee the rights of an individual natural person and the impact of Big-Data processing and management on the society and on economical markets, including in this range both the public and private scope.

In a sense, it sounds appropriate to recall firstly the Bauman's G-Local approach, in order to establish a correct balance in this innovation path, each person should think globally, but act locally, to reduce the negative effects of standardization and flattening deriving from Globalization, to use (on the contrary) its benefits in enhancing the person (Bauman 1998).

1.1 From Big-Data to Smart-Data

Firstly, some possible definitions (since its perimeter cannot be defined in a static way, due to the dynamism of the upstream phenomena that generate it) of Big-Data can be given: «Microsoft provides a notably succinct definition: *Big-data is the term increasingly used to describe the process of applying serious computing power—the latest in machine learning and artificial intelligence—to seriously massive and often highly complex sets of information*» (Ward and Barker 2013). Moreover, these another ones are interesting: «Big-data is high-volume, high-velocity and high-variety information assets that demand cost-effective, innovative forms of information processing for enhanced insight and decision-making» (Gandomi and Haider 2015) and also «Big-data is a term that describes large volumes of high velocity, complex and variable data that require advanced techniques and technologies to enable the capture, storage, distribution, management and analysis of the information» (Gandomi and Haider 2015).

It is also important to point out that we are assisting to a (natural) evolution from Big-Data to Smart-Data—i.e. Big-Data related and/or deriving from smart applications (Al Nuaimi et al. 2015)—especially in medical and pharma fields (due to a large treatment of sensitive data) in which it is fundamental to focus on the balancing between advantages and obligations, through the correct application of the "*accountability principle*" (according to Article 5 of GDPR).

The use of Big-Data (powered by Artificial Intelligence, Machine Learning, IoT, etc.) leads directly to Smart-Data which, due to their nature and customization, give an added value to what can be called "Health 4.0", i.e. that mechanism of close collaboration between the operators of the integrated health and pharmaceutical system, through the interaction of information and data.

Specifically, it must be highlighted that the aforementioned added value of Smart-Data in the pharmaceutical and healthcare fields can be traced back to the evolution of the analysis processes, passed—over time—from simple diagnostics to descriptive, predictive and finally prescriptive analytics.

The wide range of scope of Big-Data and of Smart-Data highlights the exposure to risks and possible damages for an individual personal right or interest, in case of any kind of abuse or misuse during data processing, independently to the physical location of each interested party. Such attention to the actual geographical context (and to its possible future evolutions) is paid by the European Legislator, so much so that the territorial scope of the Regulation (according to Article 3) is not only related to the place of establishment of a controller or a processor (within the European Union), regardless of whether the processing takes place in the Union or not; the second and more extensive one hinges on the "interested parties" resident in the EU, who are the object of activities of offering goods and services (profit or not) or where the behaviour is monitored, even if the processing is carried out by a controller or a processor established outside the European area.

The crucial point of the transition from Big-Data to Smart-Data is that a database selected and implemented from the outset in accordance with the principles of the GDPR (specifically in compliance with the principles of Privacy-by-Design and Privacy-by-Default) assumes an "added value"—precisely by virtue of the procedures and selection processes applied from the outset—to reduce risks to a minimum, also significantly affecting the sanctioning effect, and consequently, on the possible reputational damage deriving from it.

2 The Legal Framework of Data Protection in Health Sector

On May 25, 2018, with the entry into force and full applicability—also with sanctioning effects—of the General Data Protection Regulation (so-called, GDPR) which has standardized, homogenized and harmonized the legislative and regulatory framework of the various European countries, it deserves an appropriate reflection on those aspects related to the processing of personal data, particularly those included in "special categories"—according to Article 9 of GDPR, these are those personal data that reveal racial or ethnic origin, political opinions, religious or philosophical beliefs, or trade union membership, genetic data, biometric data for the purpose of uniquely identifying a natural person, concerning health and/or a natural person's sex life or sexual orientation—referred to patients and minors, in order to highlight how to guarantee the fundamental rights of natural persons.

For these reasons, the Regulation states as a primary objective (and assumes) that «the protection of natural persons in relation to the processing of personal data is a fundamental right» focusing attention on an issue of ethical nature, rather than a regulatory one.

As a matter of facts, a National Healthcare System (NHS) is a complex system, in which it required special attention to data processing, due to the nature of data processed (concerning the health of the interested parties) in relation to the expectation of respect of "privacy" and of confidentiality, especially in the relationship between doctors and paramedical staff and patients (Modafferi 2017). «Confidentiality is a central feature of all statements of ethical medical practice» (Jones 2003).

Moreover, the health field is one of the most delicate in the processing of data, since several and different regulations coexist, although they converge towards the respect of the inviolable rights of the data subject, and any kinds of *data breach* concerning "privacy" cannot be related only to unlawful or illicit data treatments, but also to major violations, such as cybercrimes or the violation of confidentiality deriving from professional secrecy (the term "secret" derives from the Latin "secretum", past participle of the verb "secernere" which means to separate, or to put aside).

For exhaustiveness, it should be noted that according to Article 4 of GDPR, a "personal data breach" refers to a breach of security leading to the accidental or unlawful destruction, loss, alteration, unauthorized disclosure of, or access to, personal data transmitted, stored or otherwise processed.

Especially in this context, it is important to focus on the concept of "*medical confidentiality*", that is declined in various ways.

First of all, it is appropriate to recall what is quoted in "The 1974 BMA [British Medical Association, *edit.*] handbook on Medical Ethics boldly" that «reaffirms the doctor's obligation to maintain secrecy in what appear to be most uncompromising terms: it is a doctor's duty strictly to observe the rule of professional secrecy by refraining from disclosing voluntarily to any third party, information which he has learned directly or indirectly in his professional relationship with the patient. The death of the patient does not absolve the doctor from the obligation to maintain secrecy» (Thompson 1979).

Therefore, in the patient's perspective, this can be declined in «the right of the patient to have protected and not disclosed to third parties information concerning his state of health, even about the services performed and all that he has confided to the doctor in relation to his psychophysical conditions» (Simeoni et al. 1998).

Obviously, it is clear that these affirmations derive directly from the "Hippocratic Oath" that every doctor lends as *a sine qua non* condition for the exercise of the profession, which in its traditional version states «what I may see or hear in the course of the treatment or even outside of the treatment in regard to the life of men, which on no account one must spread abroad, I will keep to myself, holding such things shameful to be spoken about».

It is important to highlight, especially introducing the interactions within the medical fields, that this confidentiality has to be respected both by doctors, but also by all the people involved in data processing, for whatever reason, regardless of their job or role, even in complex structures like hospitals, clinics, laboratories, etc.

2.1 *The Principle of «Accountability»*

The key principle of the Regulation, the most important, impacting and innovative is "*accountability*": the English term cannot be interpreted as an only *ex ante* and/or static "responsibility", but is a dynamic and ongoing "*responsibility*", which primarily invests the Data Controller with a *ripple effect* on all other persons—both natural and/or other entities—involved in the treatment process, by adopting appropriate behaviours, measures and methods to protect the rights of the individual concerned, reducing its risk exposure.

As a matter of facts, the GDPR introduces a substantial adoption of the most well-known and widespread risk management methodologies (integrated with the legal aspects of the protection of rights) in order to manage and mitigate the data processing risk exposures, and—according to Article 5, 1(f) of GDPR—aimed to «ensures appropriate security of the personal data, including protection against unauthorized or unlawful processing and against accidental loss, destruction or damage, using appropriate technical or organizational measures» i.e. *Confidentiality*, *Integrity* and *Availability*.

The lack or partial application of the privacy rules, and therefore, certainly also of the "*principle of accountability*" should not be underestimated lightly, since the sanctions are real and effective and data security is a daily and global issue.

3 Processing of "Special" Data in the Health Sector

An innovation introduced by the EU Regulation, differently to the past, states a priori that is prohibited the processing of special categories of personal data—in order to limit and avoid any unneeded risk exposure to any improper uses—with the exception in presence of an explicit consent to the processing of those personal data for one or more specified purposes, or in those cases in which processing is necessary to protect the vital interests of the data subject or of another natural person where the data subject is physically or legally incapable of giving consent.

For what the exceptions concern, the Regulation highlights how the processing of "sensitive" data in the absence of express consent, must be subjected to higher precautions, as well as to appropriate security measures (referring to the measured exposure at risks), to protect the rights and freedoms of natural persons.

In particular, it is highlighted—that in the specific sphere of personal data relating to health—data treatments carried out for reasons of public interest, a data processor cannot modify the declared purposes, as these data cannot be subject to processing by third parties (especially, by employers, insurance companies and credit institutions).

3.1 The Consent in Medical Field

The EU Regulation introduces some substantial methodological innovations, although in a logical continuity with the previous legislation, about the information to be given to the interested party, renewing the data subject's right to be informed about the methods and purposes of the processing remains, especially in medical and pharma field.

It is important to underline how the methodology to express the data treatment consent derives directly to the "informed consent" for clinical treatment, that responds, on one hand, to need and right to self-determination of the patient with regard to choices concerning his own health; on the other, «*informed consent* has become the primary paradigm for protecting the legal rights of patients and guiding the ethical practice of medicine. It may be used for different purposes in different contexts: legal, ethical or administrative. Although these purposes overlap, they are not identical, thus leading to different standards and criteria for what constitutes "adequate" informed consent» (Hall et al. 2012).

«At the end of the 1970s, a number of privacy principles emerged under the concept of Fair Information Practices and later became the foundation for the Organization for Economic Cooperation and Development (OECD) *Guidelines on the Protection of Privacy and Transborder Flows of Personal Data* adopted in 1980. Those principles, which seek to balance the "fundamental but competing values" of "privacy and the free flow of information", form the basis of most privacy legislation around the world.[…] Over the years, and especially in the context of the Internet, this system of "notice and consent", originally intended to be only one of the multiple ways through which the lawful processing of personal data can take place, has become the dominant mechanism» (Cate and Mayer-Schönberger 2013).

The "*notice and consent*" unfortunately often represent a formally correct and legal modality (compliant with the applicable provisions in force), although sometimes does not guarantee the correct "*balance of interests*" (Greenberg 1973), creating situations of "*information asymmetry*"—it «occurs when the knowledge of one contracting party is inferior to that of the other party regarding the counterparty's true intentions and planned activities» (Mas-Colell et al. 1995)—in the data subject's decision-making process, due to his/her lower "*digital awareness*" (d'Agostino Panebianco 2019a, b, c).

To contrast this negative phenomenon (i.e. information asymmetry), the EU legislator introduced (within the Chapter III*Rights of the data subject) an entire* Section called "Transparency and modalities", in which is described how the Data Controller should provide the correct information (both in terms of contents, communication and modalities) addressed to the single data subject in order to obtain a free and explicit consent and to let the exercise of the rights. It can be summarized that all communication should be "*receiver-oriented*", respecting these following features: *in a concise, transparent, intelligible and easily accessible form, using clear and plain language* (according to Article 12 of GDPR), defining an adequate storage time (according to Article 5 of GDPR), and also any notice or verification criterion of usability.

It seems interesting to highlight, for example, how the Italian Data Protection Authority in its General Decision of March 7, 2019, entitled "*Clarifications on the application of the discipline for the treatment of health data in the health field*" suggests—in order to obtain greater communication and information effectiveness—to provide the required information addressed to data subjects in a progressive manner: firstly, giving only the information relating to the treatments related to the "ordinary activities" of health services, and postponing that specific information related to some more specialistic or extraordinary activities (such as the supply of health facilities, methods of delivering medical reports online, research purposes, etc.), only at a later time, and to patients actually interested by such services and further treatment.

3.2 Exceptions to Ordinary Consent

In some circumstances, medical and/or pharma entities are called to obtain a free and explicit consent: as a matter of facts, when the status changes (i.e. the first data processing was started referring to a legal basis or a legislative measure, the patient's care goes on further, at a time when the patient/data subject is able to give consent) it is necessary to adopt the "ordinary" informed consent process (without any prejudice to the validity of the previous data processing). Moreover, this is a more sensitive case, when in the data processing are involved external private actors (such as Insurance Companies) who can finance or reimburse the related expenses deriving from the provision of the service or healthcare.

To the previous quoted cases, it is important to list some others, which are only partially provided by the Regulation, and refer to other provisions: when data subject is a minor (premises n. 38 and Article 8 of GDPR) or a patient is suffering from a disease that affects his mental faculties (so as not to allow him/her the autonomous decision) it is required to collect the consent by—respectively—in the first case, who has the right to exercise parental authority (the most delicate case concerns double or joint power), and in the second one, the "*guardian*" or the "*administrator*" appointed by the competent authority, depending on the regulatory framework of each country.

In this last quoted circumstance, it is important to highlight that even the parent, guardian and/or administrator take qualification of data subjects, and then it is important to apply the "*notice and consent*" procedure also in their favour, also paying particular attention to the verification of powers and to document it, for future and following proof of responsibility.

4 Organizational Aspects of Data Processing

Any organization is required to adopt a complex model in order to manage actively all the ongoing changes that naturally occurred daily: as a matter of facts, this does not concern only the effects of each different cause, but mainly the interactions amongst each player and factor. For this reason, it is important to have a wider range of view and perspective, that can be defined as a "*holistic complexity approach*" (d'Agostino Panebianco 2019a, b, c).

Both at the time of the determination of the means for processing and at the time of the processing itself, to answer to the required *risk-based-approach* in data processing, one of the most important aspect is the implementation and application of a pyramidal organizational model, based on differentiated authorization profiles: at the top, there is a data controller, and going down we can find one or more data processor (both internal or outsourced), all the other authorized sub-processors, and all staff involved in processing operations. To the previously quoted, it is needed to add the adoption, as well, of some other appropriate integrated technical-organizational measures, such as *pseudonymisation and data-minimization*, including also the assignment of responsibilities, awareness-raising, training, and related audits (referring to Articles 25 and 39 of GDPR).

It is important to notice that in *pseudonymisation* the processing of personal data aims to no longer be attributed to a specific data subject without the use of additional information, provided that such additional information is kept separately and is subject to technical and organizational measures to ensure that the personal data are not attributed to an identified or identifiable natural person; while in *data-minimisation* personal data shall be adequate and relevant but limited to what is necessary in relation to the purposes for which they are processed.

It is important to highlight, especially in the healthcare field (both medical and pharma) that the "sensitive nature" of processed data requires the data controller to verify the effectiveness and efficiency of the systems adopted (as a declination of the principles established by the GDPR), particularly with regard to the effective protection of rights.

4.1 The Parties Involved in Data Processing

First of all, in the aim to design the correct organizational model, it seems important to identify and determine the roles of the different "actors": starting from the definitions of the Regulation, putting them in relationship with the patient, with whom it seems appropriate to start (as "*data subject*").

4.1.1 Data Subject

According to Article 4 of Regulation (EU) 2016/679, a *Data Subject* is an identified or identifiable natural person who can be identified, directly or indirectly, in particular by reference to an identifier such as a name, an identification number, location data, an online identifier or to one or more factors specific to the physical, physiological, genetic, mental, economic, cultural or social identity of that natural person.

In the context of healthcare (medical treatment and pharma), he/she can certainly be identified as "patient", i.e. in the person who asks for, obtains and/or relies on treatment (also of a preventive nature) or is subjected to urgent health interventions for the protection of his safety or health and/or safeguarding of life. It is important to highlight, how sometimes the treatment (both medical's and data's) is made, both by public and/or private structures that—under the agreement (indirect) or in pure solvency—also through the financial intervention of third parties (like, for examples, Insurance Companies). Moreover, as previously described, some particular circumstances lead to identifying the data subject, as well as the person who exercises the power of dispose (see below).

4.1.2 The Data Controller and the Joint Controllers

The Data Controller is the first legal or natural person (even a Public Authority, agency or other body) expressly called to decline in practice the "accountability principle"—according to Article 5 of Regulation (EU) 2016/679—demonstrating the implemented organizational model's compliance and effectiveness: to the Controller (eventually with one or more joint controllers) is given the power to determine the purposes and means of the processing of personal data, as well as the adopted security solutions, are adequate in terms of protection to data subject's rights and in mitigating arising risk exposures (according to Article 24 of Regulation (EU) 2016/679).

In a National Healthcare System, a Data controller can be identified with one or more of the public or private structures to which a patient refers to ask and obtain a medical service, lato sensu.

In many cases, data processing can be performed by more than one joint data controllers (with no prejudice at all for *patient data subject's* rights, since he/she can exercise the rights against even just one of them), due to the different law structure of each National Healthcare System, and in many cases, responding to the "*principle of subsidiarity*", referring to Article 5(3) of the Treaty on European Union (TEU) and Protocol (No 2) on the application of the principles of subsidiarity and proportionality.

The principles of subsidiarity and proportionality govern the exercise of the EU's competences. In areas in which the European Union does not have exclusive competence, the principle of subsidiarity seeks to safeguard the ability of the Member States to take decisions and action and authorizes intervention by the Union when the objectives of an action cannot be sufficiently achieved by the Member States, but can be better achieved at Union level, "by reason of the scale and effects of the proposed action". The purpose of including a reference to the principle in the EU

Treaties is also to ensure that powers are exercised as close to the citizen as possible, in accordance with the proximity principle referred to in Article 10(3) of the TEU.

In many National Healthcare Systems, some medical/paramedical services, and some other activities are provided not only by public entities but even by private professionals or companies, in force of accreditation process and concessions (a sort of outsourcing): the top of the pyramidal organization is always and in any case responsible for the actions taken in the lower stages, by virtue of the principles of *culpa in eligendo* and *culpa in vigilando* (d'Agostino Panebianco 2019a, b, c).

4.1.3 The Data Processor and the "Sub-Data Processor"

The designation of (one or more) data processor is a faculty expressly provided for by the GDPR, in order to allow the data controller to adopt an organizational model that is sustainable, safe and effective: as a matter of facts, their identification should answer to the need that they provide sufficient guarantees to implement appropriate technical and organizational measures in such a manner that processing will meet the requirements of the Regulation and ensure the protection of the rights of the data subject. In this perspective, an innovation of GDPR is the possibility for the «data processor» to appoint a «sub-data processor» for the execution of specific processing activities on behalf of the Data Controller, by the prior condition of his express authorization: as a matter of facts, the Regulation states that «the processor shall not engage another processor without prior specific or general written authorization of the controller» (Article 28.2 of Regulation EU 2016/679).

4.1.4 Authorized Person

Any natural person who have access or process data, has to be expressly authorized: he/she should act only under the authority of the Controller or the Processor, following precise guidelines and instructions.

It follows that an "authorized person" is a natural person—appointed by a Controller or a Processor—that because of his/her job role provides (para)medical services or care and so processes personal data. Considering the healthcare field, most of the treated data is sensitive and belongs to particular data classification, so it is very important that data processing is performed both in compliance with present laws, but also within the purposes and methods established by the Data Controller, in the light of the results of the Privacy Risk Assessment. To do so, all the "authorized persons" have to be specifically trained.

4.1.5 The Data Protection Officer

The Regulation provides that the controller and/or the processor shall designate a Data Protection Officer in any case where: the processing is carried out by a Public

Authority or body (except for Courts acting in their judicial capacity), or the core activities of the controller or the processor consist of processing operations which, by virtue of their nature, their scope and/or their purposes, require regular and systematic monitoring of data subjects on a large-scale; or the core activities of the controller or the processor consist of processing on a large-scale of special categories of data (pursuant to Article 9 and personal data relating to criminal convictions and offences referred to in Article 10).

Considering both the public nature of Controllers and Processor and the categories of processed data (mainly sensitive), in most of the case, the designation of a DPO sounds compulsory, in most of the National legal frameworks.

But, depending on the national legislation, an appropriate evaluation of merit about the designation within affiliated and accredited subjects, and within outsourced functions, is due to evaluate whether in presence of the requirements of Article 37 of GDPR.

In general, as far as the more complex structures are concerned, it is possible to design a single DPO for different healthcare services, in coherence to the results of the Privacy Risk Assessment carried out under the responsibility of the data controller.

5 The Italian National Healthcare System (I-NHS)

The complexity of a National Healthcare System faces different problems and challenges in each different country, due to the different regional legislation and framework. The General Data Protection Regulation offers the opportunity to think about both many under-evaluated risks (especially emerging by the IT evolutions and related economical interests, and by illegal phenomena, such as cybercrimes), but also to some solutions that can standardize the trans-national approach to "particular personal data" processing.

The peculiarity and features of each national framework and Healthcare System (NHS) does not allow to create just one standard approach, but on the contrary, starting from a specific NHS it is possible to comment and highlight some important aspects, that can be useful and used in all the others.

5.1 I-NHS: A Public and Private Partnership's Model

The Italian National Health Service (I-NHS) was established in 1978, to grant universal access to a uniform level of care throughout Italy, financed by general taxation (De Felice and Petrillo 2015), transforming the previous model of a social insurance system into a national health service (Doetter and Götze 2011), strongly fragmented into more than a hundred health insurance companies. Moreover, this end-1970s reform developed a decentralization process based on regional and local authorities and autonomy (Serapioni and Duxbury 2012).

The Italian situation (it is not static, due to continuous reforms and legal adjustment) of the National Healthcare System is a regionally based one, that provides universal coverage free of charge at the point of service. «The national level is responsible for ensuring the general objectives and fundamental principles of the national healthcare system. Regional governments, through the regional health departments, are responsible for ensuring the delivery of a benefits package through a network of population-based health management organizations and public and private accredited hospitals» (Lo Scalzo et al. 2009).

The reform of the end of 1990s (Italian Legislative Decree No. 229/1999), extended the regionalization process and strengthened the role of municipalities, making clearer the division of responsibilities between levels of government, also with the results of having softened the previous shift to the market and competition, promoting cooperation among healthcare providers and partnerships with local authorities for health promotion and community care, and all the "Local Healthcare Units" (in Italian, defined as "USL, Unità Sanitaria Locale") and tertiary hospitals were transformed into autonomous bodies, aiming to a more efficient system, and a stronger cooperation between public and private partnerships, and even introducing elements of managed competition among public and private (accredited) providers (Lo Scalzo et al. 2009).

5.2 *I-NHS and Data Processing*

The new challenge of the "Italian e-medicine" expects many opportunities, but also faces many risks deriving from data protection and cybersecurity, since it is based on Citizen's Electronic Health Record, digital dossier, online disease certification and electronic prescription, which since containing many sensitive data, are desirable both from Tech-Giant and cybercriminal organizations.

So, the implementation and adoption of a correct and integrated (both management and data processing) model is one of the first and more effective measure to mitigate emerging risks and to allow a natural and safe IT development of healthcare, medical and pharma field.

Even for the exposed reason, the Italian Data Protection Authority published—on March 7th, 2019—a general position containing the guidelines of data processing in the I-NHS. First of all, referring to the aforementioned Italian Data Protection Authority's position emerges an innovative interpretation: in the context of treatments for "*care purposes*" carried out by (or under the responsibility of) a health professional subject to professional secrecy (or similar figure), it is not needed to obtain an explicit data processing consent of the patient (data subject): obviously, this regards only what concerns the necessary health treatments (which could be defined as a "principle of necessity"), regardless of the circumstance that the Data Controller operates as a freelancer or within a public or private health facility.

Firstly, one of the main focus points regards the direct relationship between the medical informed consent and data protection consent.

As a matter of facts, it should be noted that the lack of proper informed consent—if this circumstance is serious and causes prejudice to the interested party—may constitute an independent item of compensation: this has to be considered additional and distinct with respect to the pure medical liability damage (see judgment of the Italian Supreme Court n. 17022/2017).

Therefore, a parallelism can be observed between informed consent and privacy consent. In terms of "privacy", the GDPR and the Italian Privacy Code clarify the precise content of the information, as well as the areas of administrative and criminal relevance, for the non-compliance with the rules relating to the correct information of the interested party and to obtaining a free, specific, informed and unequivocal expression of will.

With regard to *informed consent* (and the related due information) in the medical field, it should be clarified that even where the doctor warns the patient of the possibility of complications, this does not eliminate tout court the responsibility for its occurrence, because of the avoidance of the fact.

The omitted or insufficient informed consent—also pursuant to Italian Law n. 219/2017 Article 1 and Article 32 of the Italian Constitution—constitutes one of the main harmful conduct carried out by the healthcare professional. A medical treatment—of any kind—in the absence of prior consent must be considered illegitimate and "*certainly unlawful, even when it is in the interest of the patient*" (see judgment of the Italian Supreme Court n. 16503/2017).

Therefore, the right to provide informed consent is an autonomous right with respect to that of receiving medical treatment, so much so that the acquisition of informed consent constitutes service on its own with respect to the execution of the medical services and is, therefore, autonomous item of non-pecuniary damage.

It is worth noting that some rulings have gone so far as to sanction also the incomplete information, reaffirming that this should be rendered in a manner "consistent with the level of scientific knowledge" of the patient (see judgment of the Italian Supreme Court n. 668/2018).

Then, from an organizational perspective, it is clear how complex is the I-NHS, made by a network of public and private "players" (see Italian Legislative decree 30/12/1992, n. 502), each one with a different role: the main ones are "Local Healthcare Units", Hospitals, Universities, Hospitalization Institutes and Care Centres operating in the prevention, diagnosis, treatment and rehabilitation activities provided by the health system, as well as administrative activities related to the aforementioned activities. To the previously described, there are some private "players" (or actors) which provide—on direct payment by patients or on behalf of I-NHS—(para)medical services, lato sensu.

Data Controller in the I-NHS—Considering the structuring (both legislative and organizational) of the Italian National Health System, the pyramidal scheme, the legislative competition between the State and the Regions, and the application of the "*principle of subsidiarity*", besides the role of single "controller" of the Central State (both for the role of coordination, financing and guarantee, and the provision by the State of IT-management tools such as, for example, the E-Health Record, the

E-Health Dossier, etc.), in the other cases, it is appropriate to consider a «joint» data processing, amongst different controllers.

Data Processors in the I-NHS—In all the other circumstances of any kind of outsourcing, and/or agreements and/or accreditation of public or private entities (facilities that provide specialist outpatient services, spa systems, private clinics and hospitals, and the entire pharmaceutical assistance system) and/or professionals (such as pharmacist, surgeon, dentist, veterinary surgeon, psychologist, nurse, obstetrician, paediatric nurse, rehabilitative health professions and other auxiliaries of the health professions, like masseurs, opticians, dental technicians, nurses), it sounds that the most appropriate designation is external "*data processor*".

5.3 Pseudonymisation of Sensitive Personal Data

A further consideration is about the conflict of the pseudonymisation with an Italian regulation about safety in adverse reactions in the field of Pharmacovigilance. «Physicians and other health workers are required to promptly report suspected Adverse Drug Reactions (so-called ADR, ed.) of which they become aware, in any case within two days as part of its activity, in a complete manner and according to the methods identified in the adverse reporting model prepared by the Italian Pharma Authority (in Italian, Agenzia Italiana del Farmaco, AIFA)» (Article 22 of the Decree of the Ministry of Health of 30 April 2015). In the *ADR reporting,* it is envisaged that the patient remains identifiable through certain information (i.e. initials of the name and surname, date of birth or age, sex, ethnic origin, weight, relevant clinical information in the specific case) and this involves the interaction of the subject with some privacy profiles, such as pseudonymisation (i.e. that treatment that allows personal data can no longer be attributed to a specific interested party without the use of additional information, stored separately). In this case, it is possible to define it as *reversibility of the personal data*, since pseudonymisation continues to allow the identification of the individual natural person, although indirectly. As a matter of facts, as highlighted by a recent experiment/survey reported in "*Estimating the success of re-identifications in incomplete datasets using generative models*" (Rocher et al. 2019) it is possible—using a specific model based on Artificial Intelligence—to reconnect the anonymized-data to data subjects (the result of the survey proved that 99.98% of Americans would be correctly re-identified in any dataset using 15 demographic attributes). The authors concluded, with concern, that the results of this survey led and suggested «that even heavily sampled anonymized datasets are unlikely to satisfy the modern standards for anonymization set forth by GDPR and seriously challenge the technical and legal adequacy of the de-identification release-and-forget model» (Rocher et al. 2019).

For example, the possible identification of patients through the «reverse» pseudonymisation can lead to remarkable legal consequences: as a matter of facts—according to Article 83 of the EU Regulation—this infringement may be subject to administrative fines up to €20mln. Moreover, this misconduct (in consequence of a Data Breach) may encourage Data Subjects to sue the Data Controller, asking for compensations: the major risk exposure for the Data Controller—considering the high number of potential claimants—is to suffer a "class action", that is a «legal procedure which enables the claims of a number of persons against the same defendant to be determined in the suit» (Mulheron 2004).

Moreover, the negative outcomes of this proceeding can generate a ripple effect that might involve also the company's reputation (and so, the damage would not only be of a short-term financial nature).

References

Al Nuaimi, E., Al Neyadi, H., Mohamed, N., & Al-Jaroodi, J. (2015). Applications of big data to smart cities. *Journal of Internet Services and Applications, 6.*

Bauman, Z. (1998). On glocalization: Or globalization for some, localization for some others. *Thesis Eleven, 54,* 37–49.

Cate, F. H., & Mayer-Schönberger, V. (2013). Notice and consent in a world of Big Data. *International Data Privacy Law, 2,* 67–73 (Oxford University Press).

Clemente, C. (2019). Presentation of the 2018 Annual Report, by the Director of the Financial Information Unit for Italy. Banca d'Italia.

d'Agostino Panebianco, M. (2019a). *Vivere nella dimensione digitale.* Rome: Themis Edizione.

d'Agostino Panebianco, M. (2019b). *Holistic complexity approach.* Statistica & Società. Retrieved November 2, 2019, from https://www.rivista.sis-statistica.org/cms/?p=710.

d'Agostino Panebianco, M. (2019c). Il trattamento dei dati nel Sistema Sanitario Nazionale Italiano alla luce del Provvedimento del Garante del 7 marzo 2019. *Ciberspazio e Diritto, 20,* 241–269.

De Felice, F., & Petrillo, A. (2015). Improving Italian healthcare service quality using analytic hierarchy process methodology. *IFMBE Proceedings, 45,* 981–982.

Doetter, L. F., & Götze, R. (2011). The Changing Role of the State in the Italia Healthcare System. *TranState Working Papers 150.*

Gandomi, A., & Haider, M. (2015). Beyond the hype: Big data concepts, methods, and analytics. *International Journal of Information Management, 35,* 137–144.

Greenberg, P. S. (1973). The balance of interests theory and the fourth amendment: A selective analysis of Supreme Court action. *California Law Review, 61.*

Hall, D. E., Prochazka, A. V., & Fink, A. S. (2012). Informed consent for clinical treatment. *CMAJ, 184*(5), 533–540.

Jones, C. (2003). The utilitarian argument for medical confidentiality: A pilot study of patients' views. *Journal of Medical Ethics, 29,* 348–352.

Lo Scalzo, A., Donatini, A., Orzella, L., Cicchetti, A., Profili, S., & Maresso, A. (2009). Italy: Health system review. *Health Systems in Transition, 11,* 1–216.

Mas-Colell, A., Whinston, M. D., & Green, J. R. (1995). *Microeconomic theory.* New York: Oxford University Press.

Modafferi, F. (2017). Privacy in Sanità, Modafferi: «Ecco come cambierà con il GDPR». agendadigitale.it. Retrieved November 2, 2019.

Mulheron, R. (2004). *The class action in common law legal systems.* Oxford: Hart Publishing.

Rocher, L., Hendrickx, J. M., & de Montjoye, Y.-A. (2019). Estimating the success of re-identifications in incomplete datasets using generative models. *Nature Communications, 10.*

Serapioni, M., & Duxbury, N. (2012). Citizens participation in the Italian health-care system: The experience of the Mixed Advisory Committees. *Health Expectations, 17,* 49.

Simeoni, E., Serpelloni, G., Crestoni, L., Spiniello, M., & Montisci, M. (1998). Segreto professionale e diritto alla riservatezza. In *HIV/AIDS Diritti e responsabilità* (pp. 51–62). Brescia: EDAS-Edizioni Surian.

Thompson, I. E. (1979). The nature of confidentiality. *Journal of Medical Ethics, 5,* 57–64.

Ward, J. S., & Barker, A. (2013). Undefined By Data: A Survey of Big Data Definitions. arXiv: 1309.5821.

Smart Tourism System in Calabria

Annarita De Maio, Daniele Ferone, Elisabetta Fersini, Enza Messina, Francesco Santoro, and Antonio Violi

Abstract We describe a Smart Tourism System called SMARTCAL, designed in the context of a R&D project aimed at supporting the tourism development in Calabria (a region in the South of Italy). The system is designed by considering Points and Events of Interest (PEOI) and their relationship with the local transport systems and infrastructures. A proactive tourist tour planner algorithm is proposed to generate optimized itineraries based on static and dynamic profiling of the users. Social media data are also taken into account for recommendations.

A. De Maio · D. Ferone · E. Fersini (✉) · E. Messina
Department of Informatics, Systems and Communication (DISCo),
University of Milano-Bicocca, Milan, Italy
e-mail: elisabetta.fersini@unimib.it

A. De Maio
e-mail: annarita.demaio@unimib.it

D. Ferone
e-mail: daniele.ferone@unimib.it

E. Messina
e-mail: enza.messina@unimib.it

F. Santoro
ITACA s.r.l., Rende, CS, Italy
e-mail: santoro@itacatech.it

A. Violi
Department of Law, Economics, Management and Quantitative Methods,
University of Sannio, Benevento, Italy
e-mail: antonio.violi@unisannio.it

P. Mariani and M. Zenga (eds.), *Data Science and Social Research II*,
Studies in Classification, Data Analysis, and Knowledge Organization,
https://doi.org/10.1007/978-3-030-51222-4_11

1 Introduction

In the era of Big Data, a massive use of the information available on the web has been registered in different sectors (Ardito et al. 2019; Del Vecchio 2018; Gajdošík 2019). In the tourism sector, travelers searching for information on the web often need to face the evaluation of a long list of possible **P**oints and **E**vents **O**f **I**nterest (PEOI) that becomes very complex and time-consuming. Furthermore, the spread of smart mobile technology enables the possibility to geo-localize and interact with lots of information sources in any location and time. Nowadays, lots of intelligent systems support our daily life, for example, suggesting the shortest path for a destination or the best restaurants, the lowest price in finding accommodation or flight. These can be considered Recommender Systems for filtering information relevant for the user to find interesting PEOI, with respect to explicit user preferences (Kontogianni et al. 2018; Khatibi et al. 2019).

In particular, Tourism Recommender Systems (TRS) have shown great developments during the last years due to the spread of web and mobile technologies. According to United Nations World Tourism Organization (2018), the total international tourist arrivals in 2017, registered an increase of +7% with respect to 2016. In this paper, we describe the TRS developed during the R&D project, funded by the Italian Ministry of Economic Development, named SMARTCAL "Smart Tourism in Calabria".

The SMARTCAL project aims at providing a strategy for strengthening the tourism offer in Calabria by creating a system that supports the tourist to plan his/her travel throughout the territory, and the decision makers to monitor and plan their business activities. These goals have been achieved by integrating multiple information sources such as user preferences, events calendar, routes, time tables, and opinions about destinations gathered from social media.

The result is a web platform and a smart mobile application to facilitate the matching between offer and demand. In the following, we describe the main architectural features of the system, focusing on the most important modules that guarantee the pro-activity and the capacity of profiling the users' preferences. In particular, we are going to characterize the following modules:

- **tourism recommender system**, that, given the explicit preferences of a user on tourism categories, provides relevant recommendations by integrating user's explicit preferences with auxiliary data obtained by analyzing social media in order to provide *accurate* and *serendipitous* recommendations.
- **proactive tourist tour planner**, that builds an itinerary for visiting a set of points of interest, considering the user preferences learned from the social networks analysis.

The paper is organized as follows: Sect. 2 briefly describes the state of the art, Sect. 3 is focused on the proposed architecture and modules for building the system, while some conclusions and ideas for future work are exposed in Sect. 4.

2 Literature Review

In order to build an effective and efficient proactive system, we have analyzed the scientific and technological literature, related to the Recommender Systems in general, and to the Tourist Recommender Systems in particular. Furthermore, we have described the best contributions related to the Tourist Tour Planner engines.

Recommender Systems (RSs) can be broadly classified in three main categories, namely *collaborative filtering*, *content-based*, and *hybrid*. Collaborative filtering (CF) recommendations are based on what people with similar tastes and preferences liked in the past. In order to provide recommendations, two different CF approaches can be used: *memory-based* and *model-based*. Memory-based CFs store the entire user-element rating matrix and perform some type of neighbor search to find the most similar users from which to select the recommendations for the new one. On the other hand, Model-based CFs learn from the rating matrix, a recommendation model is used to generate recommendations for new users. In this case, different *Machine Learning* (ML) techniques can be applied, such as: clustering (Bjelica 2010), probabilistic Latent Semantic Analysis (pLSA) (Hofmann 2004; Yin et al. 2009), matrix factorization (Bauer and Nanopoulos 2014; Zhai and Li 2015), clustering and regression techniques (Mehrbakhsh et al. 2016). A general overview is given by Sana et al. (2019).

Content-based (CB) RSs (Pazzani and Billsus 2007) recommend an item to a user based upon the item description. Features extracted from item descriptions are matched with the features of the user profile in order to make a recommendation. Also in this case, a plethora of ML approaches have been applied, such as Support Vector Machines (SVM) (Pronoza et al. 2016), Bayesian Networks (Pecli et al. 2015), Decision Trees (Alemeye and Getahun 2015), etc.

Hybrid RSs are obtained by combining collaborative filtering and content-based approaches (Geng et al. 2015; Nguyen et al. 2016; Verma et al. 2016).

Nowadays, RSs are widespread more and more in lots of different contexts: e-commerce (Linden et al. 2003), content streaming (Gomez-Uribe and Hunt 2015; Covington et al. 2016), news recommendation (Liu et al. 2010), and social networks (Ahmed et al. 2013; Rodriguez et al. 2012). A general description of the new trends is reported in Pimenidis et al. (2019) and Zhang et al. (2019). More recently, the attention has shifted to TRSs that have also a component of novelty and unexpectedness. Indeed, the great challenge is to obtain a recommender system that is able to give recommendations that are both relevant and serendipitous. Nevertheless, it is not easy to give a unique and universally accepted definition of serendipity in RSs, because it includes an emotional dimension, which is very subjective and hard to model. Kotkov et al. (2016) survey the literature of serendipity-oriented RSs and identify as serendipitous suggestions those elements that are relevant, novel, and unexpected. Unfortunately, relevance is often in contrast with novelty and unexpectedness, because a user tends to identify as relevant elements that are in his/her tastes, i.e., expected elements. Therefore, in order to improve user satisfaction, a compromise between accuracy and serendipity must be found: a very accurate recommender system creates a problem of over-personalization where there is no place for unexpected discoveries and the user will be trapped in filter bubbles. On the other hand, serendipity can be easily confused with randomness and incur again in the problem of information overload. As shown in Pan (2016), RSs used auxiliary data in several ways to improve the serendipity of recommendations. Guo et al. (2017) used auxiliary implicit feedback to overcome the data sparsity issue. Wang et al. (2018) uses sentiment analysis and social relationships to tackle the cold start problem, meanwhile Fernández-Tobías et al. (2019), faces up the same problem with the use of cross-domain information.

Tourism is an important sector for economic development and the use of RSs in this sector is becoming increasingly crucial. Indeed, several TRSs have been proposed in the last years. Most of them are based on **Multi-Agent Systems** (MASs), a set of agents that interact with each other to reach their desired goals. Agents usually cooperate, coordinate, and negotiate with each other to solve complex problems. Agent and multi-agent systems allow modeling, at a very high level, heterogeneous and distributed systems and environments.

Examples of MAS-based TRSs in the literature are *MARST* (Bedi et al. 2014), *PersonalTour* (Lorenzi et al. 2011), and *Turist@* (Batet et al. 2012). The latter, in particular, provides an easy and ubiquitous access to the desired information through a hybrid RS that considers different elements (such as the user profile and preferences, the location of the tourist, the activities and the opinions of previous tourists) and can adapt to changes in the activities and incorporate new information at execution time. Hassannia et al. (2019) proposes a MAS to recommend tour packages and establish a real-time communication between all the stakeholders (e.g., hotels, tour operators, tourists, etc.).

Other TRSs are based on knowledge-based intelligent systems. They use *ontologies* to represent the domain knowledge, in order to enable *reasoning* processes. Some example are given by *SAMAP* (Castillo et al. 2008), *SigTur* (Moreno et al. 2013), *SmartMuseum* (Ruotsalo et al. 2013), *GeOasis* (Santiago et al. 2012), and *e-Tourism* (Sebastia et al. 2008). Most of these systems define and use generic ontologies that store information which must be taken into account for a recommendation or integrate different ontologies for representing the domain knowledge. For example, Wang et al. (2008) base their TRS on two separate ontologies, the first one for the users' profile, in which demographic characteristics and personal interests are modeled, and the other one for topics concerning tourism-related information (e.g., accommodations, restaurants, transport, shopping, culture, etc.). Another example of TRS embracing this approach is *PaTac* (Ceccaroni et al. 2009), which is based on ontology mappings between various standard ontologies, such as *W3C's Time*, the *General User Model Ontology* (GUMO), the *Friend Of A Friend* ontology (FOAF). Colombo-Mendoza et al. (2017) combines ontology, and Latent Dirichlet Allocation (LDA) (David et al. 2003) to propose a context-aware software recommender system in the field of restoration. For an extensive review on TRSs, the reader can refer to Borràs et al. (2014).

Our purpose is to develop a TRS that integrates auxiliary data to increase both the serendipity and the accuracy of recommendations, by exploiting different sources of information, like user preferences and the huge amount of data available on social media. A Tourist Recommender System usually presents some functionalities related to the design of a personalized trip for the user. This activity is usually played by a particular component of the system, that is known as Tourist Tour Planner (TTP). Different contributions can be found in the literature in this field. Ardissono et al. (2003) described a real application, their INteractive TouRist Information GUidE (INTRIGUE) for the city of Torino (Italy), based on a fuzzy logic-based recommender system. Maruyama et al. (2004) introduced a P-Tour, a personal navigation device that calculates tourist routes, extended by Kinoshita et al. (2006). A complete literature overview was presented in Souffriau and Vansteenwegen (2010). Nowadays, the TTP is usually supported by another engine called Multimodal Route Planner (MMRP) which is able to provide information about how to move from one PEOI to another, using public transports, car, walking or bike in real time. The literature presents different studies in the field based on hyper-graph (Lozano and Storchi 2002) or multi-label network (Ziliaskopoulos and Wardell 2002). Moreover, the MMRP usually implements complex optimization techniques for providing a solution in a reasonable computational time. Starting from the Dijkstra algorithm (Dijkstra 1959), one of the most famous approaches for solving the shortest path problem, a lot of complex heuristics techniques were developed (Abeysundara et al. 2005; Pajo 2009). Finally, Konstantinos and Zafros (2009) presented web and mobile search engine systems for multimodal routes. Further details are described in the following section.

3 SMARTCAL as a Decision Support System

The overall architecture of the proposed system can be seen in Fig. 1, where the major components and the main interactions between them are depicted. This system can be accessed via web and mobile applications, both interacting with the core services.

A *data provider* layer is composed of several submodules, each of them specialized in gathering data from a specific source. They implement an ETL (Extraction, Transformation, Loading) subprocess for all the different sources that the platform uses (open data, API feeds, etc.) so as to read data (POIs, Events, etc.), transform them into the SMARTCAL domain and load them into the platform.

The following subsections provide a detailed description of the main modules:

- SMARTCAL Tourism Recommender System;
- Tourist Tour Planner;
- Multimodal Route Planner.

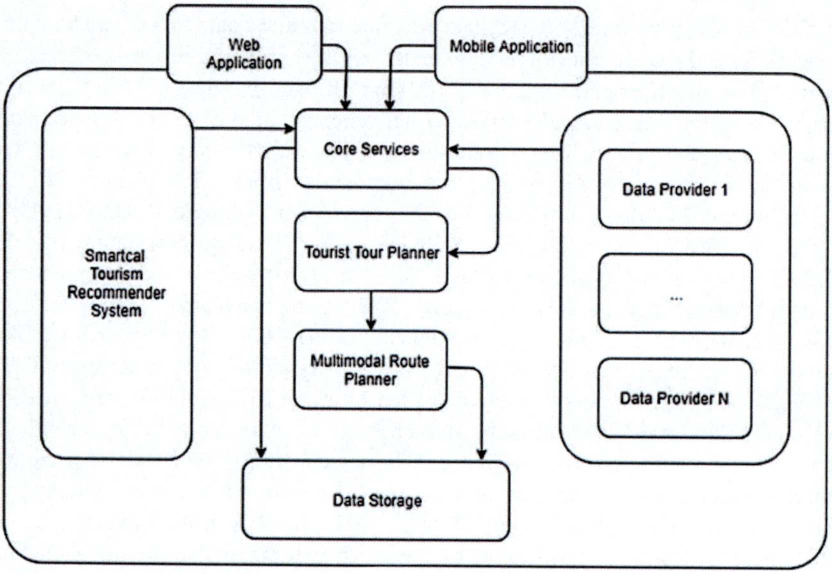

Fig. 1 Functional architecture

3.1 SMARTCAL Tourism Recommender System

The huge amount of information about travel destinations and their associated resources, such as accommodations, restaurants, attractions, museums or events, is commonly searched by tourists in order to plan a trip. Travelers are very keen on using tools that may support their decision-making processes when they are planning a trip, including the choice of destination, the selection of attractions to visit, the construction of a multi-day plan, the suggestions of appropriate restaurants and accommodations, etc. (United Nations World Tourism Organization 2018). However, the list of possible destinations may be overwhelming. In order to address this problem, the SMARTCAL Tourism Recommender System (STRS) has been designed to automatically suggest a ranked list of the most interesting tourist destinations to a given user, taking into account his/her preferences and personal interests. In particular, we propose a serendipity-oriented recommender system that uses a reranking approach exploiting auxiliary data, where relevant suggestions are evaluated also by considering PEOI reviews on social media. We use them to determine whether users are providing positive or negative emotional comments regarding specific PEOI, and use this information to enhance the prediction both in terms of accuracy and serendipity.

Given a taxonomy of tourism categories that allows us to associate each destination to one of those categories, such as *Museums*, *Abbeys Hermitages Monasteries*, *Historical Bridges*, *Restaurants*, and *Lakes*, the STRS has three main goals

- analyze the explicit preferences of the user, infer his/her **implicit preferences** and use them to obtain a category score that represents the level of interest of the user with respect to each specific category;
- extract **opinions** from reviews about PEOI to obtain the corresponding sentiment score;
- derive a **ranked list of destinations** according to the level of interest of the user with respect to the given categories (category score) and the opinion about the destinations (sentiment score).

The inferred preference of a user on a given category is based on matrix factorization and learning-to-rank approaches. Taking as input a rating matrix of explicit preferences of a user with respect to a given set of categories, latent preferences are inferred through a Probabilistic Matrix Factorization approach (Mnih and Salakhutdinov 2008). The obtained implicit preferences are then used to train a RankSVM model (Tzu-Ming et al. 2014) and to derive all the category scores. The sentiment score is derived by aggregating the reviews and rating scores specified by online users using a bayesian averaging approach.

Once the category and the sentiment scores have been obtained, they are combined into a single score through a linear combination, whose the hyper-parameter has been experimentally selected. The SMARTCAL Tourism Recommender System is depicted in Fig. 2.

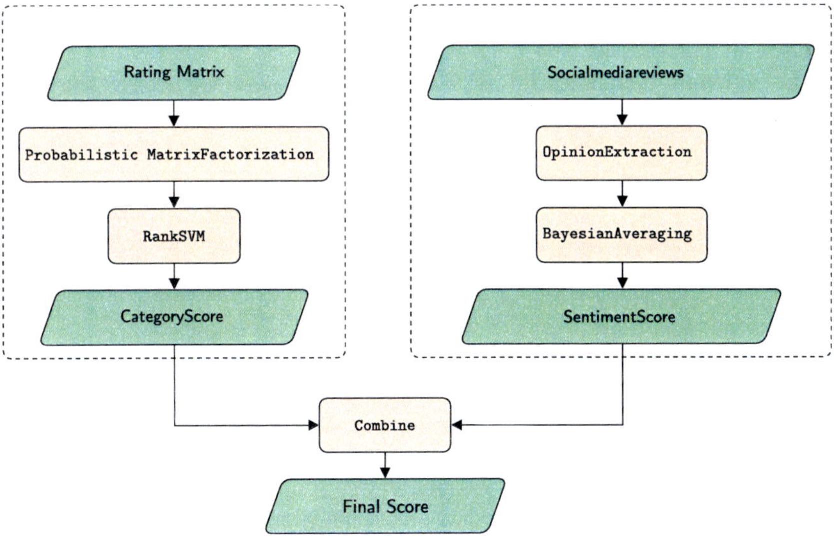

Fig. 2 SMARTCAL Recommender System

3.2 *Tourist Tour Planning Module*

In order to meet the expectations of the modern tourist, the Tourist Tour Planning module of the SMARTCAL application is based on two different main components:

- **a multimodal route planner (MMRP)**, which allows the user to book travel itineraries from a source/destination pair;
- **a tourist tour planner (TTP)**, which allows offering an optimized itinerary, based on the preferences and constraints expressed by the users, in order to discover the territory and to move by taking advantage of public transport.

3.2.1 Multimodal Route Planner

In general, a trip or route planner is a specialized search engine, used to find an optimal way to travel between two or more locations, sometimes using more than one mode of transport (in this case MMRP). Requests can be optimized according to different criteria, for example, searching for the fastest route, the shortest path, the minimum or the cheapest cost. Planning can be performed taking into account numerous constraints, such as the need to leave or arrive at a certain time, to avoid certain points, etc. A single journey can use a sequence of different modes of transport, which means that the system needs to know public transport services and transport networks for private transport in detail. Travel planners have been widely used in the travel industry since the 1970s. The growth of Internet use, the proliferation of Geo-spatial data, and the development of information technologies have generally led to the rapid development of many browser-based intermodal travel planners, such as Rome2rio, Google Transit, Yahoo, and Google Maps. In general, a MMRP implies different levels of complexity, related to the data and input representation or computational time of the algorithm.

In order to provide a solution that is competitive with respect to the market, where big and famous players are already present, the SMARTCAL architecture integrates an open-source library for managing the suggestions of multimodal trips (from one PEOI to another) into each tour, that is Open Trip Planner (OTP).[1]

OTP is an open-source library amply used in successful applications sponsored by public and private organizations around the world, thanks to its high flexibility and the big community interested in supporting its development. Furthermore, it is also used in different scientific contribution: in Liebig et al. (2017), a system for individual trip planning is studied which incorporates future traffic hazards in routing through the use of OTP, in Sierpiński et al. (2014), OTP is compared to other systems and identified among the best planners for multimodal transport, in Narboneta and Teknomo (2013) a disaster relief and recovery system is introduced based on OTP standards. OTP uses a time-dependent graph, which contains both street networks and transit networks, and exploits the A* Search algorithm with

[1]https://www.opentripplanner.org/.

Euclidean heuristics. The algorithm A* is a graph search algorithm that identifies a path from a given initial node to a given destination node. It uses an "estimating heuristic" that classifies each node through an estimation of the best route visiting that node for reaching the destination. Step by step, the best node is selected until the destination is reached. The algorithm A* is also an example of best-first research and was introduced for the first time by Hart et al. (1968). It is an extension of Dijkstra's algorithm for searching the shortest path, and so it is very flexible, like all the generic graph traversing algorithm. Indeed, it can be verified that A* does not consider more nodes than any other feasible search algorithm unless the alternative algorithm has a more accurate estimating heuristic. In this sense, A* is the most computationally efficient algorithm that guarantees the search for the shortest path in these types of networks.

3.2.2 Tourist Tour Planner Engine

The generic problem of the customized generation of tourist trip has been defined as the Tourist Trip Design Problem (TTDP). This is a routing problem which has lots of applications in logistics, tourism, and defense. Given a set of nodes, PEOIs, the decision support model aims at designing a tour visiting a subset of PEOIs. The objective of the problem is to maximize the total score of the tour while the total travel time and the total cost do not exceed some predefined thresholds related to user constraints (travel costs, time, and other attributes). The optimal solutions of the TTDP should have recommendations that correspond to the preferences of the tourist, planning the tour that can be traveled almost optimally. Different variants of the TTDP can be derived simply by considering different parameters and constraints, compared to those of the generic problem presented. The TTDP cannot be solved in polynomial time, and for this reason, all existing online applications are based on the use of efficient heuristic algorithms. The literature includes lots of modeling approaches for simplified versions of TTDP, focusing on Orienteering Problems (OP). OP is a routing problem in which the goal is to determine, if it exists, a subset of nodes to visit, and the related order, so that the total score collected is maximized or minimized based on the nature of the problem (Golden et al. 1987). OP is a well-know NP-Hard problem, for this reason, it is usually solved with heuristics approaches (Gendreau et al. 1998).

In order to build an effective engine into the project, we consider a set of PEOIs (divided into categories), featured by a score, a time window, a time for the visit, a cost. An Orienteering Problem with time windows is formulated, taking into account constraints related to the budget imposed by the user, the maximum duration of the tour and the maximum time spent for each category. The score of each PEOI is pre-computed through different techniques of Sensitive Analysis (SA) by the RS previous described. All the formulation details are described in Ciancio et al. (2018). To the best of our knowledge, the heuristic algorithms are usually the best choice for solving an orienteering problem in acceptable computational time, maintaining a good quality of the solution also when large instances are approached. For this reason, the TTDP engine is made up of a **genetic algorithm** (GA). A genetic algorithm is a search heuristic that is inspired by Charles Darwin's theory of natural evolution. This algorithm reproduces the process of natural selection, in the sense that, it selects the fittest individuals for the reproduction of the next generation at each iteration (Koza 1997). In this type of approach, each solution is represented by an entity called **chromosome**, and evaluated thought a function that considers the best features, called **fitness function.** The developed GA undertakes to evolve the solution, according to the following basic pattern:

1. random generation of the first population of solutions (generating only feasible chromosome);
2. selection of the best solutions based on the value of the fitness function;
3. generation of new solutions using different classical techniques (moves for mixing chromosome: crossover and mutation);
4. repetition of steps 2–3 for *n* iterations;
5. selection of the best found solution.

The algorithm integrates also some practical constraints, related to

- the possibility to set double-time windows for some PEOI (e.g., a church could be open during 8:00–12:00 a.m. and 3:00–6:00 p.m.);
- the specification of a time interval in which to have a break (e.g., 30 min for lunch at 12:00 a.m.–2:00 p.m.);
- imposing a maximum time allowed for each type of activities.

All the details related to the algorithm and the experimental part are described in Ciancio et al. (2018). In the following, a scheme of the algorithm is provided (figure 3).

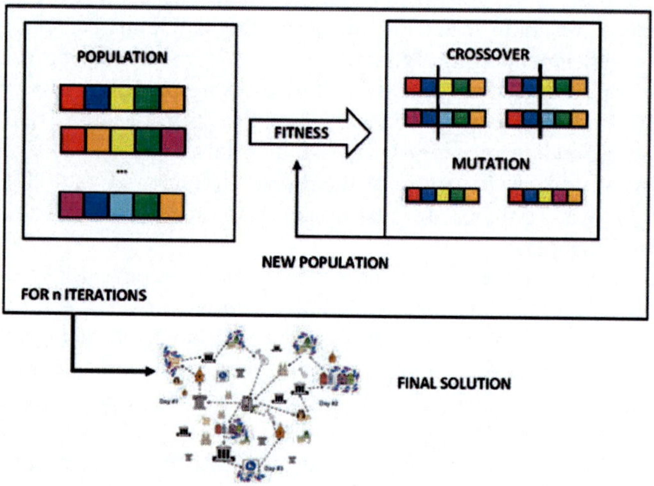

Fig. 3 Genetic algorithm scheme

4 Conclusions and Future Work

In this paper, we proposed a Tourism Recommender System that integrates explicit and implicit preferences of the user to auxiliary data present on social media. The preferences are combined with a sentiment score that represents the opinion of the community about the PEOIs. This combination allows us to improve the serendipity of the system. Starting from these recommendations, the Tourist Trip Planner computes an optimized tour and is able to give in output a personalized trip respecting a given budget, a series of time windows constraints, and mixing different typologies of PEOIs into the trip. This application presents a great potential for the touristic valorization of a particular territory, a high level of scalability and also a good opportunity for business development. Moreover, it can also be seen as a useful tool for decision makers in the tourism field, since it can allow to have a real-life perception of end-users perspectives and preferences.

Acknowledgements This work has been partially funded by MISE (Ministero Italiano dello Sviluppo Economico) under the project "SMARTCAL—Smart Tourism in Calabria" (F/050142/01-03/x32).

References

Abeysundara, S., Baladasan, G., & Kodithuwakku, S. (2005). A genetic algorithm approach to solve the shortest path problem for road maps. In *Proceedings of the International Conference on Information and Automation.*

Ahmed, A., Kanagal, B., Pandey, S., Josifovski, V., Pueyo, L. G., & Yuan, J. (2013). Latent factor models with additive and hierarchically-smoothed user preferences. In *Proceedings of the Sixth ACM International Conference on Web Search and Data Mining—WSDM '13* (pp. 385–394). ACM Press.

Alemeye, F., & Getahun, F. (2015, September). Cloud readiness assessment framework and recommendation system. In *AFRICON 2015* (pp. 1–5). IEEE.

Amin, S. A., Philips, J., & Tabrizi, N. (2019). Current trends in collaborative filtering recommendation systems. In: Y. Xia & L. J. Zhang (Eds.), *Services—SERVICES 2019*. Lecture Notes in Computer Science (Vol. 11517, pp. 46–60). Cham: Springer.

Ardissono, L., Goy, A., Petrone, G., Signan, M., & Torasso, P. (2003). Intrigue: Personalized recommendation of tourism attractions for desktop and handset devices. *Artificial Intelligence, 17*(8–9), 687–714.

Ardito, L., Cerchione, R., Del Vecchio, P., & Raguseo, E. (2019). *Big data in smart tourism: Challenges, issues and opportunities.*

Batet, M., Moreno, A., Sánchez, D., Isern, D., & Valls, A. (2012). Turist@: Agent-based personalised recommendation of tourist activities. *Expert Systems with Applications, 39*(8), 7319–7329.

Bauer, J., & Nanopoulos, A. (2014). Recommender systems based on quantitative implicit customer feedback. *Decision Support Systems, 68*, 77–88.

Bedi, P., Agarwal, S. K., Jindal, V., & Richa. (2014). MARST: Multi-Agent Recommender System for e-Tourism using reputation based collaborative filtering. In *Databases in Networked Information Systems* (pp. 189–201). Springer International Publishing.

Bjelica, M. (2010). Towards TV recommender system: Experiments with user modeling. *IEEE Transactions on Consumer Electronics, 56*(3), 1763–1769.

Blei, D. M., Ng, A. Y., & Jordan, M. I. (2003). Latent dirichlet allocation. *Journal of Machine Learning Research, 3*, 993–1022.

Borràs, J., Moreno, A., & Valls, A. (2014). Intelligent tourism recommender systems: A survey. *Expert Systems with Applications, 41*(16), 7370–7389.

Castillo, L. A., Armengol, E., Onaindia, E., Sebastia, L., González-Boticario, J., Rodríguez, A., et al. (2008). SAMAP: An user-oriented adaptive system for planning tourist visits. *Expert Systems with Applications, 34*(2), 1318–1332.

Ceccaroni, L., Codina, V., Palau, M., & Pous, M. (2009). PaTac: Urban, ubiquitous, personalized services for citizens and tourists. In *Proceedings of the 3th International Conference on Digital Society (ICDS)* (pp. 7–12).

Ciancio, C., De Maio, A., Laganà, D., Santoro, F., & Violi, A. A. (2018). A Genetic algorithm framework for the orienteering problem with time windows. *New trends in emerging complex real life problems*. AIRO Springer Series (pp. 179–188).

Colombo-Mendoza, L. O., Valencia-García, R., Rodríguez-González, A., Colomo-Palacios, R., & Alor-Hernández, G. (2017). Towards a knowledge-based probabilistic and context-aware social recommender system. *Journal of Information Science, 44*(4), 464–490.

Covington, P., Adams, J., & Sargin, E. (2016). Deep neural networks for YouTube recommendations. In *Proceedings of the 10th ACM Conference on Recommender Systems—RecSys '16* (pp. 191–198). ACM Press.

Del Vecchio, P., Mele, G., Ndou, V., & Secundo, G. (2018). Creating value from social big data: Implications for smart tourism destinations. *Information Processing & Management, 54*(5), 847–860.

Dijkstra, E. W. (1959). A note on two problems in connexion with graphs. *Numerische Mathematik, 1*, 269–271.

Fernández-Tobías, I., Cantador, I., Tomeo, P., Anelli, V. W., & Di Noia, T. (2019, January). Addressing the user cold start with cross-domain collaborative filtering: exploiting item metadata in matrix factorization. In *User modeling and user-adapted interaction*.

Gajdošík, T. (2019). Big data analytics in smart tourism destinations. A New tool for destination management organizations? *In smart tourism as a driver for culture and sustainability* (pp. 15–33). Cham: Springer.

Gendreau, M., Laporte, G., & Semet, F. (1998). A tabu search heuristic for the undirected selective travelling salesman problem. *European Journal of Operational Research, 106*(2–3), 539–545.

Geng, X., Zhang, H., Bian, J., & Chua, T.-S. (2015, December). Learning image and user features for recommendation in social networks. In *2015 IEEE International Conference on Computer Vision (ICCV)* (pp. 4274–4282). IEEE

Golden, B. L., Levy, L., & Vohra, R. (1987). The orienteering problem. *Naval Research Logistics, 34*(3), 307–318.

Gomez-Uribe, C. A., & Hunt, N. (2015). The netflix recommender system. *ACM Transactions on Management Information Systems, 6*(4), 1–19.

Guo, G., Qiu, H., Tan, Z., Liu, Y., Ma, J., & Wang, X. (2017). Resolving data sparsity by multi-type auxiliary implicit feedback for recommender systems. *Knowledge-Based Systems, 138*, 202–207.

Hart, P. E., Nilsson, N. J., & Raphad, B. (1968). A formal basis for the heuristic determination of minimum cost paths. *IEEE Transactions on Systems Science and Cybernatics, 2*, 100–107.

Hassannia, R., Barenji, A. V., Li, Z., & Alipour, H. (2019). Web-based recommendation system for smart tourism: Multiagent technology. *Sustainability, 11*(2), 323.

Hofmann, T. (2004). Latent semantic models for collaborative filtering. *ACM Transactions on Information Systems, 22*(1), 89–115.

Khatibi, A., Belém, F., da Silva, A. P. C., Almeida, J. M., & Gonçalves M. A. (2019). Fine-grained tourism prediction: Impact of social and environmental features. *Information Processing & Management*.

Kinoshita, T., Nagata, M., Shibata, N., Murata, Y., Yasumoto, K., & Ito, M. (2006) A personal navigation system for sightseeing across multiple days. In *Proceedings of the 3rd International Conference on Mobile Computing and Ubiquitous Networking (ICMU 2006)* (pp. 254–259).

Konstantinos, N. A., & Zafros, K. (2009). Solving the multi-criteria time-dependent routing and scheduling problem in a multimodal fixed scheduled network. *European Journal of Operation Research, 192*, 18–28.

Kontogianni, A., Kabassi, K., Virvou, M., & Alepis, E. (2018). Smart tourism through social network user modeling: a literature review. In *2018 9th International Conference on Information, Intelligence, Systems and Applications (IISA)* (pp. 1–4). IEEE.

Kotkov, D., Wang, S., & Veijalainen, J. (2016). A survey of serendipity in recommender systems. *Knowledge-Based Systems, 111*, 180–192.

Koza, J. R. (1997). *Genetic programming*.

Kuo, T.-M., Lee, C.-P., & Lin, C.-J. (2014). Large-scale kernel rankSVM. In *Proceedings of the 2014 SIAM International Conference on Data Mining* (pp. 812–820). SIAM.

Liebig, T., Piatkowski, N., Bockermann, C., & Morik, K. (2017). Dynamic route planning with real-time traffic predictions. *Information Systems*.

Linden, G., Smith, B., & York, J. (2003). Amazon.com recommendations: item-to-item collaborative filtering. *IEEE Internet Computing, 7*(1):76–80.

Liu, J., Dolan, P., & Pedersen, E. R. (2010). Personalized news recommendation based on click behavior. In *Proceedings of the 15th International Conference on Intelligent User Interfaces—IUI '10* (pp. 31–40). ACM Press.

Lorenzi, F., Loh, S., & Abel, M. (2011, August). PersonalTour: A recommender system for travel packages. In *2011 IEEE/WIC/ACM International Conferences on Web Intelligence and Intelligent Agent Technology* (pp. 333–336). IEEE.

Lozano, A., & Storchi, G. (2002). Shortest viable hyperpath in multimodal networks. *Transportation Research Part B, 36*, 853–874.

Maruyama, A., Shibata, N., Murata, Y., Yasumoto, K., & Ito, M. (2004). P–tour: A personal navigation system for tourism. In *Proceedings of 11th World Congress on ITS* (pp. 18–21).

Mnih, A., & Salakhutdinov, R. R. (2008). Probabilistic matrix factorization. In *Advances in neural information processing systems* (pp. 1257–1264).

Moreno, A., Valls, A., Isern, D., Marin, L., & BorríS, J. (2013). SigTur/E-Destination: Ontology-based personalized recommendation of tourism and leisure activities. *Engineering Applications of Artificial Intelligence, 26*(1), 633–651.

Narboneta, C. G., & Teknomo, K. (2013). OpenTripPlanner, OpenStreetMap, general transit feed specification: Tools for disaster relief and recovery. In *7th IEEE International Conference Humanoid, Nanotechnology, Information Technology Communication and Control, Environment and Management*.

Nguyen, H., Richards, R., Chan, C.-C., & Liszka, K. J. (2016). RedTweet: Recommendation engine for reddit. *Journal of Intelligent Information Systems, 47*(2), 247–265.

Nilashi, M., Dalvi-Esfahani, M., Roudbaraki, M. Z., Ramayah, T., & Ibrahim, O. (2016). A multi-criteria collaborative filtering recommender system using clustering and regression techniques. *Journal of Soft Computing and Decision Support Systems, 3*(5), 24–30.

Pajor, T. (2009). *Multi-Modal Route Planning*. Dissertation.

Pan, W. (2016). A survey of transfer learning for collaborative recommendation with auxiliary data. *Neurocomputing, 177*, 447–453.

Pazzani, M. J., & Billsus, D. (2007). Content-based recommendation systems. *The adaptive web* (pp. 325–341). Berlin, Heidelberg: Springer.

Pecli, A., Giovanini, B., Pacheco, C. C., Moreira, C., Ferreira, F., Tosta, F., et al. (2015). Dimensionality reduction for supervised learning in link prediction problems. In *Proceedings of the 17th International Conference on Enterprise Information Systems* (pp. 295–302). SCITEPRESS - Science and and Technology Publications.

Pimenidis, E., Polatidis, N., & Mouratidis, H. (2019). Mobile recommender systems: Identifying the major concepts. *Journal of Information Science, 45*(3), 387–397.

Pronoza, E., Yagunova, E., & Volskaya, S. (2016). Aspect-based restaurant information extraction for the recommendation system. In *Human language technology. Challenges for computer science and linguistics* (pp. 371–385). Springer International Publishing.

Rodriguez, M., Posse, C., & Zhang, E. (2012). Multiple objective optimization in recommender systems. In *Proceedings of the Sixth ACM Conference on Recommender Systems—RecSys '12* (pp. 11–18). ACM Press.

Ruotsalo, T., Haav, K., Stoyanov, A., Roche, S., Fani, E., Deliai, R., et al. (2013). SmartMuseum: A mobile recommender system for the web of data. *Web Semantics, 20*, 50–67.

Santiago, F. M., López, F. A., Montejo-Ráez, A., & López, A. U. (2012). GeOasis: A knowledge-based geo-referenced tourist assistant. *Expert Systems with Applications, 39*(14), 11737–11745.

Sebastia, L., Garcia, I., Onaindia, E., & Guzman, C. (2008). e-Tourism: A tourist recommendation and planning application. In *Proceedings of the 20th Institute of Electrical and Electronics Engineers (IEEE), International Conference on Tools with Artificial Intelligence (ICTAI)* (Vol. 2, pp. 89–96).

Sierpiński, G., Celiński, I., & Staniek, M. (2014). Using trip planners in developing proper transportation behavior. *International Journal of Architectural and Environmental Engineering, 8*.

Souffriau, W., & Vansteenwegen, P. (2010). Tourist trip planning functionalities: State–of–the–art and future. In F. Daniel & F. M. Facca (Eds.), *Current Trends in Web Engineering. ICWE 2010*. Lecture Notes in Computer Science (Vol. 6385). Berlin, Heidelberg: Springer.

United Nations World Tourism Organization. (2018). *UNWTO Tourism Highlights*.

Verma, C., Hart, M., Bhatkar, S., Parker-Wood, A., & Dey, S. (2016). Improving scalability of personalized recommendation systems for enterprise knowledge workers. *IEEE Access, 4*, 204–215.

Wang, H.-C., Jhou, H.-T., & Tsai, Y.-S. (2018). Adapting topic map and social influence to the personalized hybrid recommender system. *Information Sciences*.

Wang, W., Zeng, G., Zhang, D., Huang, Y., Qiu, Y., & Wang, X. (2008). An intelligent ontology and Bayesian network based semantic mashup for tourism. In *Proceedings of the IEEE Congress on Services—Part I (Services)* (pp. 128–135).

Yin, Z., Yueting, Z., Jiangqin, W., & Liang, Z. (2009). Applying probabilistic latent semantic analysis to multi-criteria recommender system. *AI Communications, 22*(2), 97–107.

Zhai, H., & Li, J. (2015). Refine social relations and differentiate the same friends' influence in recommender system. In *Mining intelligence and knowledge exploration* (pp. 504–514). Springer International Publishing.

Zhang, S., Yao, L., Sun, A., & Tay, Y. (2019). Deep learning based recommender system: A survey and new perspectives. *ACM Computing Surveys, 52*(5).

Ziliaskopoulos, A., & Wardell, W. (2002). An intermodal optimum path algorithm for multimodal networks with dynamic arc travel times and switching delays. *European Journal of Operational Research, 125*, 486–502.

Spatial Localization of Visitors Mobile Phones in a Sardinian Destinations' Network

Anna Maria Fiori and Ilaria Foroni

Abstract In the act of placing and receiving calls, or sending text messages, a mobile phone reports its presence to the closest cell towers and unveils its geographical position. Mobile phone providers collect in their *call data records* the information about the clients' spatial localization. The use of such data sets represents an enormous scientific opportunity to detect the structure of social networks. Quantifying and understanding network features may help to obtain deeper insight into applications of practical importance. For instance, knowing the number of visitors moving from one destination to another is an information that reveals valuable knowledge for regional tourism planning. With this aim, we investigate how visitors of Sardinia (Italy) move across the island, according to the localization tracks revealed by their cell phones. In order to study such mobility patterns, we employ aggregated call data records to construct a spatial network that we analyze with the aid of statistical tools. We also provide a comparison between the movement behaviors of national and international visitors.

1 Introduction

Mobile phones are well-known to be *tracking devices* that reveal the spatial localization of their owners. In the *call data records*, the mobile network providers register the migrations of their subscribers from one cell tower to another, and so doing, they gather information on individual mobility behaviors (Candia et al. 2008; Doyle et al. 2014). The use of such call data records provides, as affirmed in (Candia et al. 2008)

> huge scientific opportunity to uncover the structure and dynamics of the social networks at different levels (Candia et al. 2008, p. 2).

A. M. Fiori · I. Foroni (✉)
Department of Statistics and Quantitative Methods, University of Milano-Bicocca, via Bicocca degli Arcimboldi, 8, 20126 Milan, Italy
e-mail: ilaria.foroni@unimib.it

A. M. Fiori
e-mail: anna.fiori@unimib.it

P. Mariani and M. Zenga (eds.), *Data Science and Social Research II*,
Studies in Classification, Data Analysis, and Knowledge Organization,
https://doi.org/10.1007/978-3-030-51222-4_12

The scientific literature is, in fact, rich of examples where, for instance, mobile phone data have been used to capture generalized patterns of individual movements (see, in addition to the aforementioned Doyle et al. (2014) also Girardin et al. (2009), Isaacman et al. (2011), Kung et al. (2014), Mir et al. (2013), Ratti et al. (2006), Schneider et al. (2013), Toole et al. (2015) among the others). In this consideration, in our paper, we employ mobile phone data to gain some insights on how people who visit the island of Sardinia (Italy) move across its territory. Specifically, we consider a sample of data that Vodafone Italia had gathered in thirty-three different places spread across the island. These ones include twenty-seven towns located in the most touristic areas of the island, three airports and three ports. As a result of the survey, the dataset supplies for every pair of locations the number of visitors who had been present, at least once, in both the locations between September 2015 and September 2016. The investigation of such a phenomenon, which is denoted in the dataset as a *co-visit*, allows some understanding of individual mobility patterns. Specifically, being the data referred to Sardinia's visitors and to locations placed in touristic areas, it is possible to retrace how travelers move across the destinations. Such a knowledge can be particularly useful in the tourism planning of the region. In our research, we propose the creation of a network structure that embeds the information drawn from the data in order to visualize and study the mobility behavior of the visitors. In the scientific literature, it is possible to find many applications of the network analysis to study the mobility of tourists across the destinations. For instance, in Hwang et al (2006) and Shih (2006), the authors introduce a tourism network structure that emerges from the analysis of trip patterns, where locations are considered as nodes and travels between the places visited as links among these nodes. A more recent study (Asero et al. 2016), focuses on the connections among tourist destinations in Sicily (Italy) in order to define tourist networks within the region. Other examples can be found also in Candia et al. (2008), Ahas et al. (2008), Baggio and Scaglione (2017), D'Agata et al. (2013). Following this stream of literature, in our paper, we introduce a network scheme where each one of the thirty-three Sardinian locations involved in the survey is represented by a node. Moreover, we consider that two locations are connected by a link if they were both reached, at least once, by the same visitor. Two main properties characterize the network that we investigate: (1) the network is *undirected*, given the nature of the relationship that we assume to link the nodes; (2) the network is *complete* as each node is connected to any other through, at least, one co-visit. In addition, the network is weighted because, in order to quantify the intensity of the connections existing between the locations, we assigned to the link associating two nodes a weight proportional to the number of co-visits recorded between them. The paper is organized as follows. The first section provides a description of the data set we used in our study. The second section introduces the networks under study and presents the results of the network analysis. The final section draws some conclusions and gives a future research agenda.

Table 1 Sardinian locations considered in the study grouped by provinces. The expression between the parentheses indicates the denomination of the node that has been assigned to the location in the national and international networks

Provinces	Locations (Network vertices)
Cagliari	Cagliari Airport ($v02$); Cagliari ($v07$); Chia ($v11$); Muravera ($v15$); Cagliari Port ($v21$:); Pula ($v25$); Villasimius ($v33$)
Carbonia-Iglesias	Carloforte ($v09$); S. Antioco ($v26$)
Medio Campidano	Costa Verde ($v12$)
Nuoro Ogliastra	Cala Gonone ($v08$); Orosei ($v18$)
Ogliastra	Arbatax ($v20$); Santa Maria Navarrese ($v28$); Tortolí ($v32$)
Olbia-Tempio	Olbia Airport ($v03$); Budoni ($v06$); Golfo Aranci ($v13$); Olbia ($v16$); Arzachena ($v17$); Palau ($v19$); Golfo Aranci Port ($v22$); Olbia Port ($v23$); San Teodoro ($v27$); Santa Teresa ($v29$)
Sassari	Bosa ($v05$); Alghero Airport ($v01$); Sinis ($v30$); Alghero ($v04$); Castelsardo ($v10$); La Maddalena ($v14$); Porto Torres ($v24$); Stintino ($v31$)

2 The Dataset

In this paper, we employ mobile phone data that have been collected by Vodafone Italia. These data, suitably anonymized and aggregated, are publicly available on the website developed by the Sardinian regional government organization. Specifically, as mentioned in Sect. 1, the data set contains information about the number of co-visits effectuated by Sardinia's visitors. As the data distinguishe between national and international visitors, we used them to describe two separated networks, one for each group of people. For simplicity, we, respectively, denoted them as the *national network* and the *international network*. The data set spans from September 2015 to September 2016, and contains information about more than one hundred and fifty millions of co-visits that have been collected in the thirty-three locations. These ones are listed, grouped by province,[1] in Table 1. Moreover, the spatial layout of the locations forming the national and the international networks together with the arterial roads that connect the territory of the island are shown in Fig. 1. To give an idea of the role played by tourism in each one of the Sardinian provinces in the period the research refers to, we report in Fig. 2, the number of national and international arrivals (left panel) and the number of national and international overnights (right panel) recorded in 2016. The bar plots represented in Fig. 2, highlight that international tourism is far more developed in the province of Olbia-Tempio than in the other ones. Referring to national tourism, the province of Cagliari also plays an important role in this sector. Obviously, we need to take in consideration this aspect in the following analysis on visitors' mobility patterns.

[1]We notice that, differently from the actual situation, at the time of the gathering of the data the provinces in Sardinia were eight, namely, Cagliari, Carbonia-Iglesias, Medio Campidano, Nuoro, Ogliastra, Olbia-Tempio, Oristano, and Sassari.

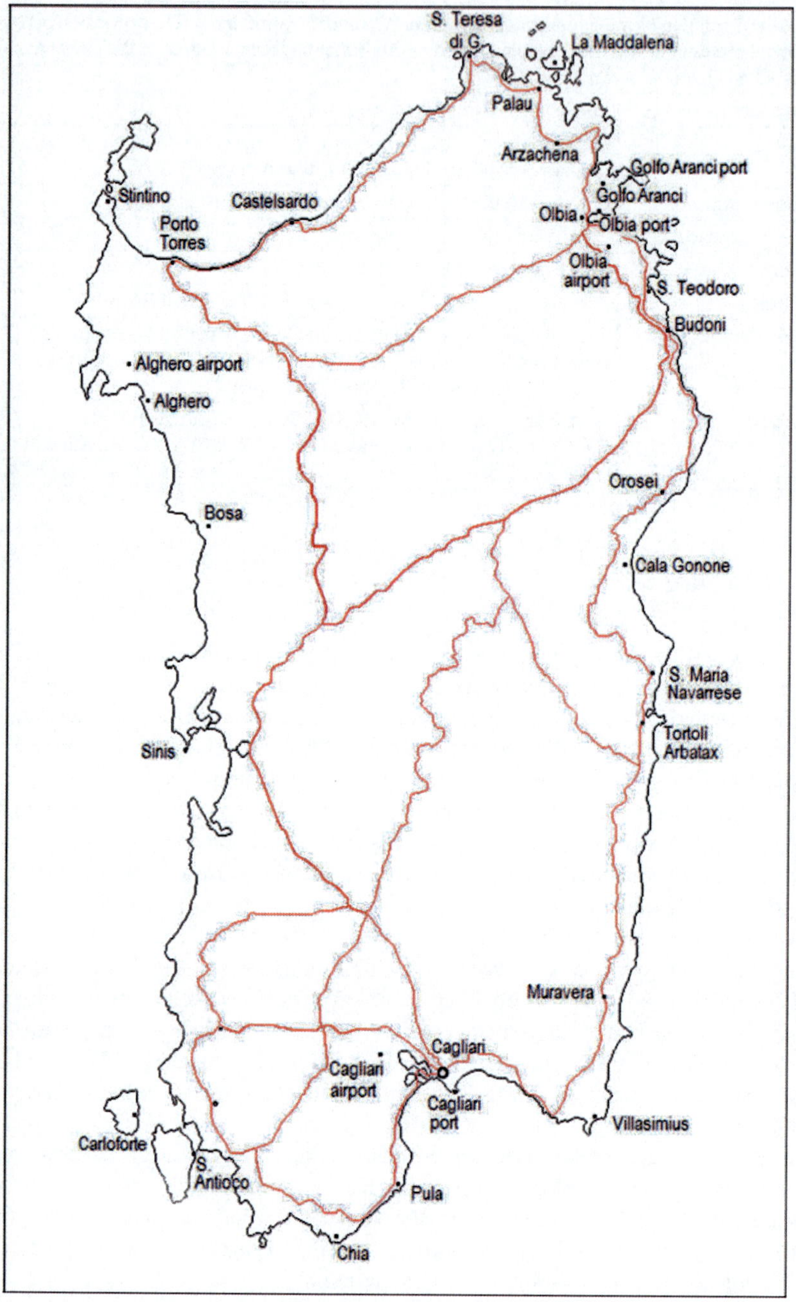

Fig. 1 The map shows the spatial layout of the thirty-three locations considered in the analysis together with the arterial roads connecting the territory

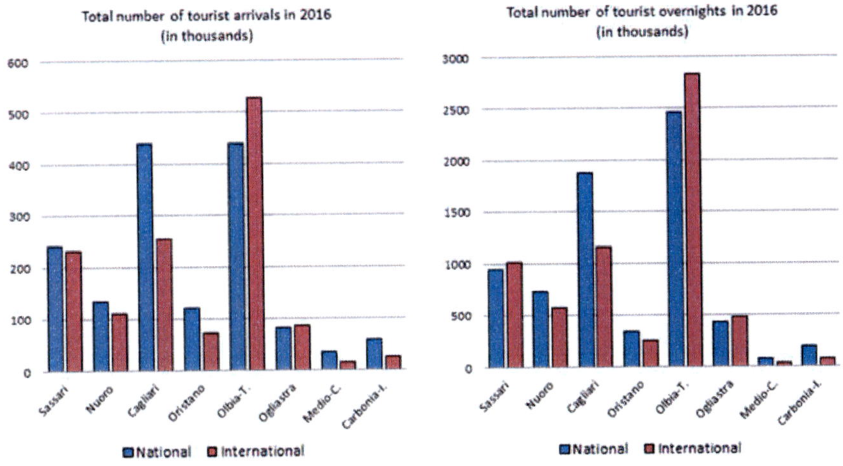

Fig. 2 National and international tourist arrivals (left panel) and overnights (right panel) in the Sardinia provinces in 2016. *Source* ISTAT census survey "Occupancy of tourist accommodation establishments" (ISTAT 2011)

3 The Visitors Mobility Networks

In the following sections, we initially formalize the structure of the networks originated from the mobile phone data, and then, by means of the methods of the network analysis, we investigate the mobility patterns of the national and international visitors.

3.1 Networks Description

From a mathematical point of view, a network is a structure consisting of a set V of N vertices (or nodes) and a set E of edges (or links) connecting the vertices. In our model, each node v_i represents a location, while an edge e_{ij} between two nodes v_i and v_j indicates the relationship introduced by the visitors' co-visits. We assigned to each edge e_{ij} a weight ω_{ij} proportional to the number of co-visits connecting v_i and v_j and normalized so as to have $0 \leq \omega_{ij} = \omega_{ji} \leq 1$. Moreover, given the meaning of the relationship, we assumed $\omega_{ii} = 0$. An important indicator of a weighted network (we refer the reader to Horvath (2011) for further details on this topic) is the node *strength* k_i that is calculated as the sum of the weights of all links attached to it. The formula of the strength for a node v_i is thus given by

$$k_i = \sum_{j=1}^{N} \omega_{ij}. \tag{1}$$

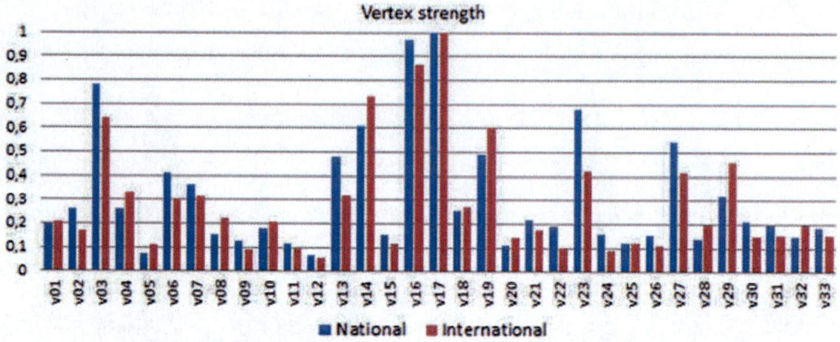

Fig. 3 Graphical representation of the scaled strength sequences associated with the national network (in blue) and to the international network (in red)

The strength of a node is a measure of the level of connectivity that characterizes it. In order to compare the national and the international networks, we consider in our study the following scaled version of the node strength

$$s_i = \frac{k_i}{k_{\max}} \tag{2}$$

where $k_{\max}$ is the maximum across the N components of the vector $\boldsymbol{k} = (k_1, \ldots, k_N)'$. Computing the *scaled strength sequence* $\boldsymbol{s} = (s_1, \ldots, s_N)'$ for the national and international networks, we obtained the results that are graphically represented in Fig. 3. The visual comparison of the national and international network scaled strength sequences shows two main aspects: the similarity between the national and international visitors' mobility patterns and the existence of some vertices that stand out with respect to the remaining ones. Indeed, in both cases, a few locations are characterized by a high level of connection (Arzachena ($v17$), Olbia ($v16$), Olbia Airport ($v03$), Olbia Port ($v23$), Palau ($v19$), San Teodoro ($v27$), and Santa Teresa ($v29$)), while the remaining ones are noticeably less connected. We notice also that, the vertices with the highest scaled strength are prevailingly located in the North-East of the island (refer to Fig. 1). Such a result reflects the fact that the majority of visitors (especially international) are concentrated in this part of Sardinia, as it can be deduced from the data reported in Fig. 2. In addition, the less connected nodes (Costa Verde ($v12$), Bosa ($v05$), and Alghero ($v04$)) are located in the area where the road network is poorly developed.

3.2 *Scaled Strength Distribution*

In the present section, we investigate the distribution of the strength sequence $\boldsymbol{s}$ because this is one of the aspects that characterizes the structure and function of a network. The probability $P(s)$ that a vertex has scaled strength s is estimated by

Table 2 Parameter estimates for the best-fit exponential curves for the national and international networks

Network	λ estimate	95% confidence interval
National	0.3297	[0.2387, 0.4853]
International	0.2977	[0.2165, 0.4353]

$$P(s) = \frac{n(s)}{N},$$

where $n(s)$ indicates the number of vertices with scaled strength s in the network. Counting how many vertices of the national and international networks have exactly scaled strength equal to s suggests that the distribution of $\boldsymbol{s}$ follows, in both cases, an exponential law. This property implies that $P(s)$ is proportional to $e^{-\lambda s}$ for some parameter $\lambda > 0$.

We estimated the best-fit exponential curves for the scaled strength sequences $\boldsymbol{s}$ of the two networks and we obtained the results listed in Table 2. To verify the goodness of fit we implemented a one-sample Kolmogorov-Smirnov test, which did not reject the exponential distribution in both networks. To visually check the validity of the test decision we present in Fig. 4, a comparison between the graph of the empirical cumulative distribution function (ECDF) of $\boldsymbol{s}$ (in blue) and the best-fit exponential curve (in red). The dashed curves represent the lower and upper confidence bounds that guarantee a coverage probability of 90%. The graphical representation shows that, with the exception of a little initial section, the best-fit exponential curve lies between the 90% confidence bounds in both networks.

As described, for instance in Gonzàlez and Barabàsi (2007), the exponential distribution typically characterizes the networks that contain a high number of nodes with homogeneously low strength and a little number of nodes more connected as it happens in the cases under study. We notice, also, that the results described in this section confirm the similarity between the national and international visitor mobility patterns.

3.3 *Assessing Network Concentration*

In this section, we quantify the geographic dispersion of the international and national visitors mobility by means of the Gini index (Gini 1912). This is a well-known measure of inequality for income and wealth distributions that had been successfully applied in network science as a coefficient of vertex strength concentration (see, for instance, Badham (2013), Goswami et al. (2018), Hu and Wang (2008), Huber (2009), Lopes et al. (2012), Oliveira et al. (2016), Reggiani et al. (2010)). To quantify the scaled strength concentration of a network, we need to calculate the Gini index in the following form

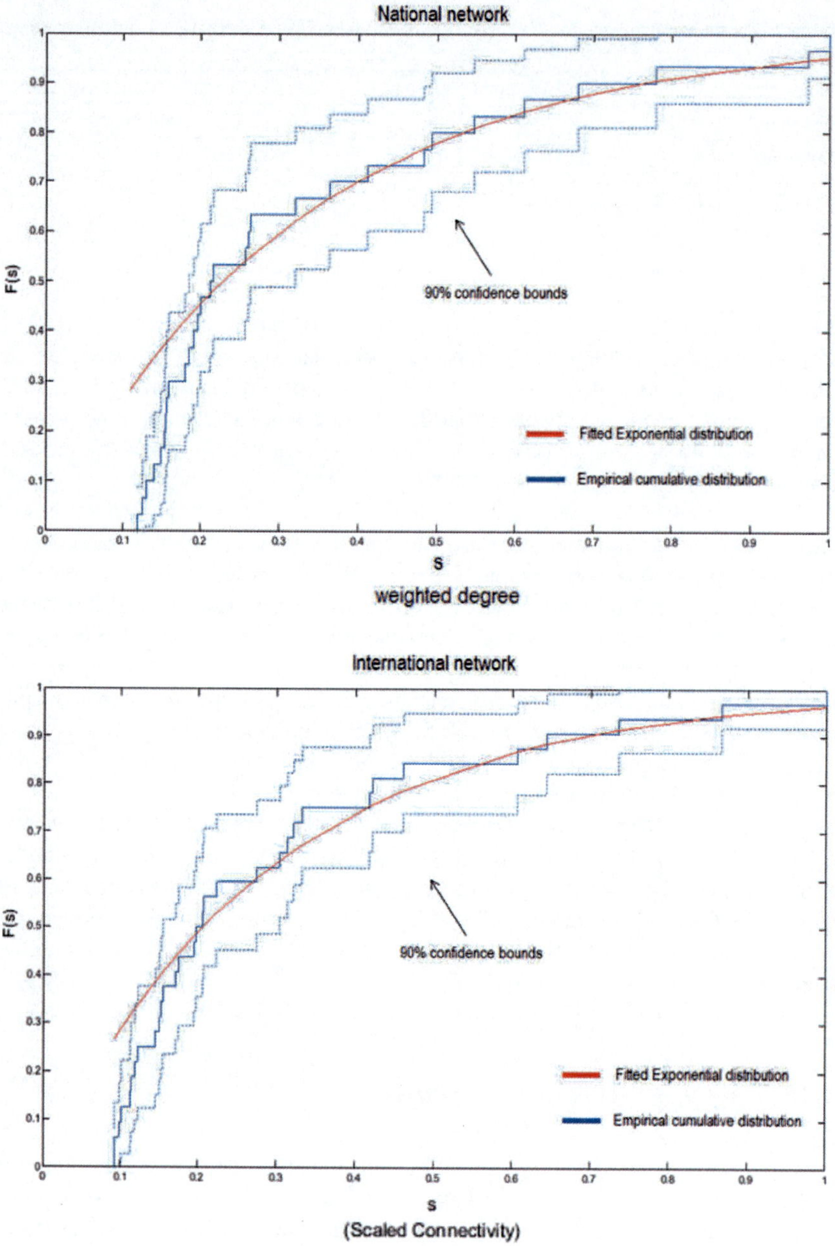

Fig. 4 Visual comparison between the empirical cumulative distribution function (ECDF) of the network strength sequence and the estimated exponential distribution curve. The dashed curves indicate the 90% confidence interval. The top panel refers to the national network and the bottom panel refers to the international network

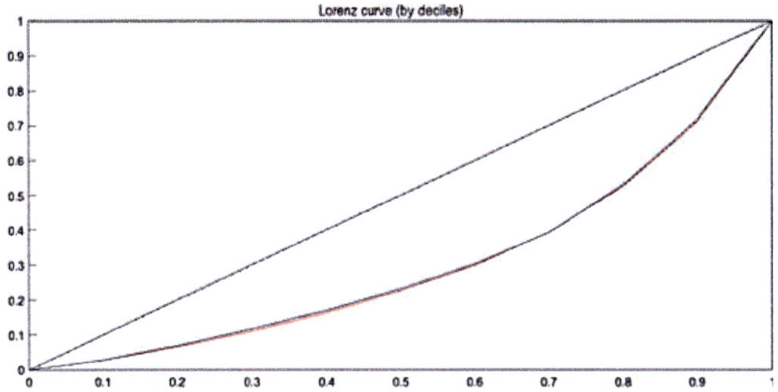

Fig. 5 The graph shows the Lorenz curves for the national network (in black) and for the international network (in red) compared to the perfect equality line

$$G = \frac{\sum_{i=1}^{n} \sum_{j=1}^{n} \left| s_i - s_j \right|}{2N^2 \overline{s}}, \tag{3}$$

where $\overline{s}$ is the average strength. In the context of strength distributions, the Gini index is formally defined as the normalized expected difference in strength between two randomly selected nodes (Badham 2013), assuming the value 0 for a completely homogeneous network and a value approaching 1 for a completely heterogeneous network. The calculation of the Gini index provided the value 0.395 for the national network and the value 0.403 for the international network. To better clarify the meaning of these outcomes, we recall the connection between the Gini index and the Lorenz curve, which provides a graphical representation of inequality (Lorenz 1905). For the strength distribution of a network, the Lorenz curve plots the cumulative proportion of the nodes ordered by strength against the cumulative proportion of the strength held by those nodes. The diagonal line, then, indicates the situation in which all nodes have the same strength. A greater distance from the diagonal line signals greater inequality. Referring to the graph of the Lorenz curve, an alternative definition of the Gini index relates to the measure of the area bounded by the Lorenz curve and the diagonal line divided by one half. The empirical Lorenz curves for the national and the international network are plotted, respectively, in red and in black in Fig. 5. The graphs underline the similarities between the two cases, in fact, they are almost overlapped and indicate a moderate level of strength concentration for both networks.

4 Conclusions and Future Research

In this paper, we used the tools of the network analysis to investigate the mobility patterns of visitors in Sardinia, derived from mobile phone data. From our study, we may draw two main results. It appears, in fact, that the mobility patterns of the national and international visitors in Sardinia, are very similar, and that in both cases the network is composed prevailingly of nodes homogeneous in strength with a little number of exceptions. The last ones correspond to nodes that are geographically located in the North-East of the island where the co-visits appeared more concentrated. This information may be particularly relevant for the governments and the policy makers because the visitor mobility may exacerbate the tourism impact and intensify the differences across locations. We lay the comparison between the mobility patterns of the residents and the visitors for future research.

References

Ahas, R., Aasa, A., Roose, A., Mark, Ü., & Silm, S. (2008). Evaluating passive mobile positioning data for tourism surveys: An Estonian case study. *Tourism Management, 29*(3), 469–486.

Asero, V., Gozzo, S., & Tomaselli, V. (2016). Building tourism networks through tourist mobility. *Journal of Travel Research, 55*(6), 751–763.

Badham, J. M. (2013). Commentary: Measuring the shape of degree distributions. *Network Science, 1*(2), 213–225.

Baggio, R., & Scaglione, M. (2017). Strategic Visitor Flows (SVF) analysis using mobile data. In R. Schegg & B. Stangl (Eds.), *Information and communication technologies in tourism 2017* (pp. 145–154). Cham: Springer.

Candia, J., Gonzàlez, M. C., Wang, P., Schoenharl, T., Madey, G., & Barabàsi, A. L. (2008). Uncovering individual and collective human dynamics from mobile phone records. *Journal of Physics A: Mathematical and Theoretical, 41*(224015), 11.

D'Agata, R., Gozzo, S., & Tomaselli, V. (2013). Network analysis approach to map tourism mobility. *Quality and Quantity, 47*(6), 3167–3184.

Doyle, J., Hung, P., Farrell, R., & McLoone, S. (2014). Population mobility dynamics estimated from mobile telephony data. *Journal of Urban Technology, 21*(2), 109–132.

Gini, C. (1912). Variabilità e mutabilità. *Studi Economico-Giuridici dell'Università di Cagliari, 3*, 1–158.

Girardin, F., Vaccari, A., Gerber, A., Biderman, A., & Ratti, C. (2009). Towards estimating the presence of visitors from the aggregate mobile phone network activity they generate. In *International Conference on Computers in Urban Planning and Urban Management*, Hong Kong.

Gonzàlez, M. C., & Barabàsi, A. L. (2007). Complex networks: From data to models. *Nature Physics, 3*, 224–225.

Goswami, S., Murthy, C. A., & Das, A. K. (2018). Sparsity measure of a network graph: Gini index. *Information Sciences, 462*, 16–39.

Horvath, S. (2011). *Weighted network analysis: Applications in genomics and systems biology*. New York: Springer.

Hu, H. B., & Wang, X. F. (2008). Unified index to quantifying heterogeneity of complex networks. *Physica A, 387*(14), 3769–3780.

Huber, H. (2009). Comparing spatial concentration and assessing relative market structure in air traffic. *Journal of Air Transport Management, 15*(4), 184–194.

Hwang, Y., Gretzel, U., & Fesenmaier, D. R. (2006). Multiplicity trip patterns: Tourists to the United States. *Annals of Tourism Research, 33*(4), 1057–1078.

Isaacman, S., Becker, R., Cáceres, R., Kobourov, S., Martonosi, M., Rowland, J., et al. (2011). Identifying important places in people's lives from cellular network data. *International Conference on Pervasive Computing* (pp. 133–151). Berlin, Heidelberg: Springer.

ISTAT. (2011). *Capacity and occupancy in collective accommodation establishments 2016*. Rome: ISTAT.

Kung, K. S., Greco, K., Sobolevsky, S., & Ratti, C. (2014). Exploring universal patterns in human home-work commuting from mobile phone data. *PloS One*, *9*(6).
Lopes, G. R., Da Silva, R., Moro, M. M., & de Oliveira, J. P. M. (2012). Scientific collaboration in research networks: A quantification method by using Gini coefficient. *IJCSA*, *9*(2), 15–31.
Lorenz, M. O. (1905). Methods of measuring the concentration of wealth. *Publications of the American Statistical Association*, *9*(70), 209–219.
Mir, D. J., Isaacman, S., Cáceres, R., Martonosi, M., & Wright, R. N. (2013). DP-WHERE: Differentially private modeling of human mobility. In *2013 IEEE International Conference on Big Data* (pp. 580–588). IEEE.
Oliveira, A. V., Lohmann, G., & Costa, T. G. (2016). Network concentration and airport congestion in a post de-regulation context: A case study of Brazil 2000–2010. *Journal of Transport Geography*, *50*, 33–44.
Ratti, C., Frenchman, D., Pulselli, R. M., & Williams, S. (2006). Mobile landscapes: using location data from cell phones for urban analysis. *Environment and Planning B: Planning and Design*, *33*(5), 727–748.
Reggiani, A., Nijkamp, P., & Cento, A. (2010). Connectivity and concentration in airline networks: A complexity analysis of Lufthansa's network. *European Journal of Information Systems*, *19*(4), 449–461.
Schneider, C. M., Belik, V., Couronné, T., Smoreda, Z., & González, M. C. (2013). Unravelling daily human mobility motifs. *Journal of the Royal Society Interface*, *10*(84), 20130246.
Shih, H.-Y. (2006). Network characteristics of drive tourism destinations: An application of network analysis in tourism. *Tourism Management*, *27*(5), 1029–1039.
Toole, J. L., Herrera-Yagüe, C., Schneider, C. M., & González, M. C. (2015). Coupling human mobility and social ties. *Journal of the Royal Society Interface*, *12*(105), 20141128.

The Role of Open Data in Healthcare Research

Carlotta Galeone, Rossella Bonzi, and Paolo Mariani

Abstract The generation and storage of data have dramatically increased worldwide in the last two decades. Computing and networking capabilities combined with openness enhance the potential impact of the accumulated data, offering society an opportunity to drive massive social, political, and economic change. Open data is a recent approach. In summary, open data can be freely used, shared, and built-on by anyone, anywhere, for any purpose. Though health open data are not regularly available, it is estimated that the value of a more effective use of data resources in the US health care sector alone could be worth USD 300 billion annually. To date, Open Government data count more than 10,000 datasets in Italy but only a few concern healthcare. In this chapter, we will try to clarify what open data are and, after having recalled the principles of Open Government, the attention will draw on open data in the health and pharmaceutical context, focusing on the state of the art in Italy and worldwide.

1 Introduction

Over the last decades, the increasing technological evolution has affected in many ways humanity, economy, and society by modifying behaviors and ways of operating in different life branches.

Alongside this huge technological development, public institutions have been endorsing public access to endless amounts of data once collected and kept protected

C. Galeone (✉) · R. Bonzi
Department of Clinical Sciences and Community Health, University of Milan, Milan, Italy
e-mail: carlotta.galeone@gmail.com

R. Bonzi
e-mail: rossella.bonzi@unimi.it

P. Mariani
Department of Economics, Management and Statistics, University of Milano-Bicocca, Milan, Italy
e-mail: paolo.mariani@unimib.it

P. Mariani and M. Zenga (eds.), *Data Science and Social Research II*,
Studies in Classification, Data Analysis, and Knowledge Organization,
https://doi.org/10.1007/978-3-030-51222-4_13

in data repositories wasting their potential to enable their real-time use by citizens, enterprises, and other public and private organizations.

The synergy between availability of data and actual technological development (the "open movement") enhances the potential value of all accumulated data and provides a great opportunity to promote a massive social, political, and economic change by spreading and consolidating other important values of our social activities, such as transparency, sharing, and teamwork (Jetzek 2016).

To date, the *open* concept has found application in software implementation, where the concept of open-source—born in 1998—has provided the freedom for developers to improve codes written by others. Moreover, open is a fundamental concept in the science and technology field, (e.g., in medical research, in particular virology) and in many phases of pharmaceutical research, where large well-established multinational companies are developing open innovation projects. It is no coincidence that we usually talk about open innovation.

In this paper, we will try to clarify what open data are, and, having recalled the principles of Open Government, we will focus on open data in the health and pharmaceutical context, in particular on the state of the art in Italy and worldwide.

2 Open Data Definition

There are various definitions of open data, one of which is provided by the Open Data Manual, i.e., a complete and constantly updated web guide explaining the basic concepts of open data, where information mainly come from the public sector. The handbook, written by the Open Knowledge Foundation, refers to open data as "data that can be freely used, reused and redistributed", the only limitation being—at most—the request for attribution and redistribution without changing its content in any way (https://opendatahandbook.org).

From this definition, it is possible to deepen the theme focusing on the key concepts of "linked data" and "source". For an open data to be defined as such a set of "data that can be freely used, reused and distributed ..." it must get a license, offering some key features: *accessibility* (the data must be accessible without legal or technological restrictions), *authorization* for use and reuse and redistribution without constraints (i.e., the license cannot oblige data users to generate open contents only), and the *possibility to quote the reference source* being the most relevant ones.

It is easy to understand how open data are comparable to an actual product, with defined features and a user license concerning their own field of application. A more recent approach on the matter aims at turning the open data concept from product to service, where the key point is what the data user needs instead of the data themselves. In other words, open data must be effectively usable and qualifiable as a data source itself (Gurstein 2013).

To date companies, institutions, and operators are paying a lot of attention to open data as they can become a reliable source of information over time, with appropriate documentation to evaluate and consequently improve the quality of data provided and to fully exploit their information potential.

The increasing use of open data will add value to the original data provided by governments and ease the creation of new friendly and personalized services, promoting the growth of new markets and businesses and encouraging citizen participation in both political and social life.

The availability of data on public policies, expenses, and investments can also increase public administration transparency and, consequently, its credibility (European Commission 2011; Kundra 2012). Because of all these benefits, many governments around the world have started to include open data in their e-government strategies and to implement open data programs, thus giving rise to the so-called Open Government Data (OGD) movement (Jetzek 2016).

3 Open Government and Open Data

The concept of Open Government was introduced in a structured way by the Obama government in 2009 and, year after year, it has become a key factor among government actions for over 70 countries in the world. Open Government is based on the principle that all Government and State Administrations activities must be open and available to foster effective actions and ensure widespread control on citizens' management of public assets.

More specifically, a government that intends to be open must guarantee:

1. transparency of information, to allow citizens access to all the necessary information to know the functioning and work of public administrations;
2. participation, in order to involve all citizens, without any discrimination, in decision-making processes and in defining policies, contributing with ideas, knowledge, and skills to the common good and the efficiency of administrations;
3. accountability, that is the obligation of governments to "give an account" to citizens of their actions and decisions, guaranteeing full responsibility for the results achieved (Jetzek 2016).

In a long-term scenario, this process will lead to a new model of interaction and collaboration between citizens and State, where the individual citizen will no longer be just a user of services provided by the Public Administration, but will actively participate in the government choices. We can then state that open data is the key tool to materialize the concept of Open Government into a real sustainable model, in which citizens can proactively evaluate decisions taken by the Public Administration.

3.1 Identifying Open Data

Companies, institutions, and operators are paying a lot of attention to open data not only as a source of information but also to the fact that they can become its source themselves over time. This is why it is worth reflecting on the fundamental features that open data, in this perspective, should have.

Open data should offer the following main features:

- *Relevance*: open data must contain measurements and statistical products that reflect current and potential users' needs.
- *Accessibility*: the information must be easily obtainable and is determined by the conditions through which users obtain the data.
- *Clarity*: the users' ease to understand the data. It is determined by the informative context in which data are presented and whether they are accompanied by appropriate metadata, illustrations such as graphs or maps or whether information on their accuracy is available. The eventual limitations of use and additional assistance from the data producer are relevant as well.
- *Accuracy*: open data must provide clear and accurate information: that is to say, there must be a high degree of correspondence between data and reality.
- *Reliability*: proximity of the initial value/estimate to the next values/subsequent estimates.
- *Comparability*: a measure of how much temporal and geographical differences are due to real variations and not to differences related to statistical concepts, measurement tools, and/or procedures.
- *Integrability*: open data should allow supporting information by integrating it with similar information coming from other sources.
- *Completeness*: concerning not only omissions or lacunae that may be present in the data provided but also the fact that very often data for some national territories are not available.
- *Timeliness*: open data must not provide data referring to periods that are too far apart in time. In other words, the time of detection should not be too far from when the data are disseminated and made available.
- *Periodicity*: the providers must periodically update the information they make available to different users.
- *Punctuality*: the period between the date of data release and the date scheduled by a calendar, by regulation or by prior agreement between partners.
- *Consistency*: the adequacy of the statistics to be differently combined and their potential for different uses.
- *Reconciliation*: the characteristic open data must have so that the statistics within the same source—related to different variables, calculated on different domains, from different sources or processes with different periodicity—can be grouped together.

Open data must be accompanied by an information provision, necessary, in terms of documentation, to evaluate—and consequently improve—the quality of the data provided and for a full and correct exploitation of all the potential information.

Besides, users must be enabled to choose the most suitable sources to satisfy their cognitive needs and to understand to what extent the available data can be used for further processing. A detailed description of the statistical methodologies used, together with measurements of sampling and non-sampling errors, presents a high information content for both the users of the data themselves and its producers.

3.2 Raw Open Data

Raw open data are data in their original raw state, promptly available in machine-readable form, underlying any application though not structured nor easily convertible in an open format.

The general attitude toward raw data, which reflects the importance of the openness of data in the global scenario, is to distribute them immediately, even in raw form, rather than not distributing them at all, making data available on the Internet in the quickest and easiest way. Then, if published data are found to be sufficiently interesting, the users or developers community will take care of converting them effectively into open data, through the so-called "data scraping" process (https://www.w3.org/TR/gov-data/).

3.3 Linked Data

Given open data and source concepts, it now is appropriate to introduce the last definitions, those of Tim Berners-Lee, the famous computer scientist best known as the inventor of the World Wide Web, who introduced in 2009 the concept of linked data (and consequently of Linked Open Data). According to Berners-Lee, linked data concerns using the web to connect related data that were previously unlinked so that a person or machine can explore the web of data at large. A step further includes the need to obtain data that are not only accessible (open) and connected in an organized way (linked) but freely accessible (i.e., interpretable by a PC) in standard formats that allow the greatest possible reusability. Open linked data are structured and linkable to each other so that we can effectively extract value and meaning from the various web sources (https://www.w3.org/DesignIssues/LinkedData.html).

Berners-Lee gives indications to software systems developers on how to make linked data, also providing a star rating. Data are rated one star if they are available on the web with an open license and up to 5 stars if, in addition to being available in an open format, they are in a non-proprietary format, they comply with specific standards (e.g., RDF) and they are interconnected (Bizer et al. 2009; https://www.

w3.org/DesignIssues/LinkedData.html). It should be noted that linked data may or may not be open (Miller 2010).

4 Open Data in Italy

In 2013, Italy has joined the "Open Data Charter of the G8", which commits the G8 Member States to adopt policies to open the information assets of their Public Administrations (https://opendatacharter.net/g8-open-data-charter/).

The inspiring strategic principles are based on the online automatic publication of all administrative data with quality and quantity characteristics, on the possibility of accessing data through digital open formats and on improving governance and promoting innovation. Also within healthcare, the availability of shared national data, such as prescribing data and health system performance data, is required.

In Italy, during the last five-year period, initiatives to open up Information Technology (IT) assets by the Public Administration (PA) have significantly increased. In 2010, the Piedmont Region was the first to launch an open data portal, containing, to date, over 800 datasets from single municipalities and provinces (www.dati.piemonte.it). The following year, the Emilia-Romagna Region has made an open data catalog available containing 390 datasets (www.dati.emilia-romagna.it).

In order to organize the PAs data-sharing process, on October 18, 2011 Agenzia per l'Italia Digitale (AGID), on behalf of the Italian government, launched the national portal www.dati.gov.it, pushing PAs toward innovation and transparency (https://www.agid.gov.it). The dedicated data store currently contains over 20,000 datasets, including geographical data, coming from 76 administrations that can be downloaded for free. To better frame the Italian position in the constantly evolving international open data landscape InfoData, the Data Blog from IlSole24ore (the reference Italian economic-financial newspaper) has compiled a list of over 2,700 open data portals (https://www.infodata.ilsole24ore.com/2018/09/23/nel-mondo-2700-portali-open-data-la-rimonta-dellitalia-2/). Information was gathered performing an accurate widespread search through opendatainception.io, the main source of information on web portals distributions, with a comprehensive list and map of open data portals worldwide (https://opendatainception.io/). This list, though incomplete, well represents the actual worldwide asset. Italy ranks 7th in the top 10 countries for open data portals availability in 2018 (https://www.infodata.ilsole24ore.com/2018/09/23/nel-mondo-2700-portali-open-data-la-rimonta-dellitalia-2/) and, albeit far from more mature realities like the United States and Britain, Italy has 56 portals on the list and ranks higher than Germany and the Netherlands. However, according to the 2018 survey of the European Data Portal (i.e., the Open Data Maturity), Italy is among the countries that have been able to implement a well-coordinated open data policy on the national territory and have a national open data portal with advanced features (https://www.europeandataportal.eu/it/news/open-data-maturity-europe-2018). This means a slow but expanding growth of the Italian movement, despite the late birth of an Italian FoIA

(Freedom of Information Act, a rule that establishes their citizens' right to make a request to access Public Administration data) which was promulgated only at the end of 2016 (https://www.funzionepubblica.gov.it/foia-7).

The datasets on the Italian national portal are sorted by theme, genre, territory of reference and releasing administration. This portal does not contain all the available Italian open data: for example, in its health section only 8 datasets from Lombardy Region are found, while on Lombardy's open data portal there are 52 datasets (www.dati.lombardia.it). To date, the regional data stores with the largest number of downloadable datasets are the Tuscany Region's one (over 1,400) and the Lombardy Region's one (over 1,200).

Despite legislation inviting all Public Administrations to release data so that it can freely be used, reused, and redistributed by anyone interested in it, open data circulation in Italy still has trouble taking off. Italian Municipalities are the main owners of public interest data, such as those on public transportation, tourism, culture, and production activities, though only 37% of them (mostly the largest municipalities) are publishing data in an open format, and, when it gets done, this is perceived as a regulatory obligation rather than a real opportunity. The difficulty to understand open data's real usefulness results in the data being low quality, not accessible, not uniform for a national level utilization. Most Italian manufacturing companies would consider strategic use of data for business, and seem to be interested in getting to know them better, but barely 4% of them use open data from PA sources. Data must be of quality, up-to-date and correct, with the reasonable certainty of being able to count on its availability in the future. On the other hand, companies must also become more aware of their huge potential as a possible predictive source to anticipate market trends and should consequently acquire a better knowledge of the professional roles for data management (https://www.osservatori.net/it_it/osservatori/comunicati-stampa/open-data-in-italia-non-decollano).

With a view to the creation and maintenance of the national portal, www.dati.gov.it, in 2011 ISTAT (Istituto Nazionale di Statistica, Italian National Statistical Institute) published a vademecum on open data, i.e., a guideline for the PAs websites on open data topics, in order to standardize methods for publishing data from various administrations. This document is structured into two macro areas: Part I) PA and open data, introducing the concepts of Open Government and Open Data, with insights on the regulatory framework; Part II) How to proceed in opening PA data, describing practical, organizational, and legal aspects to be considered before making such data available (ISTAT 2011).

According to the principles of the digital administration's new code, PAs have the responsibility to make their data available openly and digitally: the Italian Open Data License (IODL) v 2.0, a specific license agreement, has been specifically designed to manage the spreading of open data (https://opendatahandbook.org/guide/en/). In the context of the development of the information society, this procedure's intent is to make the reuse of public information easy and immediate (https://opendatahandbook.org/guide/en/; https://www.dati.gov.it/content/italian-open-data-license-v20).

4.1 Ministry of Health Datastore

With a particular focus on healthcare open data, the Ministry of Health has launched a dedicated datastore containing 44 national datasets, fewer than the 259 available in the health section of the national open data portal (which also includes those at regional or municipal level, and those belonging to AIFA, Agenzia Italiana del Farmaco) (https://www.dati.salute.gov.it/dati/homeDataset.jsp; https://www.aifa.gov.it).

The available data can be summarized by macro areas of interest: information on Ministry authorized drugs, for example, foreign medicines, master and officinal medical preparations, plant protection products, information on medical devices registered in the database and in the Ministry Directory.

The most downloaded datasets are the complete list of pharmacies open to the public (including branches, dispensaries, and seasonal dispensaries), the complete list of shops, apart from pharmacies, authorized to sell medicines to the public (Article 5 of Decree-Law 223/2006—commercial establishments) and the complete list of logistic sites authorized to distribute medicinal products and medical devices for human use on the national territory (art.100 D.Lgs. 219/2006, already Legislative Decree 538/92). Also of great concern are data sets coming from health structures such as Azienda Sanitaria Locale, Local Health Authority (ASL) and ASL's correspondence with municipalities, family counseling centers, hospitals, university hospitals, and public IRCCS (Istituto di ricovero e cura a carattere scientifico, Institute of hospitalization and treatment on a scientific basis); datasets with data on personnel such as health managers and personnel with a flexible working relationship between ASL and hospitals, divided by role.

Out of curiosity, two other useful examples are worth citing: the list of the 50 best-selling drugs available without prescription in the last semester and the data relative to accredited beds owned by the National Health System available in public hospitals, equivalent accommodation or private and accredited nursing home.

As already stated, since 2013 Italy has joined the other G8 Member States in adhering to the "Open Data Charter of the G8" (https://opendatacharter.net/g8-open-data-charter/). The governments of these countries are committed to adopting policies to open up the information assets of the PA. The five strategic principles that all member countries intend to adopt are

1. Open Data by Default, that means all administrative data must be published online automatically;
2. Quality and quantity of open data (data must be of high quality—metadata, timely, and comprehensive;
3. Universal accessibility (in both open and electronic format);
4. Use of open data to improve governance (sharing of technical skills and experience between countries, transparency on collection methods, standardization and procedures for publication of data);
5. Use of open data to promote innovation (encouraging its reuse for commercial and non-commercial purposes).

14 areas of intervention shared by all countries were identified: in particular, in the healthcare sector, the willingness to share national data, such as prescription data and health system performance data, is required (Bonelli 2013).

There are several interesting projects for the development of new applications using health open data that could ease citizens' daily life, providing new tools to make better choices. For example, the "Pronto Soccorso Lazio Ospedali" app, downloadable from Google Play and Apple Store, was the first application developed using data from the Open Data Lazio portal, which provided real-time data from various first aid units. Within this App, you can view the hospitals of Lazio Region on a map or a list, know how many people are under emergency room treatment at that precise moment, among those how many red, yellow, and green codes are issued, filtering the search by hospital name.

Other regions have recently developed similar applications. In particular, the Veneto Region app distinguishes the type of first aid unit from generic to pediatric to gynecological and of first intervention units, reporting any gynecological and pediatric fast tracks.

In addition, the electronic medical prescription fully went into effect at the beginning of 2016, allowing the automatic transmission of prescription data between general practitioners and pharmacies and already since November 2015, according to Federfarma more than 29 million prescriptions were dematerialized (https://www.federfarma.it). Of a certain interest the case of Sicily, until 2016 without an institutional open data portal (replaced by the website https://nonportale.opendatasicilia.it) and an electronic health file, but that in the same period was able to dematerialize more than 87% of medical prescriptions as well, second only to Veneto (https://dati.regione.sicilia.it). At the bottom of the ranking, we found the autonomous province of Bolzano and the region Calabria.

Videofar is a computerized drug prescription monitoring system oriented to gain an epidemiological insight into the phenomenon of "drug prescription" (https://www.epicentro.iss.it/farmaci/videofar/). It offers the possibility to analyze the volumes of different classes of drugs over time and for regional context, allowing a quick view of the prescribed dynamics over a decade. The data refer to the territorial use of prescribed medicines charged to the NHS and provided through public and private pharmacies in the period 2000–2011. The website allows an easy consultation and the possibility to filter results by drug category and sub-categories, drug consumption data, temporal evolution, and geographical area selection.

Another interesting project is the "Health Advisor" application, born with the idea of creating a web community that allows users to evaluate the performance of the National Health Service according to their own experience, even if at the moment this Apps' development may prove difficult due to a lack of open data on the subject. However, for an assessment of the performance of the Health Service, the open data of the National Outcomes Program are available, providing comparative assessments at a national level based on various indicators on the efficacy, safety, efficiency, and quality of the treatments within the National Health System.

As the number of open data on healthcare is still rather limited, though, with an increasing trend over time, it is clear that the undertaken digitization process will

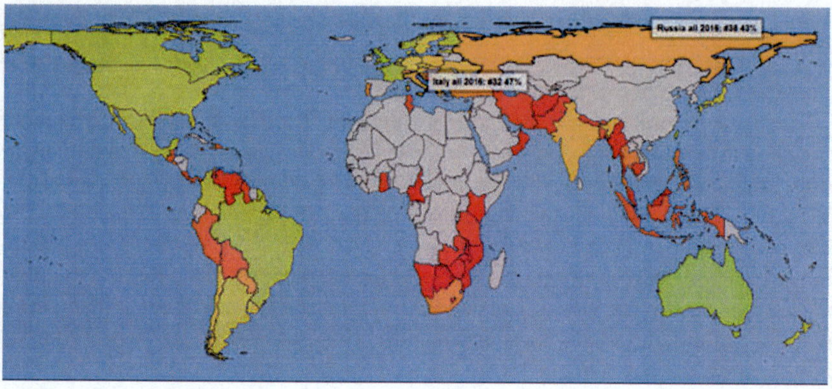

Fig. 1 Tracking the state of Italian Government open data, 2016. Place coloring represents the open data score for the currently selected dataset(s), from 0 (red) to 100 (green). *Source* https://index.okfn.org/place/

contribute to an improvement of information capacity through the integration of data. However, Italy is still struggling to start using open data massively, as the main common assessment methods for open data confirm.

The Global Open Data Index (GODI), i.e., the annual global benchmark for publication of open government data run by the Open Knowledge Network, set Italy at 17th place in 2015 (with 55% openness overall) in terms of open data number and quality, with no significant changes in the ranking position compared to previous years surveys (#21 in 2013, #25 in 2014) (https://2015.index.okfn.org).

The Italian trend for 2016 (Italy: #32, 47% open) is shown in Fig. 1 in a worldwide comparison.

Global Open Data Index methodology has changed significantly between 2015 and 2016 and that's the reason why data are not directly comparable over time. Anyway these results are qualitatively equivalent, in terms of a trend, to those from the Open Data Barometer—a global initiative defining to which extent governments publish and use Open Data for accountability, innovation, and social impact, where Italy's ranking for 2016 hits #20, with no significant change compared with previous years (https://devodb.wpengine.com/?_year=2017&indicator=ODB). By the time of this survey, this not particularly brilliant result was mainly due to the limited number of data available and reused resulting in a low impact on society and economy.

Turning to a more recent context, the Open Data Barometer-Leaders Edition released in 2017 reveals that Italy has reached a score of 50 (out of 100), 6 points up compared to 2016, the most performing sectors being datasets referring to demographic data (100), public transportation (95), and environment (80) (https://opendatabarometer.org/country-detail/?_year=2017&indicator=ODB&detail=ITA).

From a European perspective, it is recent news that Italy is gaining further positions in the European ranking, e.g., in 2017, the number of open data that the Public Administration made freely accessible by citizens and companies increased by 10%

compared to the previous year (https://www.agid.gov.it/it/agenzia/stampa-e-comunicazione/notizie/2018/11/20/open-data-maturity-report-italia-ai-primi-posti).

In addition, according to the recently published "Open Data Maturity Report", Italy hits fourth place in the 2018 European ranking for the ability to enhance its open data thus confirming itself, for the second year in a row, among the countries that have been able to implement an advanced and well-coordinated open data policy with a national open data portal with advanced functionalities (Cecconi and Radu 2018).

Despite these encouraging results, the Italian health sector remains the weak point in the open data context and much still has to be improved.

While many Italian regions are developing huge potentials on the open data front, others are not, thus leading to an overall failure of the open data network. As an example, if First Aid real-time data are not available and interconnected for all regions but only some of them, the usefulness that these data may have for inhabitants of the border provinces is nullified and it needs to be revised. The random diffusion of open data especially in the health sector makes any truly useful service unattainable. A national strategy on open data, within a coordinated development planning throughout the territory, is almost completely lacking. Indeed, raising public awareness on the potential that information assets have, especially in the healthcare sector, by putting the user back at the center of interest, should help in improving the overall local services, which, despite being territorial, should be even less bound at a regional level.

A positive sign, however, came after the launch of the web platform of the Health Ministry in 2012, where the available datasets immediately registered a significant number of downloads, the latter evidence of citizens' and developers' interest in health data. However, health data represent only a small part of the approximately 23,000 datasets released to date in Italy. For this reason, the process of opening the information carried out by the Ministry of Health can boost the publication, in an open format, of data relating to the health system (https://www.dati.gov.it/content/dati-aperti-ministero-salute).

The spending commitment is not yet sufficient to fill in the overall delay in the sector digitization. For example, as pointed out by Federsanità (the institutional subject representing local municipalities to ensure the pathways of social health and social welfare integration), the diffusion of the electronic medical record is still very limited in hospitals and territorial units, lacking integration with primary care software (e.g., the Individual Health Card, FSE). The 2020 target is the activation of the SFE in 20 regions for 70% of citizens, as, to date, 11 Regions have positively passed the interoperability tests with the central platform, the percentage of ESF services having been realized being between 90 and 100% (https://ildigitaleinitalia.it/il-digitale-in-italia-2018/il-mercato-digitale-italiano-2017-2020/consuntivi-e-previsioni-i-settori/sanita.kl).

A constant update on the developments of this interesting and dynamic topic is available to all interested users through registration to the "Spaghetti Open Data" mailing list, the longest and most active topic related Italian community (https://spaghettiopendata.org).

5 Open Data and Professional Databases

The growing number of open data and public administration portals in Italy is the litmus test of the choice of innovation, though not without obstacles, in the social action of Italian citizens. The challenge is to transform information into knowledge, evidence, in the healthcare field. The turning point will be achieved when most of the open data coming from various sources will be linkable to each other, thus creating new information summaries. These new evidences will never be able to replace those deriving from studies in the healthcare sector with ad hoc information collection, as they will never have the same level of precision and information accuracy. However, they will be complementary evidences, certainly useful for completing such a complex and multifaceted theme as public health is. The massive advent of open data will contribute to the improvement of professional databases through the integration of data (even just for validation and completion) but it will not replace them. The distinction will remain for at least three reasons:

1. in the vast majority of cases, open data are collected for administrative purposes and not for research purposes;
2. the production of open data is not subjected to professional and business logic;
3. most of the open data features that could lead them to become a source are not respected.

On the other hand, some professional databases that today have their strength in the difficulty to access the administrative data are destined to disappear or to be transformed through appropriate integrations with original and research data. Ultimately, from this brief examination, we can deduce the growing importance and usefulness of open data and the future scenario that depicts a growing interest in them by companies, institutions, and 15 citizens. However, the limits of open "sources" appear evident. Here are some examples: the possibility that the data structure changes every time an update is published; an update frequency that is not associable with a businesslike use of the data; the aim of collecting different information that barely coincides with the needs of users. Another glaring example of the difficulties encountered when exclusively relying on open data is linked to the ways of creating a connection between different data: unique and common encodings between different open data will hardly be found, thus forcing users to an expensive and sometimes impossible matching activity.

On the other hand, it is also clear that the availability of open sources has begun to revolutionize the world of data, their analysis, and their interpretation. Indeed, according to the "Open Data Charter", to which Italy has also adhered, even the prescription data could/should become open, thus modifying a more consolidated market and research scenario within the national pharmaceutical industry (https://opendatacharter.net/g8-open-data-charter/).

6 Open Data in Europe

At the end of 2012, a beta version of the European Community's open data portal was launched reaching its full operativity in the following years and which now contains over 9000 datasets (https://data.europa.eu/euodp/it/home/). The site describes many applications for reusing open data from single or multiple sources in each field, developed by institutions, agencies and other EU organizations or third parties. For example: the mapping of fishing activities, the Drought Observatory, the statistics on student mobility, and the Atlas of Road Accidents are herein reported (https://data.europa.eu/euodp/it/home/). Particularly, in January 2010 web pioneer Sir Tim Berners-Lee launched a new British government website offering free access to a huge amount of public-sector data for private or commercial reuse (https://data.gov.uk).

Denmark offers a very well implemented and easy to use web portal with statistics on the total sales of medicines in Denmark for the years 1996–2016 (https://medstat.dk/en).

The French National Address Database (BAN) containing over 25 million geocoded addresses (with no personally identifiable data) is a successful private–public partnership and a clear example of increased government efficiency thanks to open data. Users can freely access and download the addresses in BAN using its tools and geocoding services (https://adresse.data.gouv.fr/).

The PROTECT website collects drug consumption databases for European countries with constant updates. The European Medicines Agency (EMA) is the coordinator of PROTECT and the GSK is the deputy coordinator of PROTECT, managing a multi-national consortium of 34 partners including academics, regulators, small-and-medium enterprises, and the European Federation of Pharmaceutical Industries and Associations (EFPIA) companies (https://www.imi-protect.eu/index.shtml).

A very interesting infographic application is "My country in a bubble" that allows to immediately compare data of European countries through about 50 indicators (https://ec.europa.eu/eurostat/cache/BubbleChart/index.html?lg=en).

Finally, the Drugle search engine focuses on pharmacology and medicines information, listing and analyzing information on medicines available on the Internet through a simple query. The potential of this search engine is that it can be combined with other healthcare applications and systems (https://drugle.se).

7 Open Data Worldwide

The most famous and well-structured portal is the American one, born as a result of the Obama reform in 2009 (https://www.data.gov). In the following years, we have witnessed the launch of many government data-stores worldwide.

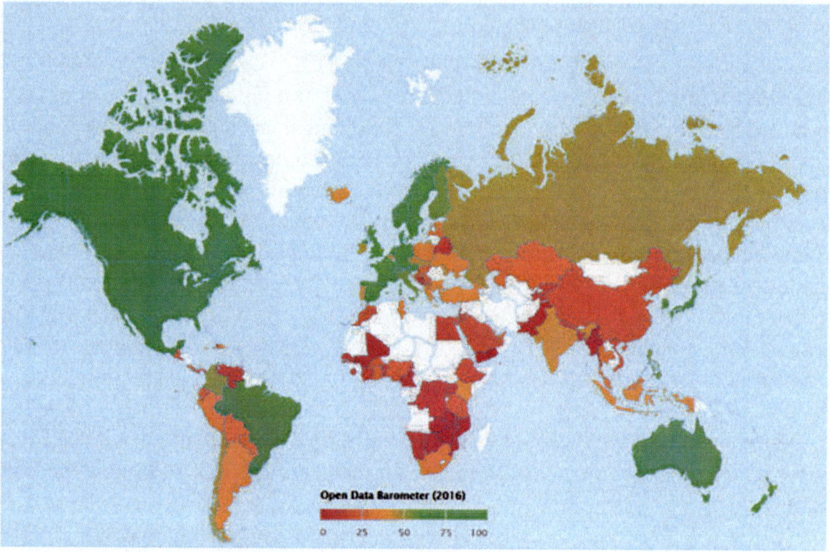

Fig. 2 Overview of countries publishing and using open data, the year 2016. *Source* https://opendatabarometer.org/4thedition/?_year=2016&indicator=ODB

The distribution and use of open data referring to the year 2016 according to the Open Data Barometer—4th Edition (the latest full edition, covering up to 115 countries) are shown in Fig. 2 (https://opendatabarometer.org/4thedition/?_year=2016&indicator=ODB).

The latest version of the website, i.e., "The Leaders Edition", referring to the period July 2016–September 2017, focuses on the 30 governments that have adopted the Open Data Charter and those that, as G20 members, have committed to G20 Anti-Corruption Open Data. The participating countries, listed from high to low-performance ranking, are shown in Fig. 3 (https://opendatabarometer.org/?_year=2017&indicator=ODB&G20=1&IODCH=1).

As shown, the top leaders in terms of overall performance are North America, Australia, France, and UK. Europe, Asia, and South America follow with acceptable results, while most African and western Asian countries are lacking information on this issue.

Political stability and government efforts in advancing wider reforms and encouraging a culture of openness can be the turning point in determining the success of open data initiatives. Countries in which political will is rendered into strong legal and political foundations, as happens in Canada, Mexico, Japan, and Korea succeed in achieving steady progress in their rankings. Otherwise, lack of government action in prioritizing open data as strongly as in the past few years, as has been happening in Nordic countries, once open data leaders, leads up to significant derailing progress.

Political changes are a big issue and strong indicators of a country's technological empowerment. To date, highly-ranking countries, like the USA, UK, and Canada,

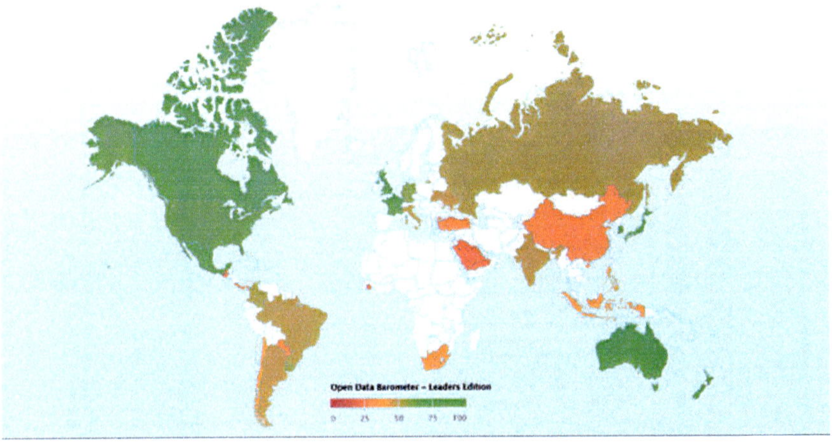

Fig. 3 Overview of countries publishing and using open data, July 2016–September 2017. *Source* https://opendatabarometer.org/?_year=2017&indicator=ODB

lack the policies needed for open data to survive: the new US administration has already removed certain key datasets from websites, leading to concerns about the future of open government data in the USA, meanwhile the UK seems to have loosened its policy commitments. At last, one of the primary reasons Canada has not overtaken UK's longstanding leadership position in the ranking is due to the restrictive licensing of several datasets, but, in spite of that, Canada will be hosting the Open Government Partnership (OGP) Global Summit in Ottawa on May 2019. The importance of political decisions is also witnessed in countries like Ukraine, Argentina, the Philippines, Burkina Faso, and Tanzania, all of which experienced remarkable improvements in their latest scores. On the other hand, a change in political assets, as seems to have happened in Costa Rica, Ecuador, and Rwanda, results in an evident ranking decrease.

Figure 4 shows the global rankings according to the Open Data Barometer—Fourth Edition (Open Data Barometer—4th Edition, 2017).

Countries that have formally adopted the Open data Charter are on average making good progress in fulfilling its principles with their performance having been improving in recent years.

In 2015, the Japanese Ministry of Land, Infrastructure, and Transport (MLIT) set up an open data site to assist the elderly and pedestrians with disabilities.

The Mexican online platform "Mejora tu Escuela", allowing users to compare over 163,785 datasets to improve educational decision-making and demanding better education for children, is a clear example of how open data could truly be used to empower and include all citizens thus making public services more effective and inclusive (https://www.mejoratuescuela.org).

Worthy of note is the result of the process of digitization of Kenya, a pioneering country on these issues in Africa and, to date, the first African country in the world

ODB 4th Edition Ranking

Rank	Score	Country	Readiness	Implementation	Impact
1	100	United Kingdom	99	100	94
2	90	Canada	96	87	82
3	85	France	100	71	88
4	82	United States of America	96	71	80
5	81	Korea	95	59	100
5	81	Australia	85	78	78
7	79	New Zealand	92	58	99
8	75	Japan	84	60	89
8	75	Netherlands	94	64	68
10	74	Norway	77	71	73
11	73	Mexico	83	58	88
11	73	Spain	81	58	88
13	71	Denmark	67	71	71
14	70	Austria	83	56	78
14	70	Sweden	87	70	47
14	70	Germany	67	69	71
17	61	Uruguay	75	64	38
18	59	Brazil	66	55	59
19	58	Switzerland	77	50	48
20	56	Italy	79	51	37
20	56	Finland	63	60	42
22	55	Philippines	58	41	76
23	53	Singapore	73	46	41
24	52	Colombia	72	42	46
25	49	Russia	60	54	27
26	47	Ireland	70	51	17
26	47	Chile	62	56	16
28	46	Israel	66	37	42
29	45	Belgium	79	38	20
29	45	Slovakia	59	43	34
31	44	Czech Republic	54	44	36
31	44	Moldova	55	54	14
33	43	India	68	32	35
34	42	Portugal	58	47	16
35	40	Kenya	57	22	58

Fig. 4 Top 30 positions of the global rankings as from the Open Data Barometer for the year 2016, originally covering 115 countries. *Source* Open Data Barometer—4th Edition, 2017

ranking issued by the Open Data Barometer, as pointed out in Fig. 4. In 2011 the Open Data Initiative, a government policy transparency operation, was launched through a project funded by the Ministry of Information and Communication. To date, the portal contains over 900 datasets, well cataloged and easy to query (https://www.opendata.go.ke).

Uruguay is doing well too, ranking at number 17 in the Open Data World, first among South American countries according to the Open Data Barometer, as highlighted in Fig. 4. Every year, in February, citizens can choose their healthcare provider, so every year 1.6 million users can potentially make a new choice and, 3 years later, the first choice can be changed. The national "ATuServicio" platform is a Ministry of Health project on public health data, integrated with data from individual hospitals to give citizens the possibility of evaluating all the services provided for a conscious choice of the provider (https://atuservicio.msp.gub.uy/). This application was considered one of the best models for the use and sharing of open data in the health sector (winner of various international awards including the Open Knowledge UK and Open Government Awards).

In the Philippines, ranking #22, community members can monitor local government budgets and engage in planning, for the first time at local level, through the Check my Barangay platform (https://www.ansa-eap.net/projects/check-my-barangay/).

8 Conclusion

In conclusion, what comes out of this overview is the constantly increasing amount and diffusion of open data worldwide and their growing availability through web portals and free consultation datasets. In spite of this expansion, the number of truly open global datasets seems to be at a standstill and the obstacles in creating new knowledge starting from the data available are not negligible. The most evident limitations are the poor and not timely updates of many open data, the possibility that the data structure could change every time an update is published and the difficulty to find unique and common coding between different open Data, requiring an expensive matching activity, were even possible.

Government-held data must be open by default and follow the principles set out in the Open Data Charter. Besides, governments must uphold their commitments to open data to avoid backsliding.

Focusing on the healthcare sector, which is of great interest to date, this bias comes true, as the availability of truly open data is still limited, both in Italy and worldwide, with no distinction concerning the economic and technological development level of the country. We have presented different and successful initiatives held by governments in countries that stand out by degree of participation and attention to the open data world, that demonstrate the potential of these data. However the healthcare data ecosystem remains mostly closed and the reasons are many and complex, to name a few: the amount of updating, the consistency between the data considered and the accuracy and completeness of the same are vital. A deficiency in one of the key data features may result in a useless service.

An improvement in quantitative and qualitative terms of open data and its use along with a deep knowledge of its limitations are the key factors to create new and useful knowledge, enabling decision-makers to understand the territory and supplying citizens with an important participatory and control tool.

References

Bizer, C., Heath, T., & Berners-Lee, T. (2009). Linked data—The story so far. *International Journal on Semantic Web and Information Systems, 5*(3), 1–22. https://doi.org/10.4018/jswis.2009081901

Bonelli, U. (2013). *Open data come politica di sviluppo economico.*

Cecconi, G., & Radu, C. (2018). Open data maturity in Europe 2018. *European Commision, Capgemini Invent*, European Data Portal.

European Commission. (2011). *Open data An engine for innovation, growth and transparent governance.* Retrieved from https://eur-lex.europa.eu/LexUriServ/LexUriServ.do?uri=COM:2011:0882:FIN:EN:PDF.

Gurstein, P. (2013). *Social equity in the network society: Implications for communities* (pp. 161–182). Policy, Planning and People: Promoting Justice in Urban Develoment.

ISTAT. (2011). *Vademecum Open Data.*

Jetzek, T. (2016). Managing complexity across multiple dimensions of liquid open data: The case of the danish basic data program. *Government Information Quarterly, 33*(1), 89–104. https://doi.org/10.1016/j.giq.2015.11.00

Kundra, V. (2012). *Digital fuel of the 21st century: Innovation through open data and the network effect.* Politics and Public Policy: Joan Shorenstein Center on the Press.

Miller, P. (2010). Linked data and government. *ePSI Platform Topic Report, 7*.

Open Data Barometer—4th Edition, G. R. (2017). *Open Data Barometer* 4th Edition Global Report.

Social Epidemiology: The Challenges and Opportunities of Worldwide Data Consortia

Carlotta Galeone, Rossella Bonzi, Federica Turati, Claudio Pelucchi, Matteo Rota, and Carlo La Vecchia

Abstract Over the last few decades, social epidemiology has developed as a solid epidemiology branch, focusing on understanding how social experiences influence population health. At the same time, growing of collaborative and interdisciplinary research led to the proliferation of multi-institutional consortia, able to assess and quantify risk-disease associations of interest with a higher degree of accuracy, to explore subgroups of the population, and to investigate interactions between environmental, genetic, and socioeconomic factors. Increasing evidence shows that low Socioeconomic Position (SEP) is a strong determinant of morbidity and premature mortality from selected non-communicable diseases, including several cancers. Thus, an accurate quantification of the impact of SEP on cancer risk is of major importance to plan public health interventions for cancer incidence and socioeconomic disparities reduction. Large data consortia as the Stomach Cancer Pooling (StoP) Project and the International Head and Neck Cancer Epidemiology (INHANCE), in which the University of Milan is proactively involved, allowed investigators to address the effects of education and household income, the main SEP determinants, on gastric and head and neck cancer, respectively, confirming the existence of a strong association between low SEP and those major neoplasms.

C. Galeone (✉) · R. Bonzi · F. Turati · C. Pelucchi · C. La Vecchia
Department of Clinical Sciences and Community Health, University of Milan, Milan, Italy
e-mail: carlo.lavecchia@unimi.it

C. La Vecchia
e-mail: carlotta.galeone@gmail.com

M. Rota
Department of Molecular and Translational Medicine, University of Brescia, Brescia, Italy

P. Mariani and M. Zenga (eds.), *Data Science and Social Research II*,
Studies in Classification, Data Analysis, and Knowledge Organization,
https://doi.org/10.1007/978-3-030-51222-4_14

1 Introduction

Social epidemiology is a relatively recent branch of epidemiology that aims to understand how social factors affect population health (Honjo 2004). One of the most important examples of sociostructural factors in social epidemiology is the study of social class in relation to Non-Communicable—or chronic—Diseases (NDCs). NDCs are diseases of long duration and generally slow progression and are the leading causes of health issues worldwide, accounting for 63% of all annual deaths globally. Cardiovascular diseases, cancers, chronic respiratory diseases, and diabetes are the main types of NDCs, whose reduction, according to the World Health Organization (WHO) member states, in terms of mortality in people aged 30–70 has been committed to be achieved by the year 2025 (WHO 2018b; Bennett et al. 2018). NDCs are the result of a combination of genetic, physiological, environmental, and behavioral factors, exposing people to health, social, and economic challenges on a daily basis. Disadvantaged people are caught in a sort of vicious circle in which poverty causes illness and illness feeds poverty (Wagstaff 2002).

Consolidated evidence shows that NCDs morbidity and premature mortality is higher in low-income and middle-income countries, and, at least in high-income countries, in people with lower socioeconomic status, making NCDs an important obstacle to reducing global and national health inequalities (Marmot 2014; Wagstaff 2002; Niessen et al. 2018).

Socioeconomic Position (SEP) reflects the availability of cultural, material, and social resources that translate into advantages in terms of decision making, social network, lifestyle habits, and also access to health services. An accurate quantification of the impact of SEP on the risk of the disease is of major importance to plan public health interventions aimed to reduce NCDs incidence and socioeconomic disparities.

Among the most commonly diffused NCDs, cancer is the second leading cause of death worldwide, and responsible for an estimated 9.6 million deaths in 2018. Globally, about 1 in 6 deaths is due to cancer and approximately 70% of deaths from cancer overall occur in low- and middle-income countries (WHO 2018a).

A precise quantification of SEP impact on cancer is difficult to point out locally, as cancer incidence, survival, and mortality are subject to large variations across countries and, within countries, across social groups. Social inequalities, as well, are continuously evolving and reshaping over time as a reflection of the economic, political, social, legislative, and technological asset of the society. Differences among social groups have a strong impact on cancer at every stage of the disease by affecting the exposure to risk factors, and hence the likelihood to develop the disease, as well as the timely access to public health measures, diagnostic and treatment facilities and health-care services.

High-income countries show higher incidence rates of all cancers than most low- and middle-income countries, mainly because of environmental and lifestyle risk factors. At the same time, low- and middle-income countries often have similar or sometimes higher mortality rates from cancer than high-income countries, mainly because of a lack of access to timely diagnosis and proper treatment. However, within almost all countries, mortality rates for most cancer types are, to a disproportionate extent, higher for groups of the population with low socioeconomic position or otherwise disadvantaged, due to poorly designed health systems or limited or even inhibited access to preventive interventions, early detection, diagnosis, treatment, or/and palliative care (Vaccarella et al. 2019; Niessen et al. 2018).

Disparities in cancer care could be resolved if the highest achievable standards in health care were attained across countries at all economic levels. Failing to translate the excellent results constantly achieved in cancer knowledge through scientific research into effective action, in terms of health infrastructure and adequate basic services, still contributes to regional, national, and international health inequities.

The advent of collaborative and interdisciplinary research framework, along with the proliferation of multi-institutional research consortia during the last two decades, markedly affected cancer epidemiology. The National Cancer Institute's Epidemiology gives the most globally recognized definition of consortium as a 'group of scientists from multiple institutions who have agreed to participate in cooperative research efforts involving activities such as methods development and validation, pooling of information from more than one study for the purpose of combined analyses, and collaborative projects. Consortia are able to address scientific questions that cannot be addressed otherwise due to scope, resources, population size, or expertise. This cooperation usually involves multiple projects over an extended time period. Consortia can also be referred to as collaboratives (NIH 2019).

A general feature and strength of consortia is the easy and prompt communication among members for an interconnected sharing of study results. Such large collaborative groups benefit in terms of dissemination of research tools and information, from the establishment of web forums, public websites, or other global means of inter-diffuse communication. Quick and fluent knowledge and data sharing are on the basis for the coordination of the scientific research whose aim is to maximize the efficiency to understand, prevent, treat, and relieve the risk and hence the incidence of diseases on the population at a global level.

Moreover, the uniquely large data set on which consortia are based permit to define and quantify, with a degree of accuracy higher than ever before, the main effects of each risk factor of interest and to adequately address associations in subgroups of the population, as well as interaction between environmental, genetic, and socioeconomic factors.

2 Definition of Socioeconomic Position

SEP is a complex concept which involves several dimensions including education, work experience, and household income, access to material resources, prestige, and social position. All of these dimensions are associated, even though each of them accounts for different aspects of the socioeconomic stratification. In a broader sense, speaking of socioeconomic status involves referring to the most common forms of inequality (Geyer et al. 2006).

The assessment of socioeconomic position in the epidemiologic research is usually performed throughout the use of a series of indicators, traditionally education, occupation, and income, though their specific use often and strictly depends on data availability.

The strengths and limitations of the selected SEP determinants are herein briefly summarized, following a measure reliability order, starting from income, the less stable measure.

Individual, or, better, household income, which may be a useful indicator in particular for women or those who may not be the main earners in the household, reflects the material component of people everyday life. People with higher incomes are more likely to experience better living conditions, social services affordability, and healthy environment than lower income groups. However, income is the SEP indicator mostly subjected to changes, also on a short-term basis, it is age-dependent and it shows the highest non-response rate in epidemiological investigations when compared to other SES measures. It has also problems in validity of reporting.

Occupation reflects the privileges related to social standing, material resources, and job-related risk factors. Occupation-based indicators of SEP are widely used in the epidemiologic research due to their large availability in many routine data sources, including census data and death certificates. The individual current and the longest-held occupation are often taken into consideration to assess adult SEP. Measures from one or several individuals belonging to the same family unit can be used to characterize the SEP of others connected to them, e.g., children, spouse, elderly, unemployed. Among the limitations, occupation indicators clearly cannot be assigned to currently unemployed or retired people, housekeepers, students, and people with informal, unpaid, or illegal jobs. Also, classification for some job categories is difficult. Moreover, the definition of occupation related to SEP may have different meanings according to individual birth date and geographical location, which consequently represents an issue in terms of international comparisons.

Education reflects the intellectual assets of individuals besides the socioeconomic conditions in childhood and adolescence and it represents people potential opportunity to access, in the future, to higher level jobs and earnings. Educational attainment is a widely used indicator of SEP. The strength of using education as a proxy for SEP in the adult population is its smaller likelihood of reverse causation (e.g., whether poor health may be cause or consequence of low SEP), which always represents a big issue of other standard SEP measures. Indeed, it is generally assumed that the cycle of education is complete, or otherwise identifiable, before health issues may occur (Galobardes et al. 2006; Shavers 2007). In many epidemiological studies where measures of income, status, and occupation are not available, educational level is frequently used as the social position indicator and it tends to be empirically associated with the other measurements (d'Errico et al. 2017). The value of this social indicator, however, varies across geographic areas and cohorts.

3 Description of Two Worldwide Epidemiological Data Consortia

The Stomach Cancer Pooling (StoP) Project and the International Head and Neck Cancer Epidemiology (INHANCE) are an example of two large data consortia, in which the University of Milan is proactively involved in. In particular, our Department is the coordinator center of the StoP project and has been promoted several investigations and statistical data analyses based on the INHANCE consortium. Brief descriptions on these data consortia are here reported.

3.1 The StoP Project Consortium

The StoP Project is a consortium of epidemiological studies on gastric cancer established in 2012; the University of Milan is among the founders of the project. Up to date, the consortium includes 33 studies for a total of 12,753 gastric cancer cases and 30,682 controls. Of the patients, 40% are from Asia, 43% from Europe, and 17% from North America; 34% are women and 66% men; the median age is 61 years (Pelucchi et al. 2015).

The main aim of the StoP Project is to examine several lifestyles, including SEP, environmental, and genetic risk factors for gastric cancer, taking advantage of a large data set with original information from various geographic areas. The statistical analyses are carried out through pooled analyses of individual-level data, after central collection and validation of the original datasets. As compared to meta-analyses, the individual-level data approach allows harmonization of information and analyses, consistency of adjustment terms and multi-variate models, and investigation of heterogeneity and interaction between covariates (Ioannidis et al. 2013).

The StoP project challenge is therefore to improve knowledge of the etiology of gastric cancer, allowing decision-makers to plan preventive strategies, and providing a contribution to its control and its impact on the health of our population (Pelucchi et al. 2015; Winn et al. 2015).

3.1.1 Definition of SEP in StoP Project Consortium

The uniquely large sample size and the access to raw patient-level data allowed the StoP consortium to accurately assess the relation of SEP with gastric cancer overall and its subsites and histological subtypes, as well as to assess the associations in subgroups of the population according to sex, age, geographic area, and macroeconomic measure of income inequality of the country.

The level of education and household income were considered as proxies for the SEP. A uniform definition of occupational position among the included studies was not available at the time of the analysis, making unfeasible the evaluation of the relationship between occupational-based social class and gastric cancer risk. Education data were standardized across studies following the International Standard Classification of Education from the United Nations Educational, Scientific and Cultural Organization (UNESCO). This international reference classification facilitates comparisons between education systems across countries worldwide. Specifically, ISCED 2011 (UNESCO 2012) was issued as reference in the StoP project consortium. Education level was divided into three categories: (i) low education level, including no education, early childhood, and primary education (ISCED 0–1); (ii) intermediate education level, including secondary education (lower and upper) and postsecondary non-tertiary education (ISCED 2–4); (iii) high education level, including tertiary vocational and higher education, often designed to provide participants with professional knowledge, skills and competencies and education leading to a university degree (ISCED 5–6).

Household income was estimated by standardizing available study questionnaires data; comparable income levels were grouped into four categories, i.e., low, lower middle, upper middle, and high (Rota et al. 2019).

3.2 *The INHANCE Consortium*

The INHANCE consortium was established in 2004 as a collaboration among international research groups and includes investigators from over 35 international studies who have pooled their data on 30,000 patients with head and neck cancer and 40,000 controls without these cancers.

The primary goal of the consortium is to address the associations of head and neck cancer with a number of environmental factors, in particular tobacco smoking and alcohol drinking (i.e., the most relevant risk factors for the disease). The large sample size achieved by pooling studies allowed to assess the role of anthropometric characteristics, nutritional factors, income, and education. Moreover, INHANCE has the sufficient sample size to investigate subtypes of head and neck cancer (specifically oral cavity, oropharyngeal, hypopharyngeal, and laryngeal cancers) and to study heterogeneity in results across studies, geographic areas, and time periods, which may help to better identify unique risk factors or vulnerable populations (Winn et al. 2015).

3.2.1 Definition of SEP in INHANCE Consortium

On the basis of its information-rich data sets, the INHANCE consortium performed a detailed study with the aim to assess the risk for head and neck cancer associated with low educational status and household income. Analyses were carried on the overall database and by age, sex, cancer subsite, geographic location, and macroeconomic measure of income inequality at country-level.

Education data were standardized across studies following the International Standard Classification of Education from UNESCO. ISCED 97 protocol was used to categorize education levels (UNESCO 1997), which were divided into three strata: (i) low education level, including no education, early childhood, and primary education (ISCED 0–1); (ii) intermediate education level, including secondary education (lower and upper) and postsecondary non-tertiary education (ISCED 2–4); (iii) high education level, which comprised further education including vocational education and higher education including university degree (ISCED 5–6).

Concerning household income, in the INHANCE consortium, data were standardized as far as possible (i.e., when in the original study questionnaire the proper categorization was addressed) by grouping comparable levels based on the strata used in the original study, starting from category 1 associated to the lowest income levels, up to the highest within category 5 (Conway et al. 2015).

4 Results

4.1 Results from the StoP Project Consortium

Findings from the StoP consortium showed that SEP, measured through education level and household income, is a strong determinant of gastric cancer.

Data on education level were available from 25 out of 33 studies participating in the StoP consortium (11 from European countries, 6 from Asian countries, 3 from North, and 5 from Central/South American countries), for a total of 10,000 gastric cancer cases and 25,000 healthy controls. Seven studies (4 from Asian countries, 2 from Brazil, and 1 from Canada) provided data on household income.

To analyze the association of education and household income with gastric cancer risk, we firstly estimated study-specific Odds Ratios (OR) and the corresponding 95% Confidence Intervals (CI) using multi-variable unconditional logistic regression models. Analyses showed that subjects with intermediate and low education levels had, respectively, 22% (pooled OR, 1.22, 95% CI, 1.01–1.48) and 65% (pooled OR 1.65, 95% CI, 1.19–2.29) increased risks of gastric cancer compared to those with higher education attainment (Fig. 1). Results were adjusted for a number of lifestyle and dietary habits, which may confound the associations of SEP with gastric cancer, including tobacco smoking, race/ethnicity, and the intake of alcohol, fruit, and vegetables. Strong positive associations were observed for both cardia and non-cardia gastric cancers, as well as for diffuse and intestinal subtypes. In addition, the positive association between education level and gastric cancer risk was evident regardless of infection with *Helicobacter Pylori* (*HP*), and in subgroups defined by age, sex, cigarette smoking, and alcohol drinking. In analyses by geographic area, strong positive associations were reported by studies from Europe and Asia, while combined results from the three North American studies indicated a non-significant positive association. Conversely, Central/South America studies (mainly Mexican

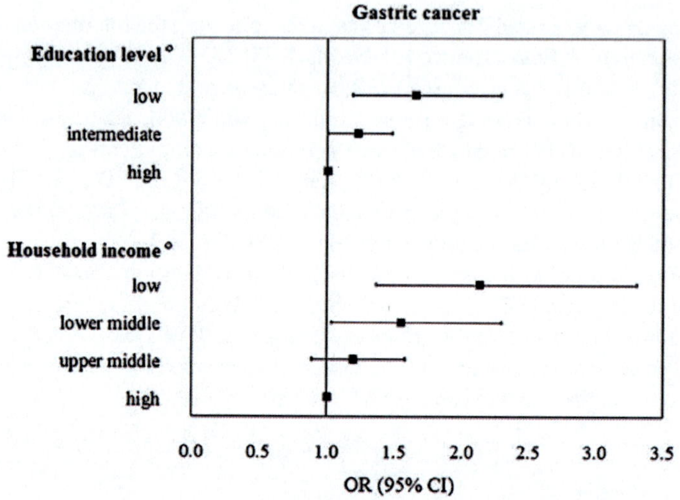

Note: ° Adjusted for age, sex, study center, alcohol drinking, tobacco smoking, race/ethnicity, fruit and vegetable consumption. Education was standardized using the International Standard Classification of Education (ISCED 2011). Low education corresponds to ISCED 0–1, Intermediate education to ISCED 2–4 and High education to ISCED 5–6.

Fig. 1 Pooled odds ratios (OR) and 95% confidence intervals (CI) of stomach cancer (Stomach cancer Pooling [StoP] Project consortium) according to education level and household income. The reference category is high level

studies) did not find any relation between education level and gastric cancer, raising concerns about the reliability of education as a proxy for the SEP in such countries. Mexico has high rates of income inequality and wealth is concentrated in a small fraction of the population, while the majority is poor and has limited access to education, and thus better living conditions. Large segments of the population still fail to achieve even basic education. Alternatively, education may be a better indicator of SEP in high- than in middle-income countries. In low- and middle-income countries, in fact, education is strongly related to social class in childhood, while in high-income countries it mainly reflects physical or psychological impairments that in the long term may influence cancer risk to a greater extent. When household income was used as proxy of SEP, a 35% increased risk of gastric cancer was observed for subjects in the lowest versus the highest income category (OR 2.13, 95% CI, 1.37–3.31) (Fig. 1) (Rota et al. 2019).

4.2 *Results from the INHANCE Consortium*

The INHANCE consortium indicated that low education level and low income are risk factors for head and neck cancer, even in the absence of the well-known lifestyle risk factor for this cancer, namely smoking, use of other tobacco products and alcohol drinking.

The estimated study-specific OR and 95% CI for the association of education and income for head and neck cancer were calculated using unconditional logistic regression based on 31 case–control studies and almost 24,000 head and neck cancer patients and 32,000 controls. The analyses indicated that subjects with low education had a more than two-fold increased risk of head and neck cancer compared to those with high education (pooled OR 2.50), after allowance for age and sex. The risk for subjects in the intermediate education category was increased by 80%. When accounting for smoking, alcohol, and selected dietary factors, the association was attenuated but still significant, with an over 30% elevated risk among subjects with low versus those with high education (pooled OR 1.34, 95% CI, 1.04–1.73) (Fig. 2). In addition, the risk remained increased by over 50% in subjects who never smoked or used other type of tobacco and never drank alcohol (OR 1.61, 95% CI, 1.13–2.31). This suggests that the association of head and neck cancer with education level is not totally attributable to these detrimental behaviors, although some degree of residual confounding could not be excluded. In addition, part of the association observed with education could be explained by *Human Papilloma Virus* (HPV) infection.

The association with low education level was observed for all head and neck cancer subsites (i.e., oral cavity, oropharynx, hypopharynx, and larynx), and was

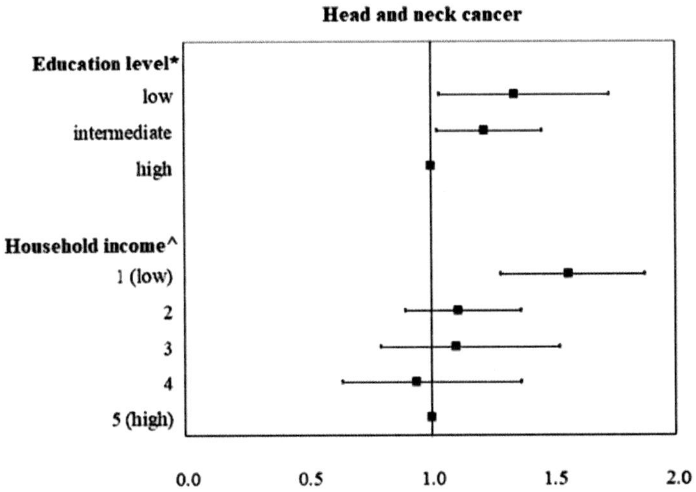

Note: * Adjusted for age, sex, study center, alcohol drinking, tobacco, fruit and vegetabl e consumption. Education was standardized using the International Standard Classification of Education (ISCED 2011). Low education corresponds to ISCED 0–1, Intermediate education to ISCED 2–4 and High education to ISCED 5–6.
^ Adjusted for age, sex, study center, alcohol drinking and smoking.

Fig. 2 Pooled odds ratios (OR) and 95% confidence intervals (CI) of head and neck cancer (International Head and Neck Cancer [INHANCE] consortium) according to education level and household income. The reference category is high level somehow stronger in North American and Central/South American populations as well as in higher income inequality countries.

The analyses on household income were based on 10 studies with available information (9 from USA and 1 from Porto Rico). Results were in line with those for education level, with an over two-fold increased risk for the lowest *vs* the highest category of income, in an analysis which takes into account age and sex. Again, the association was attenuated, but still evident, after allowance for smoking and alcohol, with an over 50% increased risk among subjects with the lower monthly income (Fig. 2) (Conway et al. 2015).

5 Conclusion

Social epidemiology is crucial to understand the sociostructural factors related to health and disease. In an era of fast inter-diffuse communication and data-sharing, large collaborative groups and data consortia are among the most effective strategies to create new social epidemiological useful evidences. In particular, data analyses of large epidemiological consortia found that SEP is strongly related to a number of cancers. Notably, the results from two large epidemiological consortia on gastric (StoP) and head and neck cancers (INHANCE) indicated that the association with SEP persists even after allowance for smoking and alcohol, i.e., unfavorable correlates of most cancer types and associated to SEP, suggesting that the SEP-cancer association follows pathways beyond these detrimental behaviors. Up to date, most industrialized countries will be challenged to identify such influencing factors and their means of operating throughout the whole of society. Reduction of socioeconomic inequalities both at national and international level is advocated to decrease the burden of cancers in deprived populations.

References

Bennett, J. E., Stevens, G. A., Mathers, C. D., Bonita, R., Rehm, J., Kruk, M. E., et al. (2018). NCD countdown 2030: Worldwide trends in non-communicable disease mortality and progress towards Sustainable Development Goal target 3.4. *Lancet*. https://doi.org/10.1016/s0140-6736(18)31992-5.

Conway, D. I., Brenner, D. R., McMahon, A. D., Macpherson, L. M. D., Agudo, A., Ahrens, W., et al. (2015). Estimating and explaining the effect of education and income on head and neck cancer risk: INHANCE consortium pooled analysis of 31 case-control studies from 27 countries. *International Journal of Cancer*. https://doi.org/10.1002/ijc.29063

d'Errico, A., Ricceri, F., Stringhini, S., Carmeli, C., Kivimaki, M., Bartley, M., et al. (2017). Socioeconomic indicators in epidemiologic research: A practical example from the LIFEPATH study. *PLoS ONE*. https://doi.org/10.1371/journal.pone.0178071

Galobardes, B., Shaw, M., Lawlor, D. A., Lynch, J. W., & Davey Smith, G. (2006). Indicators of socioeconomic position (part 1). *Journal of Epidemiology and Community Health*. https://doi.org/10.1136/jech.2004.023531

Geyer, S., Hemstrom, O., Peter, R., & Vagero, D. (2006). Education, income, and occupational class cannot be used interchangeably in social epidemiology. Empirical evidence against a common practice. *Journal Epidemiol Community Health.* doi:10.1136/jech.2005.041319.

Honjo, K. (2004). Social epidemiology: Definition, history, and research examples. *Environmental Health and Preventive Medicine.* https://doi.org/10.1007/BF02898100

Ioannidis, J. P., Schully, S. D., Lam, T. K., & Khoury, M. J. (2013). Knowledge integration in cancer: Current landscape and future prospects. *Cancer Epidemiology, Biomarkers & Prevention.* https://doi.org/10.1158/1055-9965.EPI-12-1144

Marmot, M. A., & Allen, J. J. (2014). Social determinants of health equity. *American Journal of Public Health.* https://doi.org/10.2105/AJPH.2014.302200

Niessen, L. W., Mohan, D., Akuoku, J. K., Mirelman, A. J., Ahmed, S., Koehlmoos, T. P., et al. (2018). Tackling socioeconomic inequalities and non-communicable diseases in low-income and middle-income countries under the sustainable development agenda. *Lancet.* https://doi.org/10.1016/S0140-6736(18)30482-3

NIH. (2019). Consortia to advance collaboration in epidemiologic and cancer research. Retrieved July 22, 2019, from https://epi.grants.cancer.gov/Consortia/.

Pelucchi, C., Lunet, N., Boccia, S., Zhang, Z. F., Praud, D., Boffetta, P., et al. (2015). The stomach cancer pooling (StoP) project: Study design and presentation. *European Journal of Cancer Prevention.* https://doi.org/10.1097/CEJ.0000000000000017

Rota, M., Alicandro, G., Pelucchi, C., Bonzi, R., Bertuccio, P., Hu, J., et al. (2019). Education and gastric cancer risk-an individual participant data meta-analysis in the StoP project consortium. *Internatonal Journal of Cancer.* https://doi.org/10.1002/ijc.32298

Shavers, V. L. (2007). Measurement of socioeconomic status in health disparities research. *Journal of the National Medical Association, 99*(9), 1013–1023.

UNESCO. (1997). *International Standard Classification of Education: ISCED 1997.* Paris: UNESCO Institute for Statistics.

UNESCO. (2012). *International standard classification of education: ISCED 2011.* Montreal: UNESCO Institute for Statistics.

Vaccarella, S., Lortet-Tieulent, J., Saracci, R., Conway, D. I., Straif, K., & Wild, C. P. (2019). *Reducing social inequalities in cancer evidence and priorities for research.* Geneve: IARC Scientific Publication.

Wagstaff, A. (2002). Poverty and health sector inequalities. *Bulletin of the World Health Organization, 80,* 97–105.

WHO. (2018a). Cancer—Key facts. Retrieved July 22, 2019, from https://www.who.int/news-room/fact-sheets/detail/cancer.

WHO. (2018b). Noncommunicable diseases. Retrieved July, 22, 2019, from https://www.who.int/news-room/fact-sheets/detail/noncommunicable-diseases.

Winn, D. M., Lee, Y. C., Hashibe, M., & Boffetta, P. (2015). The INHANCE consortium: Toward a better understanding of the causes and mechanisms of head and neck cancer. *Oral Diseases.* https://doi.org/10.1111/odi.12342

Identification of Opinion Makers on Twitter

Svitlana Galeshchuk and Ju Qiu

Abstract Twitter is a social platform that helps share ideas quickly and concisely. Although the network offers equal rights to post short texts, the attention these messages attract frequently depends on a user's status in the real world. Thus the tweets of real life high-profile opinion makers will *a priori* have a higher probability of spurring the interest of society than the messages from the so-called grassroots. The paper elaborates on the developed classifier that detects automatically such opinion makers on Twitter. The approach exploits the Mixed Effect Random Forests method combined with the features engineered from the Twitter data. The accuracy and the sensitivity of the proposed technique outperform the results of the other machine learning classifiers on the out-of-sample data.

1 Introduction

Social media makes a great impact on the society nowadays. Recently many social movements, political and commercial scandals found their playground on social platforms. Social platforms "stars" commence earning money by endorsing brands. When these celebrities post messages, they get through to many followers. It makes them proper opinion leaders that may influence the diffusion of stories, their magnitude, and duration of popularity. Many ideas and actions have found public support through Twitter, e.g., #metoo, #BlackLivesMatter, #ShoutYourAbortion, #WhyIStayed. Some of them started from real-world celebrities, another from grassroots but very often they involved the influencers in the news spreading. Hence, the ultimate goal of our ongoing study aims at understanding the role of social influencers in the public opinion generation and information diffusion on social networks. This paper

S. Galeshchuk (✉) · J. Qiu
Governance Analytics, University Paris Dauphine, PSL, Place du Maréchal de Lattre de Tassigny, 75016 Paris, France
e-mail: svitlana.galeshchuk@dauphine.psl.eu

J. Qiu
e-mail: ju.liu@dauphine.psl.eu

P. Mariani and M. Zenga (eds.), *Data Science and Social Research II*, Studies in Classification, Data Analysis, and Knowledge Organization,
https://doi.org/10.1007/978-3-030-51222-4_15

contributes to the social media understanding by building a method to detect social networks celebrities by means of machine learning approaches.

The focus of our study is on Twitter as this networking service has recently transformed into the global platform for opinions exchange. For this paper we define Twitter influencers as the active users who have the capacity to impact the public opinion by their tweets as they possess real life achievements and high social visibility beyond social networks. The term "grassroots" is assigned to the rest of the Twitter users. Twitter API allows retrieving necessary information on users that serves as an input for the classification model and makes the paper goal feasible.

The rest of the paper is structured as follows: Sect. 2 provides a brief review of the related literature. Section 3 expands on the data used. Section 4 describes the methodology. Section 5 presents the results of our experiments. Section 6 concludes with some observations on our findings and identifies directions of future research.

2 Related Work

Detecting Twitter influencers spurs interest among marketing experts who often sell their solutions to the companies on how to filter and target celebrities. There are number of web-service that propose identification of opinion-makers within industry. We do not know, however, their approaches. We, therefore, rely on the academic literature to set up a benchmark for our study. The paper Puigbo (2014) provides an overview of the network and diffusion-based methods for detecting the influencers in marketing strategies. Ghosh and Lerman (2010) also employ network analysis to target the influential people in the online social network Digg. However, the diffusion information techniques prone to the observation bias (see Dhamal et al. 2016) and the network approach does not comprise the information from the content of the user tweets.

The research community pays attention to the social media newsmakers also using the suite of machine learning methods to detect them. Ludu (2015) use Linear SVM to classify the Twitter users in a supervised mode using a set of gender, linguistic, age characteristics and a score taht measures the influence of Twitter users on different classes. However, the latter may be somewhat misleading for the analysis of the classifier as it characterizes accounts based on the universality of their impact and may be seen as an implicit influential score. Moreover, the approach requires many data that cannot be obtained from Twitter API (age, etc.).

Ferrara et al. (2016) try to detect promoted content by identifying different features characterizing trending campaigns discussed on Twitter. The authors apply random forest and k-nearest neighbor (KNN) with dynamic time warping. The results with the former underperformes KNN as the latter is particularly suitable for the time-series analysis. The study shows us the poor performance of the machine learning classifiers in their vanilla form with periodic data. Thus we will try to avoid time-series input switching to the aggregated values where necessary (i.e., using the mean).

Ghosh et al. (2011) implement both supervised classifiers (KNN, support vector machines) and entropy-based approach identifying differenct categories of retwiting activity on Twitter. They target semantic meaning to separate newsworthy information from the rest of tweets. Their hierarchical clustering recognizes not only information shared by humans, from the rest of retweeting but also the groupings within it based on the user's popularity. It encourages us to extract textual features like semantic and subjectivity scores from our tweets.

Lahuerta-Otero and Cordero-Gutiérrez (2016) run the descriptive analysis and multivariate regression for the tweets posted by influencers and grassroots in Japanese automotive industry. They find that Twitter opinion-makers use on average more hashtags and mentions but fewer words and links. However, the paper does not elaborate on the criteria for how the Twitter users are pre-labeled. We believe, the random-effect multivariate regression model might have provided different results and is more appropriate.

Nebot et al. (2018) use a simplistic method to identify influencers only with textual content of tweets with deep learning classifiers and the embedding methods. However, the generic approach is prone to calculate influential scores based on merely sentiment analysis that may lead to the omitted features and must be tested. Nonetheless, both Lahuerta-Otero and Cordero-Gutiérrez (2016) and Nebot et al. (2018) focus on the textual features of the tweet content as an important part of the detection problem. It motivates us to incorporate these variables in our analysis together with the user account data from Twitter API.

The literature overview proves that the existing machine learning classifiers may render high accuracy if the input features are well-engineered for the specific research question and methodological set-up. We build on the discussed papers to identify automatically the real-world opinion makers on Twitter.

3 Data

Recall that the study aims at understanding the role of influencers and grassroots in the life cycle of stories/movements on Twitter. Categorizing users on opinion makers and grassroots appears to be an important step in our research activities. We use a supervised approach to develop a classifier with manually pre-labeled users of the RepLab dataset sponsored by the EU project LiMoSINe. The accounts are assigned to 1 if the users are infuencial figures in the real life and 0 otherwise. For instance, score 1 got Twitter profiles of CEOs, important artists, established economists, etc. Additionally, each Twitter profile is manually classified into one of the following categories: Professional, Investor, Stockholder, Journalist, Public Institutions, NGO, Company, Celebrity, Sportsmen, Undecidable. Please refer to RepLab[1] for more details about each groups.

[1] http://nlp.uned.es/replab2014/.

Table 1 Input features

Feature	Comment
Mentions average	Average number of mentions per user's tweet
Retweet count average	Average number of retweets per user's tweet
Sentiment average	Average sentiment score per user's tweet
Subjectivity average	Average subjectivity score per user's tweet
Quoted average	Average number of quotes per user's tweet
Statuses average	Daily mean number of tweets posted by the user'
Favourite average	Average number of tweets that given user has marked as favorite (likes)
Friends count	Number of user's friends
Listed count	Number of public lists that this user is a member of
Replies average	Average number of replies for user's tweets
Description	If profile contains a description
User location	If user precised his/her location
Followers count	Number of user's followers
Verified	If profile is verified as authentic

Data retrieval. We collected data on 5409 RepLAb values, including 2420 Influencers. The annotation in the dataset is binary: a user is either opinion-maker or not. The dataset provides us only with the screen names and the corresponding label. Twitter API helps retrieve additional information: the account data for each screen name and corresponding 500 last tweets with a number of retweets, likes and replies. For scrapping we filtered away the retweets of user as s/he is not an original author of tweet.

Feature Engineering. We constructed a set of pertinent features which then has been used by the machine learning classifiers to learn how to detect grassroots and influencers. Table 1 summarizes those variables.

All variables create three possible groupings: (i) features defined by a user; (ii) network features; (iii) features of the tweets' content.

Features defined by a user. The first group includes information deliberately provided by an account owner. It consists of such binary variables as: a profile with description; a user declares his/her location. This information shows how open the person wants to be with readers. "Statuses Average" discloses how actively profile's page is updated with new tweets and "Mentions Average" informs about the intention of the user to get visibility and to receive feedback.

Network features. The second group aims at the user's personal popularity in the Twitter environment. A number of friends and followers show how large is a potential auditory for the user's posts. The "Listed Count" variable helps understand if the account' posts are interesting for the followers so that they include it to the particular lists to follow. It may serve as an indicator of the niche for the user's influence as lists frequently group the accounts by thematics. The variable of verification acts as a signal that the account is of public interest and is authentic. Twitter makes this verification based on the internal policy. Not every account of public interest is always an influencer and vice versa. However, the definition of the variable suggests the role of verification may be important in our experimental set-up.

Features of the tweets' content. The third group of features is retrieved from the user's posts. Again some of them are purely statistical as the average number of retweets, likes, replies per tweet. However, additional methods that measure sentiments and subjectivity facilitate the retrieval of tweet content information.

First, tweets in other languages than English (circa 18% of all corpora) are translated to English. Next step removes urls, mentions, special characters like "#" from the text. Segmentator overcomes the problem of merged words in hashtags (e.g., "metoo" transforms into "me too"). Spell checker corrects misspelled words(e.g., "looove" becomes "love"). Then we employ rule-based Valence Aware Dictionary and Sentiment Reasoner (VADER) model developed in Hutto and Gilbert (2014) to compute sentiment score for each tweet. The VADER dictionary has been built on other available solutions. Authors enriched well-established lexicons with sentiment-related acronyms (e.g., LOL), slang and emojis widely used by Twitter users. A group of 10 human experts evaluated each word in the dictionary using the scale from "[–4] Extremely Negative" to "Ludu (2015) Extremely Positive", with allowance for "[0] Neutral (or Neither, N/A)". Finally, they kept only the words with standard deviation less than 2.5 from the mean of assigned expert scores. Thus, the fine-grained lexicon we currently use consists of remaining 7500 wordings. Moreover, the method takes into consideration punctuation (e.g.,"!!!" increases the intensity of the sentiment), contrastive conjunctions (i.e., "but", "however"), intensifiers (e.g.,"extremely", "hyper" and the like). ALL-CAPITALS words are seen as more important (e.g., "the dish looks horrible but the taste is GREAT"). Three-preceding-words analysis catches flips in the polarity of the text. Finally, the sentiment score is normalized to be between -1 (most extreme negative) and +1 (most extreme positive). "Sentiment Average" represents the computed average for scrapped tweets per user.

Subjectivity average makes use of auxiliary verbs (e.g., could, would) and adverbs (e.g., definitely, maybe) to assess uncertainty (see Smedt and Daelemans (2012) for the details). Python Library "Pattern" helps estimate the subjectivity score for each tweet.[2]

Before running the classifier, additional analysis is required to understand the extent of differences between the groups of opinion-makers and non-opinion makers. The reasoning behind this test lies in the fact that the accounts were pre-labeled manually based on the capacity of a person to be an influencer in real life beyond Twitter. However, if there are no convincing distinctions between two grouping, more complex methods will be required to catch the patterns for classification. One of the

[2]https://www.clips.uantwerpen.be/pattern.

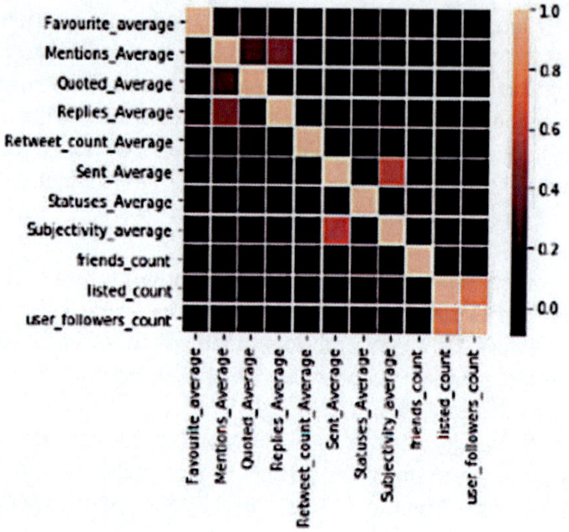

Fig. 1 Correlation matrix

Table 2 MANOVA test

Test	Value	F Value	Pr>F
Wilks' lambda	0.7114	288.4500	0.0000
Pillai's trace	0.2886	288.4500	0.0000
Hotelling-Lawley trace	0.4056	288.4500	0.0000
Roy's greatest root	0.4056	288.4500	0.0000

ways to integrate the set of continuous features is to test if two groups differ significantly from each other is to run Multivariate Analysis of Variance (MANOVA). The test measures the differences in mean scores and possible cause-effect relationships between the factors and dependent variables. The multicollinearity may distort the results as the test is linear. Hence, we construct the correlation matrix to track possible intercorrelation between independent features (see Fig. 1).

The features "Mentions Average" and "Replies Average" interrelation, and therefore we exclude the latter in our analysis. Table 2 presents the results of MANOVA: the differences between the two groups is significant.

We applied a stratified train-test split with ratio: 85:15 to train and then test our model.

4 Methodology

This section elaborates on the machine learning classifiers employed in a supervised mode to classify accounts on opinion makers (1) and non-opinion makers (0). Randomized search optimization helps us tune the models.

4.1 Conventional Machine Learning Methods

Logistic Regression (LG) is a parametric approach to estimate the class probability of instances. The method produces explainable results as it derives from linear regression outputting the logistic (sigmoid) of the regression. The functional respresentation looks as follows:

$$\sigma(t) = \frac{1}{1 + e^{-t}}. \tag{1}$$

where t is the linear regression function of the independent variables. The resulting value between 0 and 1 is a probability of the observation to be an instance of the class 1 ('Influencers'). Python Library *Scikit-Learn* is used to train the classifier.

Support Vector Machines belong to the family of versatile machine learning methods with high accuracy on non-large datasets. It tries to find the broadest possible margin between positive and negative classes. The method is sensitive to the features' scale, hence, we performed data normalization so that the values stayed within the range from -1 to 1. We use linear Support Vector Classifier (SVC) and Gaussian Radial Basis Function (RBF) in our set-up. RBF is a non-linear variant of SVM based on a similarity function between an observation and a landmark (location of each instance in the dataset). Python Library *Scikit-Learn* is used to train the classifier.

Random Forests (RF) helps overcome the disadvantages of a single decision tree by summarizing and averaging predictions over the number of trees. It is an ensemble learning approach that uses the outputs of the individual predictors as votes. If the positive class gets more votes, the method will return the corresponding result. We trained the classifier using the bootstrap with a number of maximum samples equal to the size of the set (only for training phase). Random forest looks for the most important feature to split the tree among the random set of variables. It brings lower variance and better generalization. Python Library *Scikit-Learn* is exploited to train the classifier.

In our set-up *Gradient Boosting* (GBT) method represents an ensemble of classification decision trees. Each tree sequentially joins the ensemble correcting the antecedent by fitting its residual errors. The classification error usually gets lower as trees are added to the model. The method relies on the learning rate that regulates the adaptation of the GBT each time a tree is joined. Early stopping technics permits us find the optimal number of trees. Python Library *XGBoost* is used to train the classifier.

4.2 Mixed-Effects Random Forests

So far we used the pooled models: we employed all features together as input data for the abovementioned machine learning classifiers to learn how to categorize new observations. The problem of this approach hides in the assumption that there is no difference within four available domains of users (automotive, banking, universities and music/artists) in terms of their influence. Let's take a toy example of a CEO of pharmaceutic company with 3000 employees across the country who does not have enough time for posting every day may have only 10,000 followers. We hypotetically find that the Twitter behaviour of the CEO reflects typical activity for the CEOs withing the same industry. However, journalist of local newspaper in a medium-sized city may have 15,000 followers and the established professionals in the field account for at least 30,000 followers. If we pool them together, the CEOs from pharmaceutical industry will be always seen as a grassroot and our results will generalise poorly on the out-of-sample dataset.

The Mixed-Effect Random Forests approach (MERF) developed in Hajjem et al. (2014) helps overcome the illustrated issues. Ideally, we would like to apply the method that can learn how to identify influencers using some parameter for each of 11 categories regularized by a prior derived from all the available data. These independent hyperparameters are drawn from a Gaussian distribution. Thus the user's class (Professional, Investor, Stockholder, Journalist, Public Institutions, NGO, Company, Celebrity, Sportsmen, Undecidable) represents some cluster and the mentioned approach is a part of hierarchical Bayesian clustering technics. Though such grouping is suitable more for parametric modelling as it requires some assumptions be imbibed in the functional form of the model whether it is linear or non-linear. For example, mixed-effect linear model will look as follows:

$$y = \alpha X + \beta_i Z + e. \tag{2}$$

where y is a target value, α is the fixed effect coefficiens next to X fixed effect variables, β_i is the random effect coefficiens per i cluster drawn from the same distribution multiplied by Z random effect features, e is a bias. We use same notations in the equation (1) as in Hajjem et al. (2014) to avoid confusions for those who will refer to the original paper for the in-depth understanding of MERF. The authors propose to change the original equation transforrming it into the following format:

$$y = f(X) + \beta_i Z + e. \tag{3}$$

Equation (2) looks similar to the previous one apart from the fixed effect parameters coupled influencing the fixed effect variables. Instead, we have the random forests represented by $f(.)$ - the non-linear function of the features. And each cluster i is influenced by a linear correction. Let e and β be normally distributed. Thus the set of the model's hyperparameters includes $f(.)$, β_i, σ_e prior and σ_β

prior. The approach leverages from the iterative optimization technics, in particular, expectation-maximization to learn these hyperparameters and fit the model.

We use the Python package MERF that implements methodology from Hajjem et al. (2014). The creators of the package worked closely with the authors of the approach correcting some minor errors in the original paper (see the presentation[3]). Hence, we can assume the package is fine-tuned to produce reliable results.

4.3 *The Curse of Imbalanced Data*

Recall from the previous section that we could recuperate the Twitter data on circa 5400 accounts from the RepLab list of non/opinion makers. The original dataset comprises about 8000 screen names but some of the accounts are protected from third-party Twitter API users, another may have been deleted. In the end, we obtain an imbalanced dataset with approximately 44% of data with class.

Several approaches have been employed to address the problem:

- *Oversampling* with Synthetic Minority Over-Sampling Technique (SMOTE) to increase the number of instances of minority class. SMOTE creates minority class observations by taking each minority class example and introducing synthetic examples along the line segments joining any/all of the k minority class nearest neighbors (see Chawla et al. (2002)).
- *Undersampling*. Fixed undersampling reduces the number of instances of majority class to get 50/50 proportion of classes in a dataset. Python Library *Imblearn* with default parameters is employed for over- and undersizing. Please refer to Lematre et al. (2017) for the details on the package.
- *Adjusting class weights*. The approach tends to assign larger weights to minority class so that machine learning method is not biased against it. This weight distribution follows the proportion of classes (Huang et al. (2013)). We adjust the weights during the training phase leaving testing part unchanged.

In all but one case machine learning methods with balanced class weight provide the best possible accuracy. GBT works better with SMOTE, possibly because it is the way to enlarge the dataset.

5 Results

This section discusses the classification results and evaluates the accuracy of the developed method.

[3]https://pyvideo.org/pydata-la-2018/attacking-clustered-data-with-a-mixed-effects-random-forests-model-in-python.html.

5.1 Evaluation Criteria

For this classification problem, the quality of model is measured by the proportion of correctly classified observations (accuracy). However, our dataset is skewed which means if the method is biased against the minority class, the accuracy may become a misleading performance measure. In such cases, precision and recall are better in evaluating the classifiers on test data: precision accounts for a ratio of true positives to the sum of true positives and false positives; recall measures a ratio of true positives to the sum of true positives and false negatives.

5.2 Performance Measure

We first train the machine learning classifiers (see Sect. 4.1 for details and acronyms) with a pooled set of features. Randomized search or early stopping techniques fine-tune the hyperparameters. Table 3 shows the classification performance for the test data. The ensemble methods outperform other approaches, even if the difference between non-linear SVM and RF prediction accuracy is marginal. It underlines the non-linear nature of the features' relationships in order to determine the probability of the user to be an influencer. Precision value for class 1 is the lowest in the case of LG, although its classification accuracy is better than SVC's. It demonstrates the equivocal abilities of the evaluating measure. Nonetheless, the overall results for the machine learning methods do not exceed 75%.

Next, we apply MERF model as in 4.2. Figure 2 shows the evolution MERF's accurancy on the training dataset.

Classification performance improves: accuracy approaches 81% on test data.

Table 3 Classification results

Metrics/Method	LG	SVC	RBF	RF	GBT	MERF
Accuracy (%)	68.0	66.5	69.8	73.9	74.4	80.7
Precision (0)	0.73	0.66	0.71	0.69	0.68	0.81
Recall (0)	0.68	0.83	0.78	0.85	0.92	0.86
Precision (1)	0.63	0.65	0.68	0.71	0.82	0.80
Recall (1)	0.68	0.41	0.59	0.58	0.55	0.74

Note Performance measure for class 0 is pointed out with "(0)"
Performance measure for class 1 is pointed out with "(1)"

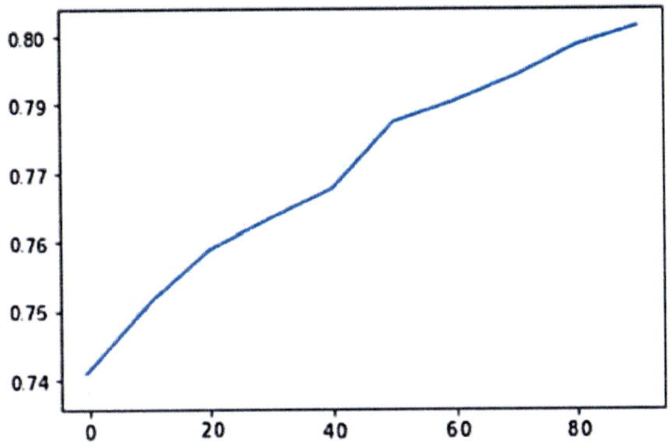

Fig. 2 MERF accuracy (y-axis) as a function of iterations (x-axis)

6 Conclusions and Further Research

In this paper, we present the supervised approach for the classification of Twitter participants on "influencer" and "grassroots" with MERF. The method proposes coupling account information (i.e., number of followers, friends, description, etc.) together with the user's activity data (i.e.,average number of posts and the like), and tweets' visibility (i.e., average number of retweets, likes per tweet) to detect whether the user is an opinion-maker.

To sum up, our experimental set-up includes the following steps:

- *Data collection*. We collected Twitter data on circa 5400 accounts pre-labelled as "Influencer" (class 1) or "Grassroot" (class 0).
- *Feature Engineering*. We processed the data to retrieve the most pertinent information that could serve as an input for the machine learning classifier. For example, sentiment and subjectivity analysis provided us with textual features from tweets. Overall three types of variables have been constructed and normalized where necessary: features defined by a user; network features; features of tweets' content.
- *Imbalanced Dataset*. We used class augmentation methods as well as weights adjustment to tackle the problem of imbalanced data.
- *Classifiers*. A suite of machine learning classifiers using the set features tried to address the challenge of users' categorization. However, the mixed effect method outperformed the accuracy of pooled modelling.

MERF provides the most promising accuracy of 80.7 % on the out-of-sample data. However, further reasearch will focus on detection of Twitter influencers independently of their status in real life. Many Twitter "celebrities" are not always prominent decision-makers beyond Twitter, yet they still have a high capacity of shaping public opinion. These celebrities gain prominence not always through the activities in real life but sometimes via simple appealing to the values, needs or preferences of the social media users. We believe, it is one of the main reasons why we could not attain better accuracy in our study. Moreover, the dataset is skewed as class 0 outnumbers class 1. Thus, we foresee preparing own dataset that will reflect the following definition of opinion-makers: "the active users who engage large audiences on the social network and have the capacity to impact public opinion by the content of their posts".

References

Chawla, N. V., Bowyer, K. W., Hall, L. O., & Kegelmeyer, W. P. (2002). SMOTE: synthetic minority over-sampling technique. *Journal of artificial intelligence research, 16*, 321–357.

Dhamal, S., Prabuchandran, K. J., & Narahari, Y. (2016). Information diffusion in social networks in two phases. *IEEE Transactions on Network Science and Engineering, 3*(4), 197–210.

Ferrara, E., Varol, O., Menczer, F., & Flammini, A. (2016). Detection of promoted social media campaigns. *In tenth international AAAI conference on web and social media.*

Ghosh, R., & Lerman, K. (2010). Predicting influential users in online social networks. arXiv preprint arXiv:1005.4882.

Ghosh, R., Surachawala, T., & Lerman, K. (2011). Entropy-based classification of 'retweeting' activity on twitter. arXiv preprint arXiv:1106.0346.

Hajjem, A., Bellavance, F., & Larocque, D. (2014). Mixed-effects random forest for clustered data. *Journal of Statistical Computation and Simulation, 84*(6), 1313–1328.

Huang, Wenhao, Guojie Song, Man Li, Weisong Hu, & Kunqing Xie. (2013). Adaptive weight optimization for classification of imbalanced data. *In International Conference on Intelligent Science and Big Data Engineering*, pp. 546–553. Berlin, Heidelber: Springer.

Hutto, C. J., & Gilbert, E. (2014). Vader: A parsimonious rule-based model for sentiment analysis of social media text. *In Eighth international AAAI conference on weblogs and social media.*

Lahuerta-Otero, E., & Cordero-Gutiérrez, R. (2016). Looking for the perfect tweet. *The use of data mining techniques to find influencers on Twitter, Computers in Human Behavior, 64*, 575–583.

Lematre, G., Nogueira, F., & Aridas, C. K. (2017). Imbalanced-learn: A python toolbox to tackle the curse of imbalanced datasets in machine learning. *The Journal of Machine Learning Research*, 18(1), 559–563.

Ludu, P. S. (2015). Inferring latent attributes of an Indian twitter user using celebrities and class influencers. *Proceedings of the 1st ACM Workshop on Social Media World Sensors*, 9–15.

Nebot, V., Rangel, F., Berlanga, R., & Rosso, P. (2018). Identifying and classifying influencers in twitter only with textual information. *International Conference on Applications of Natural Language to Information Systems*, 28–39.

Puigbo, J. Y., Sánchez-Hernández, G., Casabayó, M., & Agell, N. (2014). Influencer detection approaches in social networks: A current state-of-the-art. *CCIA*, 261–264.

Smedt, T. D., & Daelemans, W. (2012). Pattern for python. *Journal of Machine Learning Research*, 2063–2067.

Modelling Human Intelligence Using Mixed Model Approach

Thanigaivasan Gokul, Mamandur Rangaswamy Srinivasan, and Michele Gallo

Abstract In many psychometric studies, the observations may be often on longitudinal outcomes pertaining to General (G) and Specific (S) factors of human intelligence along with other covariates. Modelling human intelligence under Generalized Linear Mixed Model (GLMM) framework received the attention of psychologists in understanding the variables associated with the outcomes. In this paper, we formulate (i) a suitable GLMM model for count data of human intelligence factors and (ii) further examine the association between the outcome variables of Spearman's G and S factors of human intelligence using joint longitudinal modelling along with other covariates based on school lunch intervention data.

1 Introduction

The development of statistical model is inextricably connected to the theory of intelligence. Human intelligence is the most controversial area in psychological research as how to develop intellectual ability and interact with cognitive ability (Tourva et al. 2016). Food supplements play a vital role in improving the intelligence. School lunch programmes are introduced in order to improve the same as they may cause an impact in the school children health (Pope et al. 2018). School-aged children who suffer from severe malnutrition exhibit significantly compromised reasoning and perceptual-spatial functioning, poorer school grades, reduced attentiveness and unresponsive play behaviour, as compared to their adequately nourished peers. In addition, children suffering from mild-to-moderate malnutrition, show significant

T. Gokul · M. R. Srinivasan
Department of Statistics, University of Madras, Chennai, India
e-mail: gokultvasan@gmail.com

M. R. Srinivasan
e-mail: mrsrin8@gmail.com

M. Gallo (✉)
Department of Human and Social Sciences, University of Naples 'L'Orientale', Naples, Italy
e-mail: mgallo@unior.it

P. Mariani and M. Zenga (eds.), *Data Science and Social Research II*,
Studies in Classification, Data Analysis, and Knowledge Organization,
https://doi.org/10.1007/978-3-030-51222-4_16

deficits in intellectual and behavioural functioning. Deficits include compromised development in multiple domains, including verbal and spatial reasoning.

Research on intelligence, mainly based on correlational and factor-analytical work, with regard to development of cognitive functions, and research in cognitive psychology is similar, that is, understanding how the human being adapts to his/her own, complex environment. Tourva et al. (2016) adopted the structural equation modelling to examine the general, fluid and crystallized intelligence in children whose age ranges from 7 to 18 through the cognitive functions namely, attention, processing speed and working memory and the results revealed that the working memory alone predicts the intelligence than other two cognitive functions. Further, Martínez-Plumed et al. (2017) assessed the human intelligence through several IQ tests viz., Raven's Progressive matrices, odd-one-out and Thurstone's series to understand the role of basic cognitive functions that are needed for the intelligence tests in improving the cognitive ability among children. In addition, Coyle et al. (2018) examined the correlation between the G and S factors of human intelligence using structural equation modelling based on several cognitive tests such as short story test, reading the mind in the eyes test, etc. Furthermore, Goharpey et al. (2013) investigated the problem-solving ability in children with intellectual disability by means of the cognitive function Raven's Coloured Progressive Matrices and found to be a valid measure for assessing the intelligence in children. Inferences about the effect of dietary quality on child outcomes are more often based on observational studies in regions where mild-to-moderate malnutrition is endemic.

Generally, the cognitive measures especially intellectual in nature consist of two factors namely, G (General) and S (Specific) (Spearman 1904). All Raven's Coloured Progressive Matrices (RCPM) tests are valid measures of G. The S factor includes Verbal Meaning (VM) Test, Arithmetic Score (AS) Test and Digit Span (DS) Test in this context. The verbal comprehension measure was similar to the Peabody Picture Vocabulary Test in that the child had to select the picture matching a verbal label from a set of four pictures. The Arithmetic Test deals with logical reasoning and some maths work. The Digit Span test requires the child to repeat a series of digits after the experimenter has said them [Coyle et al. (2018) and Muniz et al. (2016)]. The present paper also in a way compares fluid intelligence, working memory and processing speed, via GLMM. The study of intelligence is often, although not always, understood as the psychometric approach of individual differences by means of standardized tests tapping cognitive or intellectual abilities. It has also been based on correlations between performance in cognitive tests, focusing on the existence of G and S factors.

Coxe et al. (2009) discussed the appropriate modelling approach for count data using Poisson regression with and without overdispersion and an alternative to the Poisson regression called negative binomial regression and concluded that the usage of Poisson model is more appropriate for analyzing count data. In addition, Fallah et al. (2011) considered a Canadian health and ageing study of participants who are 65 and above in age to predict the cognitive score changes. A comparison of the usual Poisson regression model and a proposed Poisson model with neural network is considered and from the results it is concluded that the proposed neural network

Poisson regression yields better performance and are assessed using the two well-known information criteria namely AIC and BIC. Bilker (2012) considered a Poisson predictive model to reduce the number of items in a 60-item test of Raven's Standard Progressive Matrices. The model reduces to 9-item and achieved better correlation compared with 30-item form. Graßhoff et al. (2016) investigated the Rasch Poisson-Gamma model for analyzing human intelligence by assuming the ability of a person as random component with underlying Gamma distribution and the efficiency of the model is tested by developing D-optimal designs by incorporating two binary covariates.

In longitudinal study, the data are often found to be a mixture of count, binary and other data types. Among the prevailing statistical models, an approach of random effect model is popular in analyzing longitudinal data. Breslow and Clayton (1993) extended the concept of linear mixed model by Laird and Ware (1982) to non-continuous data structure and named it as Generalized Linear Mixed Model (GLMM). The reason for considering the mixed model in this study is that it will be useful to test both fixed and random component effects in the same analysis.

Usually, longitudinal data studies the association between the different outcomes through separate analysis, whereas joint modelling allows every outcome to have its own random effects and the association can be captured in terms of the correlation between random effects. Further, Palestro et al. (2018) provides a systematic approach of fitting joint models (direct and covariance approach) for the longitudinal data and illustrated the same using psychological data involving the neural and behavioural measure of cognition. Literature is abundant in joint modelling for continuous–continuous, continuous–binary [Chakraborty et al. (2003), Thifiebaut et al. (2002), Gueorguieva (2001), Iddi and Molenberghs (2012), Molenberghs and Verbeke (2005)]. Joint modelling for different types of longitudinal outcome has been studied by Efendi et al. (2013), Njagi et al. (2013) and Horrocks and Van Den Heuvel (2009), and review on joint modelling with time-event data has been carried out by Tsiatis and Davidian (2004). Joint models for an ordinal outcome and time to event outcome is studied by Li et al. (2010).

The focus of this article is on identifying the suitable model for the longitudinal data and further examining the association between the G and S factors. For the remainder of this paper, we will focus on analyzing the psychometric concept of Raven's Progressive Matrices with cognitive studies on Arithmetic Scores, Verbal Meaning and Digit Span capacity of school children and test the relationship between the G and S factor using GLMM. The paper is organized as follows. Section 2 following the introduction and presents the dataset that are further examined in Sect. 5. The existing methodology of generalized linear mixed model is reviewed together with the joint modelling approach in Sects. 3 and 4. An application of joint model for bivariate outcome of longitudinal structure is discussed in Sect. 5. Section 6 deals with bootstrap approach for the proposed method and Sect. 7 concludes the paper.

2 School Lunch Intervention Study

School feeding programmes have been established in large parts of developing countries all over the world improving general socio-economic conditions as well as providing educational and nutritional benefits to participants. Nutritional supplements given to the school-going children plays an important role in developing the cognitive ability and intelligence. In that sense, Murphy and Allen (2003) discussed the importance of food supplements given to the children in the regions Kenya, Mexico and Egypt by comparing the quantity of micronutrients comprised in the food for improving the child mental and cognitive ability. Further, Siekmann et al. (2003) studied the impact of vitamin deficiency by providing nutritional supplements namely meat, milk, energy (small amount of milk + meat) and control (without any supplements) for the school-going children in rural Kenya. The results of Siekmann et al. (2003) revealed that the deficiency has increased significantly from the baseline for the children fed with only milk and meat supplements and there is no improvement has observed for the children fed with energy and control groups. Apart from these, recent literature such as Walingo and Musamali (2008), Kristjansson et al. (2007) and Greenhalgh et al. (2007) conclude that school feeding programmes have a significant positive effect on growth and cognitive performance in children.

A controlled Kenya school children feeding intervention study of 546 individuals was designed to test a cognitive ability among their intake of nutritional supplements. The data are collected from Neumann et al. (2003) and a complete data without missing entries (374) are considered in this study. Each nutrition group was comprised of 9 out of 12 schools with children aged 6–14. The school lunch intervention began at time $t = 0$ by adding the supplements: Meat, Milk and oil added as calories to determine the effects of human intelligence outcome measures. Here responses collected on Raven's Coloured Matrices test have been analyzed as a measure of G factor. Data were collected, at five different points of time as indicated below. Round 1 data is baseline data collected before the onset of intervention (in other words, pre-intervention). Round 2 was taken as soon as the intervention started, while rounds 3, 4 and 5 were during the second, fourth and sixth months after intervention started as indicating post-intervention scores. At the end of study, only 374 individual school children had a full-sequence data resulting from the fact that 172 individuals are missing after the first, second and fifth rounds of school lunch intervention. Total of 374 children were in this intervention study out of them 188 are boys and 182 are girls children. 96 children were given Calorie supplement, 126 children were given Meat supplement, 77 were given milk and 71 were considered as control group in this study. Table 1 gives the descriptive summaries of treatment and outcomes observed at five-time points.

In this paper, we have built a joint model for General Intelligence factor as measured by Raven's Progressive Matrices test and other three S factors assessing Verbal Meaning, Arithmetic Score and Digit Span with a view to study the association between outcomes, and how it evolves over time. There is also baseline covariate information on each subject, including gender, age, time, height and weight and

Table 1 Descriptive summary

		Minimum	Maximum	Mean	Std./Dev
G Factor	Raven's Coloured Progressive Matrices	0	33	5.49	3.09
S Factor	Arithmetic score	0	33	5.69	4.02
	Verbal meaning	0	33	8.53	8.09
	Digit span total	0	33	4.02	3.58
Covariates	Age	5	12	7.03	1.139
	Height	101.10	134.95	115.6539	5.82707
	Weight	14.30	50.15	20.2314	3.38522
	Head Circumstance	45.40	56.60	50.6230	1.42828
	Socio-economics status	36.00	167.00	84.0351	21.36984
	Read test	0	12	6.91	5.107
	Write test	0	11	5.31	4.900

socio-economic status. The study consists of developing suitable joint longitudinal model for general intelligence (G and S factors) with a set of above covariates under consideration. In this context, study on improvement of mental skills measured is correlated with the working ability and this association can only be studied if both outcomes are modelled jointly using a suitable GLMM.

3 Generalized Linear Mixed Model

Consider a linear mixed model with the outcome Y_{ij} where jth measurement for ith subject (i = 1, 2,...,N and j = 1, 2,...,n_i). In GLMM, it is assumed that the outcomes Y_{ij}, are conditioned on the random effects b_i, with densities that belong to the exponential family of the form:

$$f_i(y_{ij}|b_i, \xi, \phi) = \exp\{\phi^{-1}[y_{ij}\lambda_{ij} - \phi(\lambda_{ij})] + c(\lambda_{ij}, \phi)\} \tag{1}$$

with

$$\eta[\psi'(\lambda_{ij})] = \eta(\mu_{ij}) = \eta[E(Y_{ij}|b_i, \xi)] = x'_{ij}\xi + z'_{ij}b_i, \tag{2}$$

in which x_{ij} and z_{ij} are k-dimensional and q-dimensional vectors of known covariate values, ξ a k-dimensional vector of unknown fixed regression coefficients, ϕ a scale parameter, λ_{ij} is the canonical or natural parameter which is a function of the linear predictor η and $\psi(\cdot)$, $c(\cdot)$ that are known functions.

3.1 Joint Model for Multiple Outcomes

Consider two longitudinal outcomes, say, Y_{1ij} and Y_{2ij} denote the jth measurement on the ith subject, respectively, for the outcomes either continuous, ordinal or count in nature (i = 1,..., N, j = 1,...,n_{1i} and j = 1,..., n_{2i}). A joint model is built by describing the joint density $f(Y_{1i}; Y_{2i})$ of the first outcome vector Y_{1i} and second outcome vector Y_{2i}. This can be achieved by considering a mixed model for both outcomes where the random effects are to be correlated. Let b_{1i} and b_{2i} be the vectors of random effects for the first and second outcome. More specifically, it will be assumed that $b_i = (b_{1i}, b_{2i})'$ is normally distributed with mean zero and covariance matrix D expressed as

$$\begin{pmatrix} b_{1i} \\ b_{2i} \end{pmatrix} \sim N\left[\begin{pmatrix} 0 \\ 0 \end{pmatrix}, \begin{pmatrix} d_{11} & d_{21} \\ d_{21} & d_{22} \end{pmatrix}\right] \tag{3}$$

Especially, the following structure association within each outcome sequence separately, as well as random components to model the association between the sequences. It will also be assumed that, conditioned on b_i, the outcome vectors Y_{1i} and Y_{2i} are independent; that is, we assume that the association between the outcome vectors is completely captured by the association between the random effects. However, a general unstructured matrix D will be assumed by imposing certain conditions as well to capture the association between elements in b_i. Researchers such as Fieuws and Verbeke, (2006) and Iddi and Molenberghs, (2012) has applied this approach, but it has not been used in the context of count outcomes.

Ivanova et al. (2016) considered continuous and ordinal outcome Y_{1ij} and Y_{2ij}, respectively, to develop joint modelling for the longitudinal variables based on patient's clinical outcomes to study the effect of diabetes care in Belgium. The mathematical model can be expressed as

$$E(Y_{1ij}) = \xi_{0,1} + \xi_{1,1}t_{ij} + \xi_{1,2}X_{1,i} + \xi_{1,3}X_{2,i} + \xi_{1,4}X_{3,i} \tag{4}$$

$$Logit[P(Y_{1ij} \leq r)] = \xi_{2,0r} + \xi_{2,1}t_{ij} + \xi_{2,2}X_{1,i} + \xi_{2,3}X_{2,i} + \xi_{2,4}X_{3,i} \tag{5}$$

where Y_{1ij} denote the jth measurement (1,...,n_{1i}) on the ith subject (i = 1,...,N) for the continuous outcome and Y_{2ij} denote the jth measurement (1,..., n_{2i}) on the ith subject (i = 1,...,N) for the ordinal outcome, t_{ij} is the time point at which outcome j is measured, varying random effect component with a scale factor for four different models are considered by assuming that $b_i = (b_{1i}, b_{2i}, b_{3i})'$ is normally distributed as emerged: b_{1i} and b_{2i} are the random intercept and random slope for continuous outcome variable and b_{3i} is the random intercept for ordinal outcome variable. For example, the structure of random effects with (a) uncorrelated random intercept and random slope for scale parameter, (b) random intercept for scale parameter and (c) further with random intercepts taken to correlated is expressed as

$$\begin{pmatrix} b_{1i} \\ b_{2i} \\ b_{3i} \end{pmatrix} \sim N\left[\begin{pmatrix} 0 \\ 0 \\ 0 \end{pmatrix}, \begin{pmatrix} d_{11} & 0 & d_{13} \\ 0 & d_{22} & 0 \\ d_{31} & 0 & d_{33} \end{pmatrix}\right] \tag{6}$$

Similarly, the random effects for other three models can be obtained by simplifying the above expression and inflated factor λ also known as scale parameter is measured for different scales in the case of shared random effect. To capture the correlation between the responses, various assumptions about the distribution of the random effects are made. Joint modelling with longitudinal response becomes a massive task when the outcome variables are different. However, when the outcome variables are count in nature, joint modelling possesses equally a tough problem and are considered in the following section.

4 Joint Model for Count Outcome

Consider two longitudinal count outcome, say, Y_{1ij} denote the jth measurement (1,…,n_{1i}) on the ith subject (i = 1,…,N) for the count outcome and Y_{2ij} denote the jth measurement (1,…, n_{2i}) on the ith subject (i = 1,…,N) for the another count outcome. Following Ivanova et al. (2016), a joint model is built by describing the joint density $f(Y_{1i}, Y_{2i})$ of the vectors Y_{1i} and Y_{2i}, respectively. The mathematical model can be expressed as

$$Log[P(Y_{1ij} \leq r)] = \xi_{1,r} + \xi_{1,1}t_{ij} + \xi_{1,2}X_{1,i} + \xi_{1,3}X_{2,i} + \cdots + \xi_{1,8}X_{7,i} \tag{7}$$

$$Log[P(Y_{2ij} \leq r)] = \xi_{2,r} + \xi_{2,1}t_{ij} + \xi_{2,2}X_{1,i} + \xi_{2,3}X_{2,i} + \cdots + \xi_{2,8}X_{7,i} \tag{8}$$

where Y_{1ij} and Y_{2ij} denote the two count outcomes, t_{ij} is the time point at which outcome j is measured, $X_{i,i}$ are the covariates in this study, b_i as the vector of random effects for the first and second outcome. Usually, the random effects b_i are assumed to be normally distributed with mean zero and covariance matrix D. Also, it is assumed that association between the outcome vectors is completely captured by the association between the random effects.

Two approaches for modelling joint longitudinal count sequences: (i) separate random effects that are correlated and (ii) formulating shared random effect models with scale parameter are considered. The study encompasses the two above-mentioned approach in four different models to investigate the association between various sequences of longitudinal outcome. They are

Model 1: Model with random intercept in which the intercepts are allowed to vary with slopes as fixed and each observation are predicted by the intercept across groups

Model 2: Model with random intercept and scale parameter in which the intercepts are allowed to vary across groups and fitted with scale factor λ
Model 3: Model with uncorrelated random intercept and slope and
Model 4: Model with correlated random intercept and slope.

The mean structure of all models is of the form $\log(\mu_{1ij}) = \alpha_0 + \alpha_j T_{ij}$ and $\log(\mu_{2ij}) = \beta_0 + \beta_j T_{ij}$, where T_{ij} is a year indicator defined as $T_{ij} = 1$ if the measurement was taken in the jth year (j = 1,2,3,4) and 0 otherwise. The α and β parameters are the effect of time associated with the prevalence of G and S factors, respectively. To capture the correlation inherent in the data, a random intercept is added to the conditional mean models. Thus, let b_{1i}, b_{2i} are the random intercepts for the two outcome variables (Model 1), the following structure reveals the assumptions of random effects.

$\eta(\mu_{1ij}) = \alpha_0 + \alpha_{ij} T_{ij} + \mathrm{b}_{1\mathrm{i}}, \eta(\mu_{2ij}) = \beta_0 + \beta_{ij} T_{ij} + b_{2i}.$

More specifically, it will be assumed that $b_i = (b_{1i}, b_{2i})'$ is normally distributed with mean zero and covariance matrix D expressed as

$$\begin{pmatrix} b_{1i} \\ b_{2i} \end{pmatrix} \sim N\left[\begin{pmatrix} 0 \\ 0 \end{pmatrix}, \begin{pmatrix} d_{11} \; d_{21} \\ d_{12} \; d_{22} \end{pmatrix}\right] \tag{9}$$

Especially, following structure models the association within each outcome sequence separately, as well as the random components to model the association between the sequences.

$$\eta(\mu_{3ij}) = \alpha_0 + \alpha_{1j} T_{ij} + b_{3\mathrm{i}}, \eta(\mu_{4ij}) = \alpha_0 + \alpha_{2j} T_{ij} + b_{4\mathrm{i}}$$

$$\eta(\mu_{5ij}) = \beta_0 + \beta_{1j} T_{ij} + b_{5i}, \eta(\mu_{6ij}) = \beta_0 + \beta_{2j} T_{ij} + b_{6i}$$

Suppose, if we want to implement the model with random intercept and random slope (Model 3 & 4), the structure of random component is expressed as

$$\begin{pmatrix} b_{3i} \\ b_{4i} \\ b_{5i} \\ b_{6i} \end{pmatrix} \sim N\left[\begin{pmatrix} 0 \\ 0 \\ 0 \\ 0 \end{pmatrix}, \begin{pmatrix} d_{31} \; d_{32} \; d_{33} \; d_{34} \\ d_{41} \; d_{42} \; d_{43} \; d_{44} \\ d_{51} \; d_{52} \; d_{53} \; d_{54} \\ d_{61} \; d_{62} \; d_{63} \; d_{64} \end{pmatrix}\right] \tag{10}$$

where b_{3i} and b_{5i} are the random intercept and b_{4i},and b_{6i} are the random slope for the two outcome variables. The structure of random effects for Model 2 can be obtained in a straightforward way by simplifying the above-mentioned structure by including the scale parameter in the model. The parameters are estimated with a more specifically adaptive Gaussian quadrature, which has been implemented in the R. The order Q of the integration is determined manually by fitting the model for increasing values of Q until numerical stability is obtained in the approximated likelihood value and parameter estimates.

5 Data Analysis

The study involves four longitudinal outcome variables based on human intelligent factors measured on five-time points. To capture the relationship between the responses, various assumptions about the distribution of the random effects can be made. There is also baseline covariate information on each subject including age, gender, height, weight, head circumstance, socio-economic status, intake of food supplements such as milk, meat, calories and control, duration of the follow-up study, measurement of G factor in Analytical ability as assessed by Raven's Coloured Progressive Matrices test and S factors involving linguistic ability as assessed by Verbal Meaning, numerical ability as assessed by Arithmetic Scores and immediate memory as assessed by Digit Span. Obviously, it is expected that improvement of children cognitive skills is correlated with the nutrition supplements and this association is studied using GLMM.

In this study, we considered four different random-effects model based on (9) and (10), by assuming the nature of the data to follow Poisson (P) and Negative-binomial (NB) distribution. The following random effects are included b_{1i} is the random intercept for Raven's Coloured Progressive Matrices based on (7) and b_{2i} is the random intercept for Arithmetic Scores/Verbal Meaning/Digit Span based on (8). Similarly, for random intercept and slope, the following random effects are included b_{3i}, b_{5i} and b_{4i}, b_{6i} are the random intercept and slope for Raven's Coloured Progressive Matrices and Arithmetic Scores/Verbal Meaning/Digit Span and the scale parameter λ is fitted in the Ravens coloured progressive matrices the structure of random effects is explained in Sect. 4.

The strength of association can be studied using a suitable measure such as log-likelihood value and Pearson chi-square, and these measures are widely used in the literature. Cavanaugh and Neath (2019) discussed the properties and theoretical practices of Akaike Information Criteria (AIC) and Bayesian Information Criteria (BIC) towards the model selection process. Hence, we considered measures such as AIC, BIC and log-likelihood values to explore the objective of the study. Upon fitting the joint mixed model with four different random components, the results in studying the association between G factor in Analytical ability as assessed by Raven's Coloured Progressive Matrices test and association with three S factors namely numerical ability as assessed by Arithmetic Scores are presented in Table 2, linguistic ability as assessed by Verbal Meaning in Table 3 and immediate memory as assessed by Digit Span in Table 4 with four different nutritional supplements namely milk, meat, calories and control.

The following are the observations in the study on human intelligence:

Table 2 Joint model for G factor (RCPM) in association with S factor (AS)

Nut. Sp		Calories			Milk			Meat			Control		
Criteria		AIC	BIC	logLik	AIC	BIC	logLik	AIC	BIC	logLik	AIC	BIC	logLik
M1	P	15,739.4	15,906.4	7848.27	15,727.6	15,892.1	7834.97	15,735.9	15,901.6	7843.81	15,742	15,906.8	7848.5
	NB	19,641.5	19,819.5	9796.36	19,629	19,805.5	9782.93	19,637	19,816.7	9793.62	19,643.1	19,822.5	9798.33
M2	P	15,739.5	15,906.4	7848.64	15,732.8	15,898.9	7841.59	15,739.1	15,904.9	7847.34	15,743.6	15,909.9	7851.91
	NB	19,644	19,821.9	9797.16	19,637.6	19,814.7	9790.99	19,642.6	19,819.4	9795.58	19,646.1	19,823.7	9799.37
M3	P	15,768.9	15,933.4	7873.35	15,757.4	15,919.3	7860.88	15,764.7	15,929.5	7870.12	15,773.1	15,938.3	7876.22
	NB	19,672	19,841.3	9827.31	19,660.9	19,829.9	9815.04	19,670.3	19,837.2	9824.83	19,678.4	19,846.2	9832.03
M4	P	15,719.7	15,884	7827.22	15,713.6	15,878	7820	15,721.5	15,884.9	7826.89	15,724.5	15,888.4	7829.8
	NB	19,621	19,797.3	9774.48	19,615.1	19,791.6	9768.14	19,621.7	19,798.2	9773.87	19,624.3	19,801	9776.06

M1: model with random intercept; M2: model with random intercept and scale parameter; M3: model with uncorrelated random intercept and random slope; M4: model with correlated random intercept and slope.

Table 3 Joint model for G factor (RCPM) in association with S factor (VM)

Nut. Sp		Calories			Milk			Meat			Control		
Criteria		AIC	BIC	logLik	AIC	BIC	logLik	AIC	BIC	logLik	AIC	BIC	logLik
M1	P	16,753.5	16,920.9	8360.55	16,738.1	16,906.1	8340.07	16,745.8	16,914	8350.79	16,760.9	16,927	8368.45
	NB	20,550.3	20,728.7	10,255.8	20,536.8	20,714.8	10,237	20,543.7	20,722	10,247.1	20,557.5	20,735.1	10,264.1
M2	P	16,763.8	16,929.9	8369.07	16,754.4	16,920.6	8358.75	16,761.4	16,927.5	8364.54	16,766.8	16,933.8	8371.51
	NB	20,562.8	20,739.9	10,266.6	20,555.3	20,732.6	10,257.9	20,560.5	20,737.7	10,263.1	20,567.5	10,514	10,270.7
M3	P	16,773.4	16,939.2	8372.82	16,758.9	16,925.6	8358.4	16,769.4	16,935.8	8369.16	16,776	16,943.1	8372.27
	NB	20,571.4	20,744.8	10,270.5	20,557.6	20,732.3	10,254.1	20,569.7	20,743.2	10,268.2	20,576.5	20,750.1	10,273.8
M4	P	16,712.8	16,878.2	8316.25	16,705.5	16,871.2	8307.24	16,711.9	16,877.5	8314.35	16,716.4	16,880.3	8317.04
	NB	20,509.6	20,686	102	20,504.2	20,679.9	10,204.2	20,509.8	20,685.5	10,210.7	20,513	20,688.4	10,212.7

M1: model with random intercept; M2: model with random intercept and scale parameter; M3: model with uncorrelated random intercept and random slope; M4: model with correlated random intercept and slope.

Table 4 Joint model for G factor (RCPM) in association with S factor (DS)

Nut. Sp		Calories			Milk			Meat			Control		
Criteria		AIC	BIC	logLik	AIC	BIC	logLik	AIC	BIC	logLik	AIC	BIC	logLik
M1	P	14,644.8	14,810.4	7297.31	14,636.8	14,802.5	7292.02	14,645.3	14,811	7298.11	14,651.9	14,815.5	7302.84
	NB	17,711.1	17,887.8	8828.26	17,701.2	17,878.9	8821.51	17,709.3	17,889.1	8828.42	17,714.6	17,894.2	8833.05
M2	P	14,655.7	14,823.7	7311.83	14,642.1	14,808.5	7300.05	14,653.8	14,820.1	7306.51	14,660.4	14,826.6	7313.97
	NB	17,724.4	17,903.4	8844.2	17,710	17,887.4	8831.98	17,719.3	17,896.6	8837.25	17,727	17,902.5	8844.96
M3	P	14,677.4	14,841.1	7324.68	14,659.2	14,819.6	7307.22	14,681.1	14,834.2	7319.54	14,685.4	14,840.6	7324.4
	NB	17,744.7	17,917.2	8854.5	17,725.8	17,895.2	8835.9	17,739.7	17,910.9	8854.75	17,744	17,916.4	8859.84
M4	P	14,623.4	14,789	7276.09	14,615.3	14,780.4	7268.87	14,624.4	14,790.1	7276	14,626.9	14,791.2	7277.17
	NB	17,689.8	17,866.6	8806.21	17,680.9	17,857	8798.54	17,688.6	17,864.3	8804.48	17,690.7	17,866.2	8805.54

M1: model with random intercept; M2: model with random intercept and scale parameter; M3: model with uncorrelated random intercept and random slope; M4: model with correlated random intercept and slope.

(i) When comparing the distribution of the data between Poisson (P) and Negative Binomial (NB), Poisson GLMM turns out to be the most suitable model for all the outcome variables across the nutritional supplements on human intelligence considered in this study.
(ii) On comparing the mixed models, model 4, i.e. model with correlated random intercept and slope has outperformed the other models for the outcome variable RCPM in association with AS, in all the nutritional supplements based on the criteria values. Similar results are obtained for other outcome variables considered in the study.
(iii) The AIC, BIC and log-likelihood for the comparison of models on analytical ability by RCPM in association with numerical ability by AS are closer to each other for all the nutritional supplements but Milk supplement is the least for all the four models resulting in improvement in intelligence as compared to other supplements namely, meat, calories and control. Similarly, when comparing the analytical ability by RCPM in association with linguistic ability by Verbal Meaning and immediate memory by Digit Span total, the results revealed that Milk supplements produce smaller criteria values indicating that the supplement of milk helps in improving the intelligence on children.
(iv) lastly, in studying the association between the G–S factor, RCPM–DS showed better association than compared to other G–S factors by considering that the outcomes are independent and assuming that the association between the outcome vectors are captured completely by the association between the random effects considered.

Further, studying the association between general intelligence factor in mental abilities as RCPM and specific factor obtained as a combination of particular abilities as VM, AS and DS, we conclude that the nutritional supplement milk with smaller criteria values turns out to be better in the improvement of intelligence among children in all the models than compared to other supplements and the results are given in Table 5.

On the whole, it is evident that there exists better association between the Analytical ability as assessed by Raven's Coloured Progressive Matrices test and immediate memory as Digit Span total test than other two joint outcomes. Further, the joint mixed models with correlated random effects yielded a better fit compared to the other three models. The intervention study has shown that milk nutritional supplement is comparatively better than the other three supplements based on joint model analysis. Sakamoto (2019) stated that there will be a boundary issue in using variance components for the mixed model and the information criteria for the models considered yield more or less the same results. Hence, he concluded that bootstrap approach may be helpful in drawing inference. Therefore, the following section explains the bootstrapping technique to examine the suitable model and the best nutritional supplement in improving the human intelligence.

Table 5 Joint model for G factor (RCPM) in association with S factors (AS + VM + DS)

Nut. Sp		Calories			Milk			Meat			Control		
Criteria		AIC	BIC	logLik	AIC	BIC	logLik	AIC	BIC	logLik	AIC	BIC	logLik
M1	P	31,643.9	31,974.8	15,786.2	31,601.7	31,933.3	15,748.8	31,630.1	31,961.3	15,773	31,640.4	31,969.3	15,779.7
	NB	39,390.2	39,743.2	19,653	39,346.5	39,701.2	19,615.3	39,376	39,732.3	19,642.4	39,385.8	39,740.8	19,649.7
M2	P	31,667.5	32,002.7	15,817.1	31,648.8	31,981.7	15,796.9	31,665.1	31,997.5	15,808.8	31,674.4	32,008.9	15,820.8
	NB	39,418.8	39,775.9	19,686.8	39,399.4	39,754.4	19,667.8	39,415	39,769.4	19,679	39,426.1	29,549.5	19,690
M3	P	31,698.7	32,030.9	15,836.2	31,664.8	31,994	15,800.9	31,696.9	32,019.5	15,825.5	31,707.8	32,032	15,832.4
	NB	39,446.5	39,793.3	19,711.8	39,410.8	39,756.9	19,673.9	39,438.5	39,783.3	19,705.9	39,449.7	39,794.2	19,715.3
M4	P	31,568	31,897.6	15,710.5	31,548.8	31,878.2	15,687.5	31,569.3	31,898.2	15,708.1	31,575.4	31,902.8	15,711.7
	NB	39,313.9	39,666.5	19,577	39,294.1	39,645.4	19,555	39,313.6	39,666.6	19,575	39,320.1	39,671.1	19,578.5

M1: model with random intercept; M2: model with random intercept and scale parameter; M3: model with uncorrelated random intercept and random slope; M4: model with correlated random intercept and slope.

6 Bootstrap

Bootstrap technique often requires fewer assumptions and yields greater accuracy than many statistical methods. Mohanraj and Srinivasan (2015) considered the resampling technique for the longitudinal repeated measures data and the same method of resampling is adapted in this paper. The accuracy and efficiency of the proposed modelling strategy are evaluated through the bootstrap method for complete data with varying sample sizes.

The study considers the intervention data assuming the nature of the data is count with varied sample sizes (10, 20 and 30 %) for the bootstrap process with 1000 runs. The longitudinal data were analysed by fitting a joint model with correlated random intercept and slope via ML method of estimation. Hence, the performance of the bootstrap samples towards the proposed joint model is assessed by the model information criteria AIC, BIC and the log-likelihood and the results are presented in Table 6.

The conclusions based on varying bootstrap samples of varied percentage (10, 20 and 30 %) are very closer to each other. However, the numerical values of AIC, BIC and log-likelihood from bootstrap study shows that the nutritional supplement Milk consistently performs well in all the bootstrap samples. Also, the joint model fitted for factor RPCM–DS yield better performance than the other factors.

7 Conclusions

Longitudinal studies are common in many psychometric studies particularly on cognitive ability of school children. Literature is abundant on studying the association between the G and S factors of human intelligence. But many of the research studies deal with analyzing each response separately. Here, we considered four different models with varying random effect terms to capture the association between each response variable and the covariates. Also, we have focused on modelling bivariate longitudinal sequences by assuming mixed models for both outcomes with separate random effects that are correlated and formulating shared parameter model, perhaps with scale parameter. Emphasis is placed on two count sequences. Among animal-source foods, milk is believed to play a unique role in promoting children's growth and development.

The Kenya intervention study suggests that the supplementation of milk has shown improvement gradually on cognitive performance of school children compared to other supplements. Digit span is a reliable and valid measure of attention assessing "S" factor of human intelligence. Based on the log-likelihood value of the bivariate analysis, the influence of "G" is quite likely to be felt more on attention (Digit Span) immediate memory shown better association than the other two "S" factors deals with higher order cognitive functions involving reasoning abilities on linguistic

Table 6 Bootstrap results of joint model for G factor (RCPM) in association with S factors (AS + VM + DS)

Nut. Sp		Calories			Milk			Meat			Control		
Criteria	%	AIC	BIC	logLik	AIC	BIC	logLik	AIC	BIC	logLik	AIC	BIC	logLik
RPCM- AS	10	19,695.5	19,941.6	13,750.8	16,687.1	16,730.6	12,073.6	18,929.9	18,982.5	13,232.9	21,127.6	21,179.2	14,369.1
	20	22,959.7	23,016.2	15,277.4	19,588.2	19,634.8	13,523.8	21,871.5	21,933.6	14,737.2	24,051	24,115.2	15,853.1
	30	25,545.8	25,609.7	16,575.5	22,163.4	22,216.4	14,810.2	24,455.9	24,525.8	16,034.8	26,643.9	26,716.2	17,157.4
RPCM- VM	10	22,499.8	22,554.3	16,107.9	18,964	19,013.9	14,303	21,411.9	21,467.1	15,578.3	23,760.4	23,824.5	16,809.1
	20	25,971.4	26,031.5	18,067.9	22,354.1	22,407.3	16,167.1	24,852	24,915.4	17,514.4	27,188.8	27,254	18,716.8
	30	29,000.5	29,068.3	19,740.3	25,372.3	25,432.6	17,825.8	27,881	27,952.3	19,187.3	30,220.8	30,294.1	20,394.4
RPCM- DS	10	14,186.2	14,218.6	12,153.2	10,429.9	10,459.7	10,247.3	13,053.9	13,089	11,601.8	15,538.9	15,577.1	12,874.9
	20	17,013.4	17,055.1	13,907.3	13,199.1	13,231.3	11,931.6	15,860.5	15,899.2	13,333.4	18,336.9	18,380.8	14,589.1
	30	19,265.1	19,312.8	15,188.4	15,442	15,479.6	13,202.2	18,112	18,156.4	14,614.6	20,590.5	20,640.6	15,873

and numerical domains and the joint mixed models with correlated random effects yielded a better fit compared to other models.

On the whole, Poisson model is the suitable model in studying the association between the variables in this study. Further, from the bootstrap technique of varied sample sizes (10, 20, 30 %), it is clear that the milk supplement has paved a way to increase the intelligence factor among school children. However, in practice, there are other factors (covariates) such as stress, depression, anxiety, parental behaviour which affect the improvement of intelligent factor among the school children. Thus, this analysis can be extended with more covariates related to the study and could be extended to handle more appropriate statistical procedure in the improvement of nutritional supplements.

References

Bilker, W. B., Hansen, J. A., Brensinger, C. M., Richard, J., Raquel, E., & Gur, R. C. (2012). Development of abbreviated nine-item forms of the Raven's standard progressive matrices test. *HHS Public Access, 19*(3), 354–369.

Breslow, N. E., & Clayton, D. G. (1993). Approximate inference in generalized linear mixed models. *Journal of the American Statistical Association, 88,* 9–25.

Cavanaugh, J. E., & Neath, A. A. (2019). The Akaike information criterion: Background, derivation, properties, application, interpretation, and refinements. *Wiley Interdisciplinary Reviews: Computational Statistics, 1460,* 1–11.

Chakraborty, H., Helms, R., Sen, P., & Cohen, M. (2003). Estimating correlation by using a general linear mixed model: Evaluation of the relationship between the concentration of HIV- RNA in blood and semen. *Statistics in Medicine, 22,* 1457–1464.

Coxe, S., West, S. G., & Aiken, L. S. (2009). The analysis of count data: A gentle introduction to poisson regression and its alternatives. *Journal of Personality Assessment, 91*(2), 121–136.

Coyle, T. R., Elpers, K. E., Gonzalez, M. C., Freeman, J., & Baggio, J. A. (2018). General intelligence (g), ACT scores, and theory of mind: (ACT)g predicts limited variance among theory of mind tests. *Intelligence, 71,* 85–91.

Efendi, A., Molenberghs, G., Njagi, N. E., & Dendale, P. (2013). A joint model for longitudinal continuous and time-to-event outcomes with direct marginal interpretation. *Biometrical Journal, 55*(4), 572–588.

Fallah, N., Mitnitski, A., & Rockwood, K. (2011). Applying neural network Poisson regression to predict cognitive score changes. *Journal of Applied Statistics, 38*(9), 2051–2062.

Fieuws, S., & Verbeke, G. (2006). Pairwise fitting of mixed models for the joint modeling of multivariate longitudinal profiles. *Biometrics, 62,* 424–431.

Goharpey, N., Crewther, D. P., & Crewther, S. G. (2013). Research in developmental disabilities problem solving ability in children with intellectual disability as measured by the Raven's colored progressive matrices. *Research in Developmental Disabilities, 34*(12), 4366–4374.

Graßhoff, U., Holling, H., & Schwabe, R. (2016). Optimal design for the rasch poisson-gamma model. In Kunert J., Müller C. H., & Atkinson A. C. (eds.), mODa 11—Advances in model-oriented design and analysis, pp. 133–141. Springer.

Greenhalgh, T., Kristjansson, E., & Robinson, V. (2007). Realist review to understand the efficacy of school feeding programmes. *BMJ, 335*(7625), 858.

Gueorguieva, R. (2001). A multivariate generalized linear mixed model for joint modelling of clustered outcomes in the exponential family. *Statistical Modeling, 1,* 177–193.

Horrocks, J., & Van Den Heuvel, M. J. (2009). Prediction of pregnancy: A joint model for longitudinal and binary data. *Bayesian Analysis, 4,* 523–538.

Iddi, S., & Molenberghs, G. (2012). A joint marginalized multilevel model for continuous and binary longitudinal outcomes. *Journal of Applied Statistics, 39*(11), 2413–2430.

Ivanova, A., Molenberghs, G., & Verbeke, G. (2016). Mixed models approaches for joint modelling of different types of responses. *Journal of Biopharmaceutical Statistics, 26*(4), 601–618.

Kristjansson, B., Petticrew, M., MacDonald, B., Krasevec, J., Janzen, L., Greenhalgh, T., Wells, G, A., MacGowan, J., Farmer, A. P., Shea, B., Mayhew, A., Tugwell, P., & Welch, V. (2007). School feeding for improving the physical and psychosocial health of disadvantaged students. *Cochrane Database of Systematic Reviews*, 24(1), Art. No.: CD004676.

Laird, N., & Ware, J. (1982). Random-effects models for longitudinal data. *Biometrics, 38,* 963–974.

Li, N., Elashofi, R. M., Li, G., & Saver, J. (2010). Joint modeling of longitudinal ordinal data and competing risks survival times and analysis of the NINDS rt-PA stroke trial. *Statistics in Medicine, 29,* 546–557.

Martínez-Plumed, F., Ferri, C., Hernández-Orallo, J., & Ramírez-Quintana, M. J. (2017). A computational analysis of general intelligence tests for evaluating cognitive development. *Cognitive Systems Research, 43,* 100–118.

Mohanraj, J., & Srinivasan, M. R. (2015). Selection of the best covariance structure in longitudinal data based on bootstrap. *Assam Statistical Review, 29*(1), 1–29.

Molenberghs, G., & Verbeke, G. (2005). *Models for discrete longitudinal data.* New York: Springer.

Muniz, M., Gomes, C. M. A., & Pasian, S. R. (2016). Factor structure of Raven's coloured progressive matrices. *Psico-USF, 21*(2), 259–272.

Murphy, S. P., & Allen, L. H. (2003). Nutritional importance of animal source foods. *The Journal of Nutrition, 133*(11), 3932S-3935S.

Neumann, C. G., Bwibo, N. O., Murphy, S. P., Sigman, M., Whaley, S., Allen, L. H., et al. (2003). Animal source foods improve dietary quality, micronutrient status, growth and cognitive function in Kenyan school children: Background, study design and baseline findings. *The Journal of Nutrition, 133,* 3941S-3949S.

Njagi, N. E., Molenberghs, G., Verbeke, G., Kenward, M. G., Dendale, P., & Willekens, K. (2013). A exible joint-modelling framework for longitudinal and time-to-event data with overdispersion. *Statistical Methods in Medical Research, 25*(4), 1661–1676.

Palestro, J. J., Bahg, G., Sederberg, P. B., Lu, Z.-L., Steyvers, M., & Turner, B. M. (2018). A tutorial on joint models of neural and behavioral measures of cognition. *Journal of Mathematical Psychology, 84,* 20–48.

Pope, L., Roche, E., Morgan, C. B., & Kolodinsky, J. (2018). Sampling tomorrow lunch today: Examining the effect of sampling a vegetable-focused entrée on school lunch participation, a pilot study. *Preventive Medicine Reports, 12,* 152–157.

Sakamoto, W. (2019). Inference on variance components near boundary in linear mixed effect models. *WIREs Computational Statistics, 1466,* 1–10.

Siekmann, J. H., Allen, L. H., Bwibo, N. O., Demment, M. W., Murphy, S. P., & Neumann, C. G. (2003). Kenyan school children have multiple micronutrient deficiencies, but increased plasma vitamin B-12 is the only detectable micronutrient response to meat or milk supplementation. *The Journal of Nutrition, 133*(11), 3972–3980.

Thifiebaut, R., Jacqmin-Gadda, H., Chene, G., Leport, C., & Commenges, D. (2002). Bivariate linear mixed models using SAS PROC MIXED. *Computer Methods and Programs in Biomedicine, 69,* 249–256.

Tourva, A., Spanoudis, G., & Demetriou, A. (2016). Intelligence cognitive correlates of developing intelligence: The contribution of working memory, processing speed and attention. *Intelligence, 54,* 136–146.

Tsiatis, A., & Davidian, M. (2004). Joint modeling of longitudinal and time-to-event data: An overview. *Statistica Sinica, 14,* 809–834.

Walingo, M. K., & Musamali, B. (2008). Nutrient intake and nutritional status indicators of participant and nonparticipant pupils of a parent-supported school lunch program in Kenya. *Journal of Nutrition Education and Behavior, 40*(5), 298–304.

An Analysis of the Impact of Requirements on Wages Within Sectors of the Tourism Industry

Paolo Mariani, Andrea Marletta, Lucio Masserini, and Mariangela Zenga

Abstract The definitions of professional roles have changed quickly in the last few years due to several factors, such as the spread of new technologies. The recruitment process represents a way to evaluate the skills that a candidate needs to have in the workplace. This paper aims to evaluate the requirements for new hires in the tourism sector. In particular, we analysed the profiles of 1.526 workers recruited in 2017 by The Adecco Group in Italy. In the first phase, a conjoint analysis is performed to evaluate skills in the tourism sector, while in the second phase, a multinomial logistic regression is carried out to obtain more in-depth knowledge regarding the most selected (or preferred) profiles by employers, among those evaluated.

1 Introduction

In social and economic systems, jobs play a fundamental role, both in terms of production factors and aspects related to employers. In particular, access to jobs is important for the job market supply and demand. On the demand side, the position of knowledge, abilities and attitudes requires an evaluation of models and educational paths for their creation and implementation. On the other hand, on the supply side, the economic situation and the effects of technological progress have started to change the models for qualifications and have created difficulties in defining short-term scenarios.

P. Mariani · A. Marletta (✉) · M. Zenga
University of Milano Bicocca, Via Bicocca Degli Arcimboldi, 8, Milano, Italy
e-mail: andrea.marletta@unimib.it

P. Mariani
e-mail: paolo.mariani@unimib.it

M. Zenga
e-mail: mariangela.zenga@unimib.it

L. Masserini
University of Pisa, Via Cosimo Ridolfi, 10, Pisa, Italy
e-mail: lucio.masserini@unipi.it

P. Mariani and M. Zenga (eds.), *Data Science and Social Research II*,
Studies in Classification, Data Analysis, and Knowledge Organization,
https://doi.org/10.1007/978-3-030-51222-4_17

The development of the job market has shown acceleration during the last few years in terms of the definition of professional roles. Some new roles have appeared in working environments. The classification of these new roles is not simple using older definitions and the process of new interpretation is slower than the quantitative growth of the new positions.

Even when new job profiles conserve some of the general features of the traditional ones, the new positions update the definition of competence given the transformations occurring in the market.

The nomenclature of the professional roles is often no different from the past, i.e. the changes are in company requests and actions. The beginning of the digital era and the presence of new technologies are factors that determine a bigger gap with respect to the past. These elements have an effect in other spheres, generating interconnected processes and organisational levels in companies.

These aspects have increased the difficulties in finding a balance between demand and supply in the job market ISTAT (2017). This effect could be explained by repeated moments of economic crisis, both in terms of the mismatch in requested competencies and professions OCSE (2017). For the firms that have decided to exploit opportunities to meet workers, their motivation is a need for new and different competencies that are not available in the business. These meetings have taken place using numerous channels: direct candidacy, word of mouth, newspapers and press, temporary employment agencies, industry associations, the Internet, company databases and employment exchanges Mariani et al. (2018).

In regard to temporary employment agencies, the activity could be represented as a relationship between three stakeholders, namely the job agency, the selected employer and the company. This model represents one of the external processes used by companies to solve temporary fluctuations in demand through adjustments in their manpower. Generally, these adjustments are converted into short-term contracts and constitute important phenomena for an analysis of some possible trends in job demand. They are also related to job offers because they influence the study of the moments of insertion in the activity structure.

Data associated with short-term hires generates a high flow of activations and suspensions, representing a moderate proportion of the total amount of working days. Moreover, it is necessary to analyse the activations and suspensions by the same employer during a defined timeframe. The average number of working days for short-term contracts is equal to 12 Mariani et al. (2018). An analysis of age shows that extreme time slots occur more frequently. For the youngest people, the extension of their studies limited the impact of the crisis, and irregular careers were created with more difficulties in finding their first employment. For the eldest workers, there was a postponement of the retirement age, which prolonged careers or increased the risk of being exposed to a long break before being eligible for pension.

These developments led to research being carried out on the requirements of companies that tied themselves in the matching phase to various professional roles, investigating the knowledge, skills, attitudes, and more generally skillsets, by using as evidence the actions meeting supply and demand. The analysis was based on the research proposed by The Adecco Group on new hires starting from 2017. The data was made comparable at the European level by using the European Skills/Competences, Qualifications and Occupations (ESCO) international classification, which is the basis for the creation of a national system for the permanent observation of professions and related needs. Information regarding goodwill, albeit with a managerial and administrative slant, provides a source of knowledge structured based on the criteria that companies adopt in their choices of workers who apply for job positions in their companies. The aim of this work is to measure the monetary evaluation of the skills that are useful for receiving a job offer using an a posteriori analysis of the hired candidates.

The structures of this paper are following: Sect. 2 contains some definitions about the recruitment process and the profile of workers, Sect. 3 contains the description of the proposed methodology, Sect. 4 is devoted to the application and Sect. 5 includes final remarks and discussions.

2 The Recruitment Process

The recruitment process represents a large resource utility for a company and the choice of the right person for a vacancy has an essential relevance in terms of savings and satisfaction. Otherwise, a wrong choice during the process could provoke negative consequences, not only from a monetary point of view but also in terms of loss of time. The basic requirement for obtaining good results in the recruitment process is to measure the attributes and competencies of the candidates for a job vacancy. Only by means of an objective and impartial evaluation of the candidates can a selection process yield satisfactory results Farnham and Stevens (2000), Humburg and der Velden (2015), Rowe (1995), Taylor and Bergmann (1987). The recruitment process consists of the following steps:

- Identify the role covered by a new recruit;
- Recruit a shortlist of candidates using job advertisements and then analyse their CVs;
- Evaluate/assess the competencies, attributes and attitudes of the candidates through psychological tests; and
- Select the best profile after conducting job interviews.

The first step is to identify exactly what the company truly needs, that is, the duties of the new recruit, starting with a description of the job vacancy in the job advertisement. An error during this step could have very serious consequences, such as the hiring of someone with inappropriate attributes or the exclusion of the right person from the shortlist of candidates. The second step consists of primary screening of all the applicants to reduce the number of possible interviewees. This screening could involve eliminating applicants who do not possess the attributes required for the vacancy. The measure of competencies is the most complicated part of the recruitment process. Candidate assessment can involve psychological tests and interviews. The tests aim to measure aspects of personality, predispositions and inclinations of the applicants, and the interviews can be conducted in groups (assessment centre) or individually. During the interviews, candidates can be assigned problem-solving tasks or be asked about generic skills and/or experiences mentioned in the CV. Sometimes, the human resources manager is assisted by a specialist from the business area to evaluate the specific competencies by asking technical questions regarding the job vacancy. The last phase is the selection of the best profile and the integration of the nominee into the job vacancy. After evaluating all candidates, by reading CVs and interviews, the applicant is chosen on the basis of some particular objective and subjective criteria that match the expectations of the company at that time. A period of integration and orientation is needed for the training of new recruits.

The requested requirements of companies during the recruitment process can be divided into three categories:

- Knowledges: the set of structured principles and theories useful for the correct implementation of the profession;
- Abilities: the procedures and processes that define the capabilities to accomplish the professional tasks; and
- Attitudes: the cognitive features affecting the professional development and execution of job activities.

2.1 A Focus on the Tourism Sector

In our work, we analysed the tourism sector, in particular the role of travel consultants and clerks, who provide information about travel destinations, arrange travel itineraries, make travel and accommodation reservations and register passengers at check-in and departure.

Tasks include:

- obtaining information about the availability, cost and convenience of different types of transport and accommodation, ascertaining customers requirements and advising them on travel arrangements
- providing information about local and regional attractions, sightseeing tours, restaurants and providing maps and brochures
- making and confirming reservations for travel and accommodation
- preparing itineraries and issuing tickets, boarding passes, vouchers, bills and receiving payments
- helping customers in obtaining necessary travel documents
- verifying travel documentation and registering passengers and luggage at check-in and departure.

According to Isfol (2017), skills more required for this professional figure are customer and personal services and foreign language. The first has to be intended as knowledge of principles and processes for providing customer and personal services. This includes customer needs assessment, meeting quality standards for services and evaluation of customer satisfaction. The second one included knowledge of the structure and content of a foreign language including the meaning and spelling of words, rules of composition and grammar and pronunciation.

Examples of the occupations classified here, namely Airline ticket agent, Check-in attendant, Ticket issuing clerk, Travel agency clerk, Travel consultant, Travel desk clerk, Tourism information clerk.

3 Methodology

In this paper, we use the conjoint analysis Krantz (1964) that is applied in this case for the study of the models of choice Street and Burgess (2007) of the companies, starting from the preference expressed by the companies with respect to different possible configurations of requirements related to the professional profiles. The value of the level of satisfaction obtained by a company with respect to the obtained requirements is indicated as Utility.

The Utility function assigns a level of satisfaction to each requirement considered, in particular, in the form:

$$U = f(X), \tag{1}$$

where U is the utility level and X are the characteristics of the requirements.

It is necessary to introduce some terminologies related to this methodology:

- Factors or attributes: these are the requirements under examination, i.e. the variables that the researcher controls in a conjoint experiment to measure the effect on the consumer's usefulness;
- Levels: these indicate the different ways in which the attributes are manifested (the categories of the attributes); and
- Profile or stimulus concerns a specific combination of the attribute levels.

The profile is assigned by assigning a level to each attribute; the number of the profile depends on the number of levels and attributes. In this analysis, the attributes are represented by the requirements requested by the companies. For the conjoint analysis, the preference and utility are in a biunivocal correspondence: the more a candidate meets the requirements of a company, the more his/her use will lead to usefulness. The preference can be interpreted as the function of the levels of the characteristics of a candidate. Subsequently, based on the preferred choice of the company, the partial utilities are calculated. They represent the importance associated with each level of the attributes and are called part worth. Finally, the total utility is analysed as the sum or the product of partial utilities. From an analytical point of view, this modelling is expressed as follows Luce and Tukey (1964), Luce and Krantz (1971):

$$U_j = \sum_{l=1}^{L}\sum_{k=1}^{K} u_{jkl} * x_{jkl} + e_j \tag{2}$$

where:

- U_j the utility of the j-th profile,
- u_{jkl} the partial utility referred to the l-th level of the k-th attribute,
- x_{jkl} a dummy variable that assumes value 1 if the level l of the attribute k is present in profile j, and assumes value 0 otherwise,
- e_j is the random error.

For this case, since the p_{ij} derives from a conditional logit model Dagsvik (1998), this choice requires the construction of all candidate profiles a priori as a combination of all the attributes and levels. Among these, the only one represented the choice of the company is the one related to the profile of the candidate launched. The method used to estimate partial utilities in this context is the logistic regression McFadden (1973), as a model with qualitative predictors Louviere et al. (2010).

In particular, the probability that the i-th company chooses the j-th profile is given by:

$$p_{ij} = \frac{exp(u_{jkl} * x_{jkl})}{\sum_{l=1}^{L} \sum_{k=1}^{K} exp(u_{jkl} * x_{jkl})} \tag{3}$$

It is possible to obtain the overall utility in correspondence of all the profiles, simply by applying the linear combination of utilities. In the context of the conjoint analysis, it is also possible to evaluate the relative importance of the attribute, in order to make partial utilities comparable and to reach the importance values of the factors comprised between 0 and 1, or in percentage, through the following formula:

$$I_k = \frac{\max(u_k) - \min(u_k)}{\sum_{j=1}^{J} [\max(u_k) - \min(u_k)]}, \tag{4}$$

where $u_k = [u_{k1}, \ldots, u_{kL}]$ is the vector containing the partial utilities of the k-th attribute. The percentage of importance for the k-th attribute is obtained by comparing the difference between the maximum value and the minimum value of utility relative to the attribute itself, to the sum for all the attributes of that difference. The more the change in the levels of an attribute affects utility, the greater the importance of that factor. The relative importance for each attribute can be used to obtain an assessment of the change in the remuneration of new hiring, associated with the modification of the simultaneous combination of several attributes describing the skills and characteristics of the professional profile examined. As discussed by Mariani et al. (2018), we introduce the following notation:

- b is the current profile, which is referred to as status quo, of the requirements considered;
- i, with $i = 1, ..., n$, is the alternative profile, which differs from b by attribute level i;
- U_b indicates the sum of the partial utilities associated with the status quo of the requirements under consideration;
- U_i denotes the sum of the utility scores associated with the i-th alternative profile.

Let now M_i indicate the ratio that is obtained by dividing the difference between the total utility of the i alternative and the status quo, divided by the total utility of the status quo. Formally,

$$M_i = \frac{U_i - U_b}{U_b} \tag{5}$$

assuming that U_b is different from 0. The term M_i indicates whether the change in the status quo generates a loss or gain in terms of total utility. It is evident that $M_i = 0$ represents the indifferent situation between loss and gain in terms of total utility. Let M_{ik} be

$$MI_{ij} = M_i * I_k. \tag{6}$$

Through 6, it is possible to evaluate the variation of the total salaries generated by assuming a change in the status quo profile. Given the Gross Annual Salary (GAS) associated with the status quo profile, the economic revaluation coefficient is expressed as follows:

$$V_{ik} = MI_{ik} * GAS, \tag{7}$$

where V_{ik} is the amount of the variation of the GAS. It is obtained by assuming that the monetary attribute referring to the requirement varies in proportion to the change in the total usefulness of this requirement. The evidence regarding the requisites is reported with respect to the industry and the set of professions, which, from a quantitative perspective and an information completeness aspect, have been analysed by the proposed methodology.

4 Application

In this paper, the data is sourced from the 2017 Adecco Group database, where the statistical unit is represented by a candidate receiving a job offer and the explanatory variables are the mandatory requirements needed to pass the recruitment process. The job positions are made comparable using the ESCO international classification. Information about job offers generates knowledge about the criteria used for the selection of the best candidate. In 2017, there were more than 120.000 job positions divided into the following 9 industries: Information Technology and digital, engineering, medical, finance, tourism, Human Resource, commercial, food services and production. In this paper, the work positions analysed are in the tourism sector, so the sample size is $n = 1.526$. The job profiles included are hotel concierge, airport baggage handler and travel consultant. The results are presented only for the sub-sample related to the travel consultant professional role. The number of the analysed job offers is $n = 626$.

4.1 A Monetary Revaluation Based on Conjoint Analysis

A large amount of information is available for each new hire, but for the purposes of this work, we analysed 7 skills selected from 26 competences collected by The Adecco Group, including information regarding previous work experience, knowledge of the English language and education level. The Adecco Group's Human Resources tested the 26 skills of the new hires during the recruitment phase. In the end, the requirements for each new hire were as follows:

- Quality orientation (1 = 'Yes', 0 = 'No');
- Teamworking (1 = 'Yes', 0 = 'No');
- Participation and responsibility (1 = 'Yes', 0 = 'No');
- Problem-solving and analysis (1 = 'Yes', 0 = 'No');
- Communication (1 = 'Yes', 0 = 'No');
- Self-control (1 = 'Yes', 0 = 'No');
- Customer orientation (1 = 'Yes', 0 = 'No');
- Previous work experience (1 = 'Yes', 0 = 'No');
- English knowledge (1 = 'Yes', 0 = 'No');
- Education level (1 = 'Up or equal to secondary school', 0 = 'Otherwise');

To build the conjoint plan, we needed to consider the possible profiles as a priori combinations of all of the attributes' levels: in this case, the number of generated profiles was $2^{10} = 1.024$. Among these, there is a profile corresponding to the profile of the new hire. Therefore, for each new hire, the dataset now contains 1.024 rows, at the total rows in the dataset are $1.526 \times 1.024 = 1.562.624$ ($626 \times 1.024 = 641.024$ for travel consultant sub-sample). The method used to estimate partial utilities in this context is the multinomial logistic regression; the dependent variable is represented by a set of dummy variables related to the 1.024 generated profiles: it is equal to 1 if the profile is that of the new hiring, or it is equal to 0 if the profile doesn't match to the profile of the new hiring. Data manipulation and the conjoint analysis were done using R and mlogit package Croissant (2012).

For the choice of the GAS value, we considered the Assohandlers Contracted Grossing Costs (level 4) through which the differential reference value is defined.

In Table 1, the relative importance index of the attribute I_k and rank position is displayed for tourism sector and travel consultant professional figure. The entire sector enhances requirements as problem-solving and analysis, self-control and quality orientation with an I_k bigger than 15%. In a central position, there are teamworking, communication, customer orientation and participation and responsibility. Finally, education level and previous experience are classified as requirements very widespread and less appreciated.

The second column of the Table 1 reports values and rank of I_k for travel consultants and clerks. The two rankings for the entire sector and the selected role are positive correlated, but now the most important skill is teamwork with $I_k = 20.1\%$, followed by problem-solving and analysis and quality orientation, respectively, with 19.1 and 17.3%. About the bottom of the ranking, there is a confirmation for education level and previous experience with values for I_k very close to 0.

Table 1 Relative importance of the attribute I_k and rank position in brackets for tourism sector and travel consultant figure—Italy, %, 2017

Requirements	Tourism sector (%)	Travel consultant (%)
Problem-solving and analysis	17.6 (1)	19.1 (2)
Self-control	15.2 (2)	12.6 (4)
Quality orientation	15.1 (3)	17.3 (3)
Teamworking	11.7 (4)	20.6 (1)
Communication	10.8 (5)	6.7 (6)
Customer orientation	10.5 (6)	5.8 (8)
Participation and responsibility	9.4 (7)	8.7 (5)
English knowledge	7.8 (8)	6.0 (7)
Education level	1.0 (9)	2.4 (9)
Previous experience	0.9 (10)	0.8 (10)

Source Elaboration on AdeccoGroup data

Since values for I_k are computed, it is necessary to attach a benchmark value for GAS in order to obtain the monetary re-valuation $V_i k$ for the tourism sector and the travel consultant and clerk roles. According to the proposed hypothesis, the minimum GAS provided by the Assohandlers contracted grossing costs for the entire sector is equal to € 23.000,00. The average GAS provided by Job Pricing for the industry sector is equal to € 33.000,00, generating a ΔGAS equal to € 10.000,00.

The estimates of ΔGAS (with standard error for the partial utility associated) are presented in Table 2. Moreover, a p-value on the significance of the single partial utility is reported. Problem-solving and analysis, self-control and quality orientation are the most important requirements with an estimated ΔGAS over € 1.000,00. For requirements in a central position, the values of ΔGAS are between € 500,00 and € 1.000,00. The monetary evaluation for the last requirements is very close to € 0,00 (see education level and previous experience).

The total amount of estimated ΔGAS is equal to € 8.467,44. It appears to be very close to the real $\Delta GAS =$ € 10.000,00 and this represents a satisfactory result for the proposed approach.

The same approach could be extended to the single professional role of the travel consultant. For the entire tourist sector, the minimum GAS provided by the Assohandlers contracted grossing costs is equal to € 23.000,00, while the average GAS provided by Job Pricing is equal to € 27.000,00, producing a ΔGAS equal to € 4.000,00.

The estimates of ΔGAS are presented in Table 3. Teamwork is the most important requirement with an estimated ΔGAS equal to € 2.292,03. The second most appreciated skill is problem-solving and analysis, which was in the first position for the entire sector. With respect to the previous table, its ΔGAS is stable, at about € 2.000,00. The same results are found for quality orientation with a ΔGAS close to € 1.500,00. A similar scenario is present for last requirements such as education level and previous experience with a very residual component estimated at ΔGAS.

Table 2 Estimated ΔGAS for requirements for tourism sector—Italy, €, 2017

Requirements	Estimated Δ GAS	Std. Err	P-value
Problem-solving and analysis	€ 2.040,27	0.176	<0.001
Self-control	€ 1.533,98	0.139	<0.001
Quality orientation	€ 1.499,16	0.136	<0.001
Teamworking	€ 902,74	0.098	<0.001
Communication	€ 769,22	0.090	<0.001
Customer orientation	€ 722,95	0.088	<0.001
Participation and responsibility	€ 586,48	0.080	<0.001
English knowledge	€ 400,33	0.071	<0.001
Education level	€ 6,70	0.052	<0.001
Previous experience	€ 5,61	0.051	<0.001
Total	**€ 8.467,44**		

Source Elaboration on AdeccoGroup data

Table 3 Estimated ΔGAS for requirements for travel consultant figure—Italy, €, 2017

Requirements	Estimated Δ GAS	Std. Err	P-value
Teamworking	€ 2.292,03	0.708	<0.001
Problem-solving and analysis	€ 1.978,88	0.579	<0.001
Quality orientation	€ 1.616,29	0.449	<0.001
Self-control	€ 861,92	0.239	<0.001
Participation and responsibility	€ 408,91	0.146	<0.001
Communication	€ 239,06	0.117	<0.001
English knowledge	€ 195,29	0.109	<0.001
Customer orientation	€ 179,28	0.107	<0.001
Education level	€ 31,09	0.084	<0.001
Previous experience	€ 3,37	0.080	0.006
Total	**€ 7.806,12**		

Source Elaboration on AdeccoGroup data

The total amount of estimated ΔGAS is equal to € 7.806,12. This estimate is very close to that of the entire sector, but since the real value for ΔGAS is € 4.000,00, our amount seems to overrate this difference in monetary terms.

4.2 An Alternative Approach Based on Multinomial Regression

In a further step in our analysis, namely, a multinomial logistic regression Greene (2012) was carried out in order to obtain a more in-depth knowledge of the most selected (or preferred) profiles by employers, among those evaluated. In particular, the following four profiles of candidates were included: candidates without any specific characteristics (Profile 1); candidates with a degree or a high school diploma (Profile 2); candidates with previous experience (Profile 4); and candidates with both a degree or a high school diploma and a previous experience (Profile 5). For the purposes of our analysis, Profile 1 was taken as the base profile and served as the reference group against which all the other profiles were compared. Furthermore, a set of covariates was used to explain the choice of the profile. Some of these refer to the characteristics of the employers as follows:

- Two binary variables for identifying the enterprise's sector of economic activity: Fashion, Show and Events (not = 0; yes = 1); and Tourism, Tour Operator and Travel Agencies (not = 0; yes = 1)
- Two binary variables for identifying the macro-region of Italy where the enterprise is located: North (not = 0; yes = 1) and Centre (not = 0; yes = 1);
- Two binary variables for identifying the size of the enterprise: Small (not = 0; yes = 1) and Medium (not = 0; yes = 1); and
- A quantitative variable that quantifies the length of the employment contract (expressed in months): Duration of the employment contract.

Moreover, there are also two binary variables that refer to features of the candidates, as identified by employers during the selection phase as follows:

- A binary variable that distinguishes more motivated candidates: Motivation (not = 0; yes = 1); and
- A binary variable that identifies candidates that could have greater adaptability: Adaptability (not = 0; yes = 1).

Table 4 shows the maximum likelihood estimates of the multinomial logistic regression model that allows identifying the variables that affect the choice of candidates' profile by employers. By raising the regression coefficients to an exponent, the model's results can also be interpreted in terms of relative risk ratios. That is, the relative risk ratio tells us how much the probability of being in a certain category of the response variable relative to that of the referent group is expected to change for a unit change in the predictor variable, given that the other variables in the model are held constant. After estimation, the small p-value from the Likelihood Ratio (LR) test ($p < 0.001$) leads us to conclude that at least one of the regression coefficients in the model is not equal to zero. However, McFadden's Pseudo R-Squared is rather low (0.0611), and therefore the contribution of the model's covariates is limited.

As regard Profile 2 which refers to candidates with a degree or a high school diploma, the only variable that influences the choice is the size of the enterprise. In

Table 4 Maximum likelihood estimates of multinomial logistic regression model

	Estimate	**Std. Err**	**P-value**
Profile 1 (base outcome)			
Profile 2			
Intercept	–0.733	0.538	0.173
Fashion, show and events	–0.477	0.392	0.223
Tourism, tour operator and travel agencies	0.720	1.062	0.498
North	–0.688	0.486	0.156
Centre	0.184	0.499	0.713
Small	–0.287	0.648	0.658
Medium	0.762	0.339	0.024
Duration of the employment contract	–0.001	0.001	0.154
Motivation	0.038	0.615	0.950
Adaptability	0.947	0.589	0.108
Profile 4			
Intercept	–2.707	0.802	0.001
Fashion, show and events	–0.623	0.408	0.127
Tourism, tour operator and travel agencies	1.232	1.302	0.344
North	1.603	0.794	0.043
Centre	0.670	0.881	0.447
Small	–0.327	0.830	0.694
Medium	0.818	0.304	0.007
Duration of the employment contract	–0.001	0.004	0.751
Motivation	0.439	0.484	0.365
Adaptability	–0.807	0.983	0.412
Profile 5			
Intercept	–0.280	0.382	0.464
Fashion, show and events	–0.452	0.298	0.129
Tourism, tour operator and travel agencies	1.840	0.807	0.023
North	–0.797	0.366	0.029
Centre	–0.860	0.423	0.042
Small	–0.074	0.499	0.883
Medium	1.132	0.261	0.000
Duration of the employment contract	0.001	<0.001	0.002
Motivation	–0.987	0.581	0.089
Adaptability	0.960	0.479	0.045

Source Elaboration on AdeccoGroup data

particular, medium-sized enterprises are those more interested in selecting graduates; for these enterprises the relative risk of choosing a candidate with a degree or a high school diploma over candidates without any specific characteristics (Profile 1) is 2.14 relative to that of other enterprises.

On the other hand, for Profile 4, which refers to candidates with previous experience, there are two influential variables. The first one is still the size of the enterprise. As with Profile 2, medium-sized enterprises prefer this kind of candidate; the relative risk of 2.27 means that the probability of choosing candidates with a previous experience over candidates without any specific characteristics (Profile 1) is more than twice that of the enterprises of other sizes. The second one is the macro-region where the enterprise is located. For enterprises located in the north of Italy, the probability of selecting candidates with a previous experience over candidates without any specific characteristics (Profile 1) is 4.97 times higher than that of enterprises located in other macro-regions.

Finally, for Profile 5, which refers to candidates who have both a degree or a high school diploma and a previous experience, there are six variables affecting the choice of employers. Again, medium-sized enterprises prefer this kind of candidates, with a relative risk of 3.10. However, the macro-region also shows relevant differences; enterprises located in the north and centre of Italy have a lower probability of choosing these candidates, with a relative risk of 0.451 and 0.423, respectively. Moreover, enterprises operating in the sectors of tourism, tour operator and travel agencies are particularly interested in recruiting these candidates, with a probability of selection over candidates without any specific characteristics (Profile 1) that is 6.30 times higher than that of enterprises in other sectors of economic activity. As a final point, two further variables influence the choice. The first one is adaptability, with a relative risk of 2.96 which indicates that for this profile, employers also require a greater ability to adapt from chosen candidates. The second one is the length of the employment contract since for this profile selected candidates tend to have, on average, contracts with a longer duration.

5 Discussion and Final Remarks

Since job access represents a crucial area in the study of demand and supply in the job market, an a posteriori analysis on the requirements of candidates has been carried out using job offer data from Italy in 2017. The objective was to investigate the knowledges, skills and attitudes required for the various professional roles in the matching phase.

From a methodological point of view, this objective has been pursued using two approaches both of which use logistic regression. The first one is a Choice-Based Conjoint Analysis used in combination with an economic index of re-valuation. The second one uses multinomial logistic regression carried out to obtain a more in-depth knowledge of the most selected profiles by employers.

The first approach measured the monetary evaluation of the requested requirements; in particular, it evaluated how the difference between an average salary and the minimum one provided by the Assohandlers engineering contracted grossing costs could be distributed among these requirements. In the tourism sector, the attitude to solve problems has been identified as the requirement with the highest values of revaluation. Self-control and quality orientation were very important requirements, too. Other standard requirements, such as education level and previous experience, show an almost null revaluation. Moreover, when all the monetary revaluations are summed up in a total amount, this is close to the real difference obtained as the benchmark value. The application of the same model to a single professional role in the tourism sector (the travel consultant) leads to similar results, both in terms of the most valued requirements and the proximity of the total amount to the real value.

The results derived from the alternative approach were analysed with specific profiles of candidates, in conjunction with a set of covariates, such as the dimension of the enterprise, the sub-sector of tourism, the geographic area and the length of the contract. This approach focused more on the hard skills; that is, knowledge of the English language, education level and previous experience. Even if these were penalized in the choice-based analysis model by a low monetary evaluation, they were still able to identify different ways of thinking by some specific categories of enterprises. In particular, medium-sized companies prefer profiles with hard skills, while if a firm is located in the north of Italy, a significant effect is found for previous experience.

In conclusion, it is possible to state that one the one hand, that soft skills have a bigger influence in terms of the gross annual salary; however, on the other hand, hard skills are still considered crucial and represent a discriminating factor in the choice of a new resource for some categories of enterprises.

References

Croissant, Y. (2012). Estimation of multinomial logit models in R: The mlogit Packages. R package version 0.2-2. http://cran.rproject.org/web/packages/mlogit/vignettes/mlogit.pdf.

Farnham, D., & Stevens, A. (2000). Developing and implementing competence-based recruitment and selection in a social services department–A case study of West Sussex County Council. *International Journal of Public Sector Management*, *13*(4), 369–382.

Dagsvik, J. K. (1998). Random utility models for discrete choice behavior. An introduction. In *Statistics Norway research department*. Norway. https://www.ssb.no/a/histstat/doc/doc199815.pdf.

Greene, W. H. (2012). *Econometric analysis* (7th ed.). Upper Saddle River. NJ: Prentice Hall.

Humburg, M., & der Velden, R. (2015). Skills and the graduate recruitment process: Evidence from two discrete choice experiments. *Economics of Education Review*, *49*, 24–41.

ISTAT. (2017). Il mercato del lavoro. Verso una lettura integrata, Roma.

Isfol. (2017). Ministero del lavoro—Classificatore delle professioni. http://fabbisogni.isfol.it.

Krantz, D. H. (1964). Conjoint measurement: The Luce-Tukey axiomatization and some extensions. *Journal of Mathematical Psychology*, *2*, 248–277.

Louviere, J. J., Flynn, T. N., & Carson, R. T. (2010). Discrete choice experiments are not conjoint analysis. *Journal of Choice Modelling*, *3*, 57–72.
Luce, R. D. (1964). Tukey J W Simultaneous conjoint measurement: A new type of fundamental measurement. *Journal of Mathematical Psychology*, *1*, 1–27.
Luce, R. D., & Krantz, D. H. (1971). Conditional expected utility. *Econometrica*, *2*, 253–271.
McFadden, D. (1973). Conditional logit analysis of qualitative choice behavior. In P. Zarembka (Ed.), *Frontiers in econometrics* (pp. 105–142). New York: Wiley.
Mariani, P., Zavanella, B., Mussini, M., Crosato, L. (2018). Length of searching and job matching: public employment services and firm recruitment in Italy. In Le preferenze degli imprenditori lombardi nella scelta dei neolaureati Mariani, P., Marletta, A., Zenga, M. PKE, Milano.
OCSE. (2017). Strategia per le Competenze dell'OCSE—Sintesi del Rapporto Italia, Paris.
Rowe, C. (1995). Clarifying the use of competence and competency models in recruitment, assessment and staff development. *Industrial and Commercial Training*, *27*(11), 12–17.
Street, D. J., & Burgess, L. (2007). *The construction of optimal stated choice experiments: Theory and methods*. New York: Wiley.
Taylor, M. S., & Bergmann, T. J. (1987). Organizational recruitment activities and applicants reactions at different stages of the recruitment process. *Personnel Psychology*, *40*(2), 261–285.

Big Data and Economic Analysis: The Challenge of a Harmonized Database

Caterina Marini and Vittorio Nicolardi

Abstract The real challenge that in the nowadays society needs to be scientifically faced is to accurately handle the enormous flow of information that in an IT world can be tremendously powerful to analyse the social and economic changes. The huge flow of data that private organizations and public administrations are storing in their databases is a precious and important source of information to complete the official statistics yielded by the National Statistics Institutes but not exempt from obstacles and issues that need to be solved. The dimension of private/public databases has to be considered in the Data Science scenario and involves that set of problems related to the so-called Big Data. This chapter provides a first scientific successful attempt to merge administrative databases and official statistical data in the field of research referred to the real estate economy that still suffers the consequences of the dearth of a complete and harmonized data warehouse.

1 Introduction

The problem of dealing with enormous databases is undoubtedly the new challenge that the scientific world needs to face to delineate and analyse reality in continuous and fast development whose data availability is rapidly growing due also to the explosive increase of Internet usage over the last decades. The network voluntary generation of data, as Pyne et al. (2016) assert, persuades scientists to confront not only such obvious issues as data volume, velocity and variety but also data veracity, individual privacy and ethics. And in this sense, Reiter (2012) discussed in detail the statistical approaches to ensure everyone right to personal privacy and, consequentially, protect the confidentiality of data, which are widely used, without affecting

C. Marini (✉) · V. Nicolardi (✉)
Department of Economics and Finance, University of Bari Aldo Moro, Largo Abbazia S. Scolastica, 70124 Bari, Italy
e-mail: caterina.marini@uniba.it

V. Nicolardi
e-mail: vittorio.nicolardi@uniba.it

P. Mariani and M. Zenga (eds.), *Data Science and Social Research II*, Studies in Classification, Data Analysis, and Knowledge Organization,
https://doi.org/10.1007/978-3-030-51222-4_18

the statistical analyses. Therefore, the challenge is ambitious for statisticians and computer scientists, though it is quite reductive to limit the Big Data topic to the network exchange of information that individuals normally yield whenever they use the Internet to interact with people or buy/sell something or conduct some research activity. The concrete revolution inside the so-called Big Data is centred on the increased availability of new sources and types of data that were not previously available to scientists and not necessarily derive from online activities or social network use (Connelly et al. 2016). The problem of handling and treating a massive set of data is, in fact, not new in the scientific context where the development of new technologies and research laboratory equipment allow scientists to analyse very complex biological, clinical and physical phenomena. And without being poor in the exemplifications that follow, in the scientific debate it is well recognized how the biomedical big data availability has opened new opportunities to enhance the understanding of disease heterogeneity in humans (Hamada et al. 2017), and the benefits and challenges that the big data approaches of analysis experience in the cell and molecular biology as discussed by Dolinski and Troyanskaya (2015) or the CERN's expertise in big data, which will be shared with the biomedical community in the near future. Moreover, the problem of handling and treating massive sets of data is not also new to the private organizations and public administrations because, since the beginning of the IT transformation of business and management in both contexts, they are generating huge amounts of data very often sensitive and stored in enormous databases that are intricate to manage and analyse. In this sense, although huge administrative private/public databases may not apparently hold all the characteristics that commonly describe the Big Data (e.g. velocity, variety and volume), they are considered part of them. In fact, as Hadford (2014) asserts, the definition of Big Data is still nowadays vague and many other characteristics have to be added to complete their meaning and define their different typologies. Kitchin (2014a) highlights that some huge databases may hold even a single Big Data characteristic or a totally different set of characteristics but they can still be considered in the Big Data framework. Nevertheless, as suggested by Karr and Reiter (2014) and Wachter and Mittelstadt (2019), it is extremely important that the administrative data are treated for protecting the confidentiality before their storage because the security of everyone privacy is nowadays no longer guaranteed. Therefore, independently of the scenario in which the so-called Big Data is positioned, one of the main issues that regard the Big Data topic is the way by which data need to be handled and managed considering also that the conventional and traditional statistical and computer tools are significantly inappropriate. The problem mainly arises because Big Data are normally unstructured. In fact, they frequently derive from a variety of data sources that can indifferently include structured, semi-structured and unstructured data and very often be independent of each other (Pusala et al. 2016). The latter is, for instance, the case of information provided by the Public Administration (PA, hereafter): the entirety of the administrative data on the phenomenon analysed requires very often the merging of two or more databases that belong to two or more PA offices. Furthermore, the limitations of tools normally used in the socio-economic analysis of the various phenomena depend also on the primary purpose by which the public and private databases are fulfilled, which is very often not statistical unlike the conventional databases and data warehouses traditionally yielded by the National Statistics Institutes (NSIs, hereafter) to support any kind of analysis. Therefore, the questions of the way of dealing with massive sets of data, mainly when they are referred to administrative information, are still far to be resolved. The crucial issues related to the utilization of the administrative data to integrate the official statistical information are widely discussed in the literature (Nordbotten 2010; Kitchin 2015; Thomsen and Holmoy 1998) and some of them have been already described, but the NSIs and all the worldwide Statistical Research Offices are already dealing with the related obstacles to guarantee the integration of such information (Calzaroni 2008; Di Consiglio and Falorsi 2015).

In this paper, the analysis is focused on the Italian real estate phenomenon and how the administrative data are powerful in adding new information on the phenomenon in terms of both volume and value comparing with the limited evidence that normally arises from the official statistics yielded by NSIs. The importance of the analysis yielded in this work is unique and original in its attempt to describe an economic phenomenon that, in Italy and in many European countries, still suffer the consequences, not only on the real estate business but also on the public management, of the dearth of a complete and harmonized data warehouse. In fact, a unique administrative database perfectly aligned with the Italian NSI Census database has been created starting from huge independent databases managed by autonomous Italian PA offices. In this sense, as suggested from Kitchin (2014b) that emphasizes the need to produce a taxonomy of Big Data with detailed examples of particular data types, the very large databases involved in this work are considered in the Big Data framework. Therefore, the Big Data analytic practice and the GIS processes have been necessary to guarantee the exact matching of data and depict the real estate territorial framework in detail.

Furthermore, as well recognized in the Big Data literature, the availability of a huge amount of data, when appropriately summarized into a comprehensible format, can help private organizations in the decision-making process (Laha 2016; Japec et al. 2015; Olszak 2016) and be also appreciable for the PA. Therefore, as a counterpart of the creation of the full information harmonized real estate database an economic indicator, conveniently graphically referred, has been yielded to provide policymakers and business managers with an instrumental measure to affect the real estate market and the public fiscal policies.

2 The Data

In the scientific context, the real estate economy has been extensively analysed from innumerable points of view that involve almost all the actors in the economic panorama, households and private sector on one side, and central and local PAs on the other. Nevertheless, almost all works in the literature refer to databases that comprise their own full information that is, however, partial in providing an overall view of the phenomenon. The main scientific obstacle encountered when omni-comprehensive information on socio-economic phenomena is necessary to deeply analyse their development, trend and correlated relations is connected to that set of problems that involve the merging of official statistical databases as yielded by NSIs and the administrative or differently public databases. The problems normally encountered when it is necessary to work with administrative databases are several and, among them, their accessibility where the confidentiality of individuals is disclosed, and the typology of data and the corresponding quality that they contain. Without erring on the intent of avoiding discussing the problem of administrative database accessibility, it is intention of this work to depict the enormous potentialities that, in the study of the real estate economic phenomenon, a harmonized and omni-comprehensive administrative database experiences to support/partially alter the related economic dogma, still theoretically valid, when the accessibility of each composing database is resolved. Therefore, the attention is focused on the components of each database that have been used in this work, the typology and the quality of data. In fact, the latter are both important because the type of data is fundamental to plan the analytical approach, and the quality guarantees the reliability of the outcomes. Both issues clearly require a great attention when databases are fairly big to be considered in the Big Data/Data Science scenario. In fact, it is always important to remind that the opportunities relying on the major availability of information risk to be a weakness for the purposes of the studies and, in this sense, one of the most delicate phases in the study of huge databases regards their cleaning and management. Therefore, in this work, it has been decided to independently work on each database to pre-process data and select the key features of each database to finally proceed with the merging action. The analysis is restricted to the territory of the city of Bari, in South Italy, because that is part of a national research project, but the outcomes generated can be perfectly replicated in any dimensional geographical area.

More specifically, four independent administrative databases normally managed by two independent PA offices have been used to create the full information complete real estate administrative database.

Three of the four databases belong to the Real Estate Registry (RER, hereafter) and contain all information related to the real estate. Although the PA office is the same, the three databases are independent and autonomous in providing the corresponding information. Therefore, the Italian RER has complete information on real estate though utilises that in a roundabout way that complicates its same use. The main database is named Real Estate Units (REU, hereafter) and includes a list of records in which all the technical and economic real estate information of each unit is recorded. REU provides items of valuable information referred to the real estate category to identify the various typologies of units such as, for instance, dwelling or shop or office, the corresponding council value, i.e. the economic value to calculate the council tax, and the size of each unit. The size of the database is remarkable: 283,217 records, without duplication, referring to all units but 20,240 records lack council value because they belong to units of a particular real estate category without income (i.e. the F real estate category). The other two RER databases are functional to build the final database. The first is named Cadastral Identifiers (CI, hereafter) and comprises other real estate information, mainly the Urban Section and the corresponding Sheet, Subordinate and Parcel. The CI dataset includes 421,324 records, a number much higher than REU records because of both duplications, caused by some administrative change, and the presence of some real estate unit whose record has not been deleted though the building was demolished and is not really anymore existing. The last RER database is named Cadastral Addresses (CA, hereafter) and comprises the toponyms of each real estate unit. Toponyms are important to identify the exact localization of each unit on the urban territory. The size of the CA dataset is 668,302 records and, likewise the previous database, many duplications occur because of the modifications of some toponym and/or building number.

The fourth database used in this work belongs to the Real Estate Italian Observatory (REIO, hereafter) of the Italian Revenue Agency (IRA, hereafter). In Italy, this source of data is the main and one of the most reliable to analyse the real estate monetary value dynamics. The REIO real estate value data are calculated based on the trade price per square meter of the properties. They are open data on biannual basis referred to the minimum and maximum price for all the different types of real estates at the level of the council territory. In order to use a univocal REIO value in the analysis, the midrange value for each record has been calculated. Furthermore, the council territory is split in homogeneous areas that experience the same economic and socio-environmental characteristics, i.e. the REIO zones. All zones of the same city are then grouped in five territorial districts that delineate precise geographical portions of the urban space: Centre, Near-centre, Outskirts, Suburbs and Extra-Urban. In this work, the REIO dataset of the city of Bari for the years 2015 and 2018 has been used. Biannual data are referred to the 14 typologies of the real estates that exist in Bari, for a total of 7,814 records.

The four databases, therefore, experience different sizes because of technical reasons due to the administrative information they report, and the quality of the data they include is affected by material and human errors, such as duplication, missing values and erroneous information, that complicate the merging of the data.

3 The Method

The method that in this work has been used to build the full information harmonized database is a stepwise procedure.

The first phase has consisted of cleaning the RER databases because of the abovementioned errors, without statically affecting the analysis of the phenomenon. Therefore, a first cleaning action has been necessary to delete the uncomplete records in the REU database to guarantee a homogeneous set of information reminding that it includes more than 20 thousand records lack the council value. The cleaning action has been clearly also indispensable for homogenizing information in CI and CA databases to be aligned with REU contents. The cleaning of the CI and CA databases regarded duplications and erroneous data and has been yielded by means of, respectively, the Protocol Number field and the Sequential field that report the several

Table 1 Database size details

Database	Original size		Final size	
	Fields	Records	Fields	Records
Real estate units	29	283,217	9	262,977
Cadastral identifiers	13	421,324	6	262,977
Cadastral addresses	5	668,302	5	262,977
Real estate Italian observatory	24	7,814	6	7,814
Italian NSI census sections	4	82,576	4	82,576
Final harmonized database	–	–	26	262,977

modifications that involved the real estate units over time. Finally, all RER databases are equal in size. In the case of REIO database, the quality of data is perfectly suitable for the analysis because data are already statistical values. Therefore, the cleaning action has not been necessary. Table 1 shows the size of all the databases used in this work in terms of fields and records, original size and final size after cleaning when occurred.

Once the numeric and structural homogeneity of the RER database size has been obtained, the successive step is to merge them. The Cadastral Office Real Estate Identification Code (COREIC, hereafter) has been identified as the only plausible merging field because that is the sole in common between the three databases. Therefore, the Thorough Real Estate Registry (TRER, hereafter) database has been created by means of COREIC.

The purpose of this work is the creation of a full information harmonized database of the real estate phenomenon that includes all the economic and administrative information. Therefore, TRER and REIO databases need to be perfectly aligned. In fact, it is important to highlight that the two databases previously described are unlinked and not directly connectable through any field although they are referred to same object. To merge the two databases, it is crucial to solve two technical problems that involve the procedure.

The first and most important issue is related to the specific territorial context that is differently defined in each database. In particular, the territorial context is the single real estate Parcel in TRER, while in REIO the geo-context is the REIO zone. Therefore, to surmount the obstacle a GIS procedure has been yielded by means of three additional databases that in this work appear to be instrumental on the way to the final objective. The first is the Italian NSI Census Section database that contains, among all data, the official statistical geo-localization information that are used to yield all the official territorial statistics. The Italian NSI database has allowed to precisely locate all TRER data per each real estate unit within the census sections through the second instrumental database that belongs to RER and includes all the geo-localization real estate data. The last instrumental database belongs to IRA and includes the geo-localization data of the REIO zones. The GIS procedure has created two distinctive maps that have been overlapped to obtain a new database that has been named BRIDGEDB. BRIDGEDB connects the REIO zones with the Italian NSI census sections. In other words, the procedure allows to link the Italian NSI census sections, and indirectly the real estate units, with each REIO zone. Figure 1 depicts graphically BRIDGEDB in detail, namely the Italian NSI census sections, the city neighbourhoods and the REIO zones.

The second problem in the alignment of TRER and REIO is related to the real estate typologies because they are differently classified in the two databases. It has been, therefore, decided to build a Transformation Matrix to relate the two different classifications and surmount this second obstacle.

Finally, the utilization of the Transformation Matrix and BRIDGEDB has allowed to assign the REIO real estate midrange value to each real estate unit for each category and compute the market value through the real estate size of each real estate unit.

The final database includes, therefore, all the harmonized administrative and market data for each real estate unit.

4 An Economic Evidence

The extraordinary potentialities of the complete and harmonized real estate database are very large and can involve many aspects of the PA activities on one hand, and the household/private business on the other. In fact, the alignment of council and market values, which can refer to entire city areas or portions of neighbourhoods or even single buildings, is without precedent in the literature.

In this work, it has been decided to calculate, for instance, the distance that exists between the two values to analyse the economic evidence that derives from the utilization of omni-comprehensive information. Therefore, monetary differentials between the real estate market monetary values and the real estate council values have been calculated and graphically depicted. The only two outcomes displayed in the text have been selected because they represent two opposite socio-economic realities that characterized two areas of the city of Bari, still nowadays.[1] Figures 2 and 3 depict the percentage average differentials of respectively the Economic Dwellings in Murat neighbourhood and the Villas and Detached Houses in Carbonara neighbourhood, and compare 2015 data with 2018 data for both. The maps in Fig. 2 highlight that in the Murat neighbourhood the real estate council value is always much lower than the market value, up to 80% in some cases. Furthermore, time comparison shows a significant increase in the differentials between 2015 and 2018 underlining the effects of requalification actions that involved the Murat historical area of the city. In the neighbourhood of Carbonara, contrariwise, the opposite is the case. In fact, the

[1]All the outcomes are not represented because of space limits and are available from the authors.

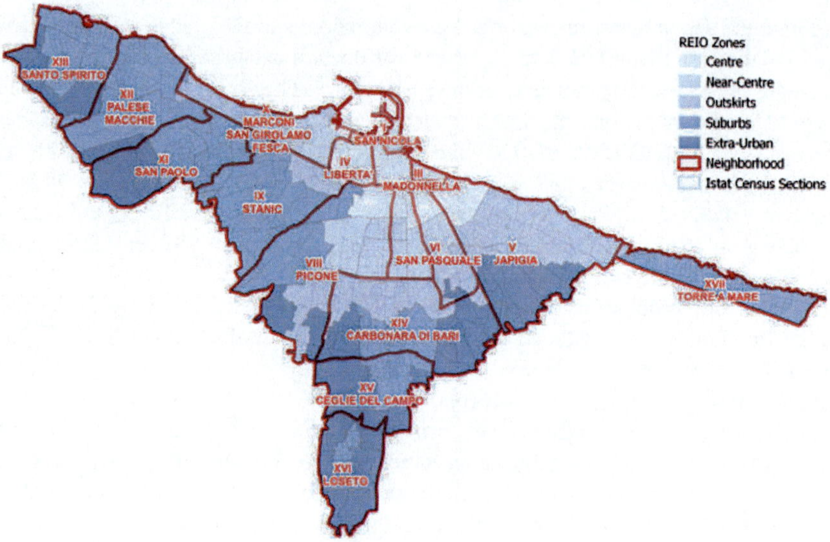

Fig. 1 BRIDGEDB: Neighbourhoods, REIO Zones and Italian NSI Census Sections of the City of Bari

maps in Fig. 3 show positive values that underline how the real estate market values are lower than the council value in the great part of the area. Time comparison of 2015 and 2018 data highlights that the distance between the two values increased pointing out that it is less appealing to live in Villa or Detached Houses in this part of the city because they are located in Outskirt and Suburb areas that are suffering a gradual socio-economic decline.

The reported outcomes are suitable for several interpretations depending on the point of view that it is used. In other words, it is possible to evaluate the PA advantage/disadvantage to fiscally profit from more fairly redefining the council tax rates on the base of a more exact economic evaluation of the socio-economic characteristics of the city areas. At the same time, it is possible to consider aligning the quality of public services based on the tax burden that households and private companies sustain. Furthermore, for households and private enterprises the analysis can be an opportunity to overall evaluate the economic profitability of an investment in the real estate market and its incomparable return in terms of business based on the location.

5 Final Remarks

The availability of a complete harmonized database that includes all information on the phenomenon analysed is a precious analytical instrument very often unavailable because of issues related to important obstacles that involve both statistical/computer

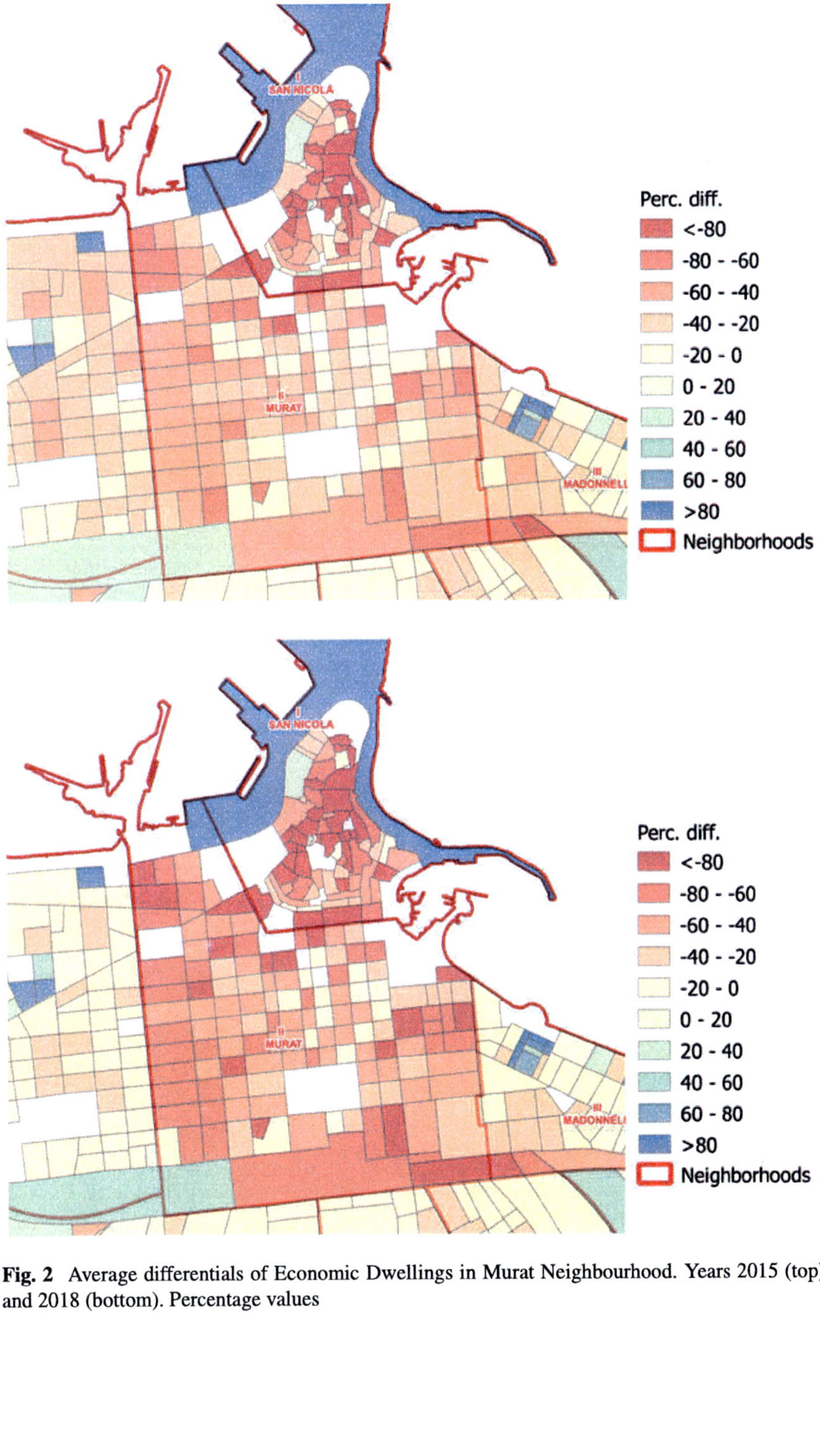

Fig. 2 Average differentials of Economic Dwellings in Murat Neighbourhood. Years 2015 (top) and 2018 (bottom). Percentage values

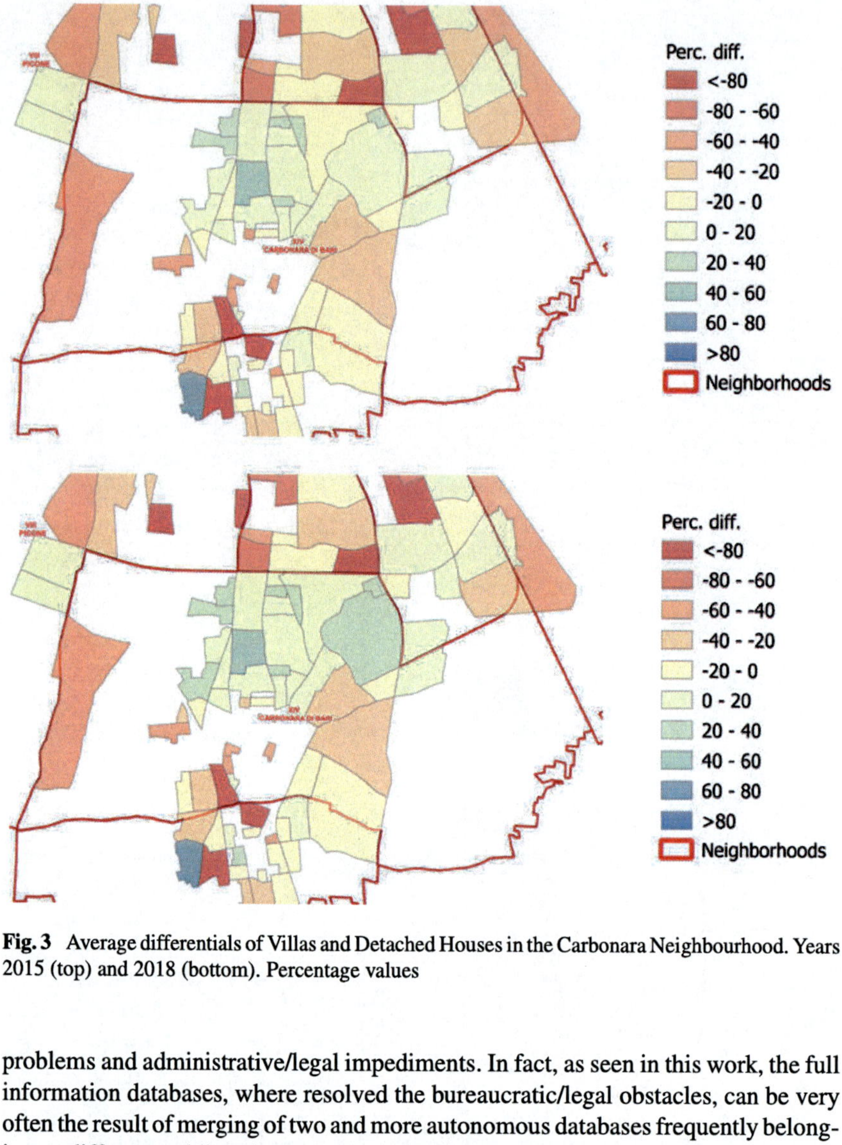

Fig. 3 Average differentials of Villas and Detached Houses in the Carbonara Neighbourhood. Years 2015 (top) and 2018 (bottom). Percentage values

problems and administrative/legal impediments. In fact, as seen in this work, the full information databases, where resolved the bureaucratic/legal obstacles, can be very often the result of merging of two and more autonomous databases frequently belonging to different public/private entities and, therefore, huge in their dimension. This evidence means that the Big Data/Data Science analytic practices are necessary to synthesize all information within a procedure that can guarantee the integrity of data representing the phenomenon. And the state-of-the-art statistical practice partially relies upon graphical approaches to ensure the exact matching of data and depict the phenomenon in detail. The importance of the analysis yielded in this work is unique and original in its attempt to describe the phenomenon of the real estate economy that still suffers the consequences of the dearth of complete information because of the practical non-existence of a full information harmonized data warehouse. In fact, in literature there is not any attempt in this sense and this work provides the first important contribution to face and solve the issue. The complete and harmonized real estate database yielded through the basic Big Data analytic practice and the GIS processes has been used to depict the value discrepancy between the real estate council value and the corresponding market value that is de facto considered as a known issue but never numerically quantified. The analysis, furthermore, shows valuable results that would support policymakers and households/business managers to preciously assess the return on investment depending on all-encompassing information. Finally, the full information harmonized database is perfectly suitable for being integrated with any geo-localized information and any statistical and administrative datasets provided by the NSIs and the PA and private sector. Therefore, the potentialities of its application in the most variegated socio-economic analyses are noteworthy, and forecasting models involving the detailed territorial information provided can be implemented to evaluate the real estate phenomenon over time.

Acknowledgements The authors are grateful to the PA offices of the city of Bari for providing RER databases.

References

Calzaroni, M. (2008). Le fonti amministrative nei processi e nei prodotti della statistica ufficiale, in Atti della Nona Conferenza Nazionale di Statistica.

Connelly, R., Playfordv, C. J., Gayle, V., & Dibben, C. (2016). The role of administrative data in the big data revolution in social science research. *Social Science Research, 59*, 1–12.

Di Consiglio, L., Falorsi, D.P. (2015) Different contexts for the statistical use of administrative data. In: Proceedings of Statistics Canada Symposium 2014 on Beyond traditional survey taking: adapting to a changing world.

Dolinski, K., & Troyanskaya, O. G. (2015). Implications of Big Data for cell biology. *Molecular Biology of the Cell (MBoC), 26*(14), 2575–2578.

Hadford, T. (2014). Big data: a big mistake? *Significance, 11*, 14–19.

Hamada, T., Keum, N., Nishihara, R., & Ogino, S. (2017). Molecular pathological epidemiology: new developing frontiers of big data science to study etiologies and pathogenesis. *J Gastroenterol, 52*(3), 265–275.

Japec, L., Kreuter, F., Berg, M., Biemer, P., Decker, P., Lampe, C., et al. (2015). Big Data in Survey Research. *Public Opinion Quarterly, 79*(4), 839–880.

Karr, A. F., & Reiter, J. P. (2014). Using Statistics to Protect Privacy. In Julia Lane, Victoria Stodden, Stefan Bender, & Helen Nissenbaum (Eds.), *Privacy, Big Data, and the Public Good: Frameworks for Engagement* (pp. 276–95). New York: Cambridge University Press.

Kitchin, R. (2014a). Data, new epistemologies and paradigm shift. *Big Data & Society., 1*, 20539517145228481.

Kitchin, R. (2014b). *The Data Revolution: Big Data, Open Data.* Data Infrastructures and Their Consequences: Sage Publications, London.

Kitchin, R. (2015). The opportunities, challenges and risks of bigdata for official statistics. *Statistical Journal of the IAOS, 31*(3), 471–481.

Laha, A. (2016). Statistical Challenges with Big Data in Management Science. In S. Pyne, B. Rao, & S. Rao (Eds.), *Big Data Analytics* (pp. 41–55). New Delhi: Springer.

Nordbotten, S. (2010). The Use of Administrative Data in Official Statistics - Past, Present, and Future - With Special Reference to the Nordic Countries. *Journal of official statistics*, 205–223.
Olszak, C. M. (2016). Toward better understanding and use of business intelligence in organizations. *Information Systems Management*, *33*(2), 105–123.
Pusala, M. K., Amini, Salehi M., Katukuri, J. R., Xie, Y., & Raghavan, V. (2016). Massive Data Analysis: Tasks, Tools, Applications, and Challenges. In S. Pyne, B. Rao, & S. Rao (Eds.), *Big Data Analytics*. New Delhi: Springer.
Pyne, S., Prakasa Rao, B. L. S., & Rao, S. B. (2016). Big Data Analytics: Views from Statistical and Computational Perspectives. In S. Pyne, B. Rao, & S. Rao (Eds.), *Big Data Analytics* (pp. 1–10). New Delhi: Springer.
Reiter, J. P. (2012). Statistical approaches to protecting confidentiality for microdata and their effects on the quality of statistical inferences. *Public Opin Q*, *76*(1), 163–181.
Thomsen, I., & Holmoy, A. M. K. (1998). Combining Data from Surveys and Administrative Record Systems. *The Norwegian Experience. International Statistical Review*, *66*(2), 201–221.
Wachter, S., & Mittelstadt, B. (2019). A Right to Reasonable Inferences: Re-Thinking Data Protection Law in the Age of Big Data and AI. *Columbia Business Law Review*, *2019*(2),

ROC Curve in GAMLSS as Prediction Tool for Big Data

Andrea Marletta

Abstract During the latest years, Big Data appears as one of the most innovative and growing scientific areas of interest. In this field, finding reliable methods to make accurate predictions represents one of the most inspirational challenges. In the following paper, the use of ROC (Receiver Operating Characteristic) Curve, a binary tool, often used for medical tests, has been used to make predictions. In particular, the attention is focused on the implementation of the ROC Curve in GAMLSS (Generalized Additive Models for Location Scale and Shape), semi-parametric models suitable for huge and flexible datasets.

1 Introduction

Big Data analysis represents a new fascinating challenge for the researchers interested in information mining from data matrices with a huge number of observations. The term 'Big Data' seems only concern a large quantity of data, but actually, it refers to a new way to focus the attention on the quality of these data. For this reason, it is not sufficient to have millions of records to classify a dataset as 'Big Data.' Data have to own some features as Volume, Velocity, Variety, Value, Veridicity, and Validity Liberati and Mariani (2016). These characteristics show that data have both to contain a lot of observations. Secondly, they have to be subject to changes and updates without losing the properties of truth and efficiency. Social media seems to be the typical area where it happens; this is the reason why the term Big Data is often associated with the words 'Internet of Things.' Moreover, it is a field in which to measure the possibility of interaction among profiles is one of the most interesting objectives to focus.

Social networks are identified as an online informative system allowing the realization of virtual social interactions. They are websites or technologies permitting to share textual contents, images, videos, and interactions among users Finger (2013).

A. Marletta (✉)
University of Milano-Bicocca, Milan, Italy
e-mail: andrea.marletta@unimib.it

P. Mariani and M. Zenga (eds.), *Data Science and Social Research II*,
Studies in Classification, Data Analysis, and Knowledge Organization,
https://doi.org/10.1007/978-3-030-51222-4_19

Social media data are data collected from social networks. Among social media, Twitter is one of the most spread and well known. Different from Facebook or Instagram, Twitter has been used to share news, official contents about economics and political issues. For this purpose, Twitter data have been provided and statistical units are represented by Twitter records. In particular, here each record represents a tweet made by a Twitter user. A tweet is a written post on Twitter with a maximum of 280 characters.

The paper is organized as follows: after a brief introduction on Big Data and Twitter, in Sect. 2, Generalized Additive Models for Location Scale and Shape will be presented. In Sect. 3, the proposal to implement the ROC Curve in GAMLSS will be introduced in order to make predictions for Twitter data; in Sect. ROC Curve in GAMLSS, data and the application of the proposed approach will be shown, while Sect. 5 will be devoted to the discussion and future works. All the analyses in this paper are implemented using the *R* statistical environment.

2 GAMLSS

Generalized Additive Models for Location Scale and Shape (GAMLSS) were introduced by Rigby and Stasinopoulos Rigby and Stasinopoulos (2005). GAMLSS are defined as semi-parametric models. Actually, besides requiring a definition of a parametric distribution for the response variable, it is possible to add non-parametric smoothing functions for each parameter considered in the model specification. The authors presented GAMLSS as a way to overcome some limitations of GLM (Generalized Linear Models) and GAM (Generalized Additive Models). GLM were introduced by Nelder and Wedderburn (1972) and represent a generalization of the linear regression model in which it is possible to use as response variable probability distribution different from the normal. A further generalization is represented by GAM introduced by Hastie and Tibshirani (1990) as an extension of GLM where a non-parametric smoothing component is considered. In comparison with GLM and GAM, the basic features of GAMLSS are two. Firstly, the exponential family distribution assumption for the response variable is replaced by a more general distribution family. Moreover, GAMLSS allow expanding the modeling to scale and shape parameters as skewness and kurtosis too. For these reasons, they are particularly flexible and suitable to model data in which the response variable shows some of these features.

GAMLSS assume independent observations y_i for $i = 1, 2, \ldots, n$ with probability density function $f(y_i|\theta^i)$ conditional on $\theta^i = (\theta_{1i}, \theta_{2i}, \theta_{3i}, \theta_{4i}) = (\mu_i, \sigma_i, \nu_i, \tau_i)$ a vector of four distribution parameters. Each parameter can be a function of the explanatory variables. The first two parameters μ_i and σ_i represent location and scale parameters, while the remaining ν_i and τ_i refer to the shape parameters (skewness and kurtosis).

The original formulation of GAMLSS is given by

$$g_k(\theta_k) = \eta_k = X_k\beta_k + \sum_{j=1}^{J_k} Z_{jk}\gamma_{jk}$$

where for $k = 1, 2, 3, 4$, $g_k(.)$ are monotonic link functions relating the distribution parameters to explanatory variables, X_k is a known design matrix of order $n \times J'_k$, $\beta'_k = (\beta_1, \dots, \beta_{J'_k})$ is a parametric vector of length J'_k and $Z_{jk}\gamma_{jk}$ the non-parametric additive terms.

Expanded formulation of GAMLSS is:

$$\begin{cases} g_1(\mu) = \eta_1 = X_1\beta_1 + \sum_{j=1}^{J_1} Z_{j1}\gamma_{j1} \\ g_2(\sigma) = \eta_2 = X_2\beta_2 + \sum_{j=1}^{J_2} Z_{j2}\gamma_{j2} \\ g_3(\nu) = \eta_3 = X_3\beta_3 + \sum_{j=1}^{J_3} Z_{j3}\gamma_{j3} \\ g_4(\tau) = \eta_4 = X_4\beta_4 + \sum_{j=1}^{J_4} Z_{j4}\gamma_{j4} \end{cases}$$

In this way, each distribution parameter can be modeled as a linear function of explanatory variables and/or as linear functions of random variables.

Other alternative formulations of GAMLSS could be considered.

The population probability (density) function $f(y|\theta)$ is left general with no explicit conditional distribution form for y. The only restriction that the R implementation of GAMLSS has for specifying the distribution of y is that function $f(y|\theta)$ and its first derivatives with respect to each of the parameters of θ must be computable. We shall use the notation:

$$y \sim D\{g_1(\theta_1) = t_1, g_2(\theta_2) = t_2, \dots, g_p(\theta_p) = t_p\}$$

to identify uniquely a GAMLSS, where D is the response variable distribution, $(g_1, \dots, g_p)$ the link functions, $(t_1, \dots, t_p)$ the model formulae for the explanatory terms in the predictors $(\eta_1, \dots, \eta_p)$.

There are two basic algorithms used for maximizing the penalized likelihood in GAMLSS. The first, the CG algorithm, is a generalization of the Cole and Green algorithm Cole and Green (1992) and it uses the first derivatives and the expected values of the second and cross-derivatives of the likelihood function with respect to $\theta = (\mu, \sigma, \nu, \tau)$ for a four-parameter distribution. However, for many probability distribution functions $f(y|\theta)$ the parameters θ are orthogonal. In this case, the second, the RS (Rigby–Stasinopoulos) algorithm is more suited. The RS is a generalization of the algorithm for fitting MADAM (Mean and Dispersion Additive Models). Essentially the RS algorithm has an outer cycle that maximizes the penalized likelihood with respect to the fixed and random effects in the model for each θ_k. At each iteration, the current updated values of all the quantities are used. This algorithm is not a special case of the CG algorithm because in the RS the diagonal weight matrix W_{kk} is computed within the fitting of each parameter θ_k, whereas in the CG all weight matrices W_{ks} are evaluated after fitting all θ_k.

The aim of both algorithms is maximizing a penalized likelihood function l_p given by

$$l_p = l - \frac{1}{2}\sum_{k=1}^{p}\sum_{j=1}^{J_k}\lambda_k\gamma'_{jk}G_{jk}\gamma_{jk}$$

where $l = \sum_{i=1}^{n} log f(y_i|\theta^i)$.

This is achieved in two steps: firstly, the first and second derivatives of the aforementioned equation are obtained to give a Newton–Raphson step for maximizing it with respect to β_k and γ_{jk}; moreover, each step of the Newton–Raphson algorithm is implemented by using a backfitting procedure cycling through the parameters and through the additive terms of the k linear predictors.

Each GAMLSS parametric model can be assessed by using its fitted global deviance GD given by $GD = -2l(\hat{\theta})$, where $l(\hat{\theta}) = \sum_{i=1}^{n} l(\hat{\theta}^i)$. Two nested models M_0 and M_1 may be compared by using the test statistic $\Lambda = GD_0 - GD_1$ which has an asymptotic χ^2-distribution under M_0 with degrees of freedom $d = df_{M_0} - df_{M_1}$. For comparing non-nested GAMLSS the GAIC (Generalized Akaike Information Criterion) Akaike (1973) can be used. GAIC is obtained by adding a fixed penalty term for each effective degree of freedom used in the model. The model with the smallest value of GAIC will be selected.

For each model M, the normalized randomized quantile residuals of Dunn and Smyth (1996) are used to check its global adequacy of M and the distribution component D. These residuals are given by $\hat{r}_i = \Phi^{-1}(u_i)$, where Φ^{-1} is the inverse CDF of a standard normal variate with $u_i = F(y_i|\hat{\theta^i})$ if y_i is an observation from a continuous response, whereas u_i is a random value from the uniform distribution on the interval $[F(y_i - 1|\hat{\theta}^i), F(y_i|\hat{\theta}^i)]$ if y_i is an observation from a discrete integer response variable, where $F(y|\theta)$ is the CDF. The true residuals r_i have a standard normal distribution if the model is correct.

3 The ROC Curve

Receiver Operating Characteristic (ROC) curve is one of the most used tools to measure the accuracy of a binary medical test. Let D be the dummy variable to indicate the presence of disease and Y the result of the diagnostic test ($Y = 1$ positive test for disease, and $Y = 0$ negative test for disease). A binary medical test is informative if it is able to predict the disease better than randomly. For this reason, in the presence of a dichotomous outcome and a binary prediction, four different situations can appear:

- True Positive (TP) when you have a disease and your prediction test is positive;
- True Negative (TN) when you have not a disease and your prediction test is negative;

Table 1 Definition of 2-by-2 table for ROC curve

	$D = 0$	$D = 1$
$Y = 0$	TN	FN
$Y = 1$	FP	TP
Total	TN + FP	FN + TP

- False Positive (FP) when you have not a disease and your prediction test is positive;
- False Negative (FN) when you have a disease and your prediction test is negative.

Arranging the outcomes in a 2-by-2 table, if D is used for the disease and Y for the test result, we will have the following Table 1:

The accuracy of a test could be computed as the sum of the main diagonal $(TP + TN)$ over the n total number of subjects. Two important factors that characterize a binary test are sensitivity and specificity. Sensitivity measures the proportion of subjects that are correctly predicted when a disease is present, so it is defined as $TP/(TP + FN)$. On the other hand, specificity measures the proportion of subjects that are correctly predicted when the outcome is negative, defined as $TN / (TN + FP)$. Sensitivity is also called True Positive Rate (TPR) or True Positive Fraction (TPF), while specificity is also named True Negative Rate (TNR) or True Negative Fraction (TNF). Most of the time, TNF is expressed as the difference between 1 and the False Positive Fraction $(1 - FPF)$. An ideal test supposes all patients correctly predicted with $TPF = 1$ and $TNF = 1$ and all observation in 2-by-2 table will be on the main diagonal.

For a binary test, ROC curve is a graphical plot of sensitivity versus (1 − specificity), i.e., (TPF) versus (FPF), where each point of the curve represents a different value for the cutoff to classify a subject as diseased or non-diseased. Since specificity and sensitivity are ranged between 0 and 1, this curve is always included in a square of dimensions (0,1) x (0,1). The point (0,0) represents $TPF = 0$ and $FPF = 0$ which predicts all subjects to be negative, while the point (1,1) represents $TPF = 1$ and $FPF = 1$ which predicts all subjects to be positive. When all subjects are correctly classified for all cutoff points then the ROC curve is just a broken line following the points (0,0), (0,1), and (1,1), where the first value is on the horizontal axis and the second value is on the vertical axis. On the contrary, a completely random test would give a diagonal line from the left bottom to the top right corner. So every test, whose curve is above the diagonal line, is an informative test. Consequently, the closer to the upper left corner is the curve, the better is the test (see Fig. 1).

Another way to check if a medical test is informative is to compute the Area Under a ROC Curve (AUC). This index is the most commonly used method for summarizing a diagnostic test's overall accuracy. It ranges from 0 to 1 (perfect classification) and takes value 0.5 for a random test. Hence, the higher above 0.5 the AUC is, the more informative is the test.

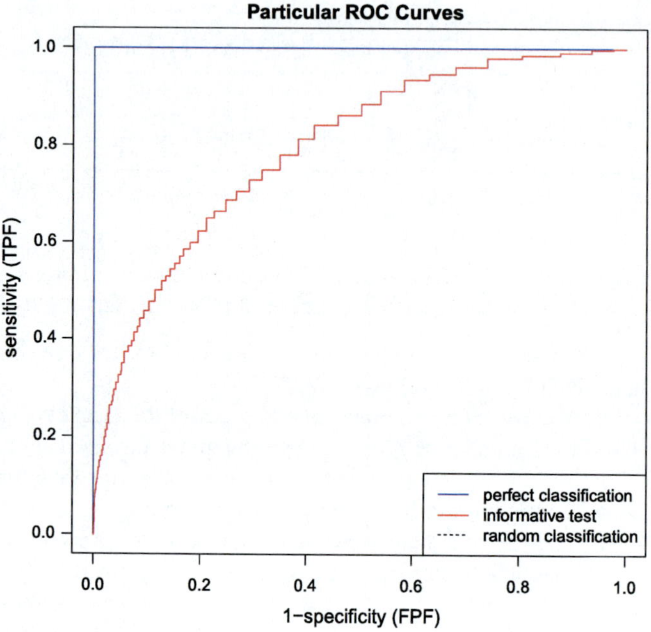

Fig. 1 ROC curve examples

ROC curves are also present in a binary regression framework Pepe (2003); Alonzo and Pepe (2002); in fact, it is possible to draw a ROC curve starting from a 2-by-2 table generated from the fitted model $\hat{p}$ and the true binary classification D. A cutoff or threshold value $0 < t < 1$ is chosen and set $Y = 1$ if $\hat{p} \geq t$, while $Y = 0$ if $\hat{p} < t$. The 2-by-2 table is a frequency cross-tabulation of $Y = 0, 1$ against $D = 0, 1$. All the points of the curve are obtained as FPF and TPF corresponding to different values of t. For each t, a 2-by-2 table is generated with resulting values for sensitivity and specificity of prediction. All these values are plotted in a square of dimensions (0,1) x (0,1) creating a binary regression ROC curve.

4 ROC Curve in GAMLSS

In the previous section, the ROC Curve has been shown as a tool used for a binary test in prediction. Using our dataset, the first issue is that our response variable is not binary but discrete, and hence, the distribution to fit the data is discrete too. Actually, we could dichotomize the variable of interest leading to a great loss of information. This categorization will lead to a ROC curve for our dataset, but following this approach, it is necessary to fit statistical models just using a binary distribution for the response, coming back to a logistic regression model.

For this reason, a new approach is proposed, in which it is possible to use the ROC curve starting from a model with continuous response variables.

ROC curves are suitable to binary data because in logistic regression FPF and TPF are computed starting by fitted values of $\hat{p} = P(Y = 1)$ in a range (0,1). The difference between logistic regression and GAMLSS is that, fitted values for data is not ranged in (0,1) but in $(0, \infty)$, so it is necessary to calculate $\hat{p} = P(Y > k)$, where k is the threshold chosen for dichotomization.

It is made possible by considering the density function of the chosen distribution for GAMLSS. In the proposed approach, for values in the estimates (0,1), $\hat{p} = P(Y > k) = 1 - P(Y \leq k) = 1 - F(k|\mu = \hat{\mu}, \sigma = \hat{\sigma}, \nu = \hat{\nu}, \tau = \hat{\tau})$ are obtained using the difference between 1 and the density function of BCPEo distribution at an established cutoff (2 m/s), where parameters are the fitted values computed for GAMLSS model. Using this approach, a direct correspondence between each observation y and a probability $\hat{p}$ that lies in (0,1) has been obtained. Then we can use these n probabilities to derive the ROC curve. Using a ROC curve in GAMLSS has a double aim: firstly, to justify the use of this approach compared with the standard logistic regression and secondly, to compare distributions by using the same method with other distributions for the response variable.

In order to obtain the ROC curve as a prediction tool using these data, it is necessary to split up the dataset in two subsets: the training and the validation set.

A selected GAMLSS model represents the starting point for estimating the ROC curve. This model was fitted on the complete dataset with different weights for training ($w = 1$) and validation ($w = 0$) individuals. Predicted values $\hat{\mu}, \hat{\sigma}, \hat{\nu}, \hat{\tau}$ were extracted for this weighted model. Predictor values for each parameter are included in $1 - F(cutoff)$ where F is the density function for BCPEo and the selected cutoff is k.

$$\hat{Y} \rightarrow \hat{p} \qquad\qquad \hat{p} = 1 - F(k|\mu = \hat{\mu}, \sigma = \hat{\sigma}, \nu = \hat{\nu}, \tau = \hat{\tau})$$

Step procedure to implement ROC curve predictions in GAMLSS are

- Split observations in the training and validation set
- Fit a GAMLSS weighted model
- Extract predicted values $\hat{\mu}, \hat{\sigma}, \hat{\nu}, \hat{\tau}$ for each y and evaluate $\hat{y}$
- Transform $\hat{y}$ to $\hat{p}$ only for observations in the validation set
- Compute specificity and sensitivity and draw ROC curve using the true indicator of cirrhosis D and the fitted probabilities $\hat{p}$ for each observation y in the validation set.

A vector of probabilities $\hat{p}$ has been obtained and now it is possible to use the same procedure used in binary logistic regression to compute accuracy, sensitivity, and specificity of the prediction for the validation set.

Training and validation sets are used for prediction, the results could be affected by the sampling; for this reason, the sampling procedure has been repeated 50 times to make more accurate predictions and to obtain more robust results.

As seen in the previous section, the use of this approach needs to be validated comparing it with other statistical models. For the comparison, we will select other GAMLSS models in which the response variable distribution is discrete.

Two possible ways of comparing different statistical models are possible using ROC curves. The first one is a graphical comparison, where different ROC curves are drawn in order to identify the higher curve. The higher the curve, the better the prediction. Secondly, the AUC index can be computed for all models; the model with a higher AUC index will be better.

5 An Application on Twitter Data

The application here showed concerns the tweets of the official account of the Monza circuit from 1 January 2016 to 31 December 2016. Each tweet represents a statistical unit and the total number of observations is equal to 737. The response variable is the number of received likes. The statistical model evaluates explanatory variables described by special characters in the text corpus. Specifically, these variables have been considered: the hashtag type (#) showing the presence of a topic or a keyword of the tweet, the tag type (@) allowing the user to mention another user in the text and the string <<http>> identifying the existence of a link containing generally pictures, videos or web pages. The basic idea is to verify the existence of a relationship between these variables using GAMLSS taking into account not only the mean but also the other moments of the response variable. In particular, since these models give back a not easy interpretable output, the attention will be focused on the predictive capability of these models compared to other statistical models.

As mentioned in the previous section, ROC curves will be used as a tool to measure the prediction and the relative classification error. The procedure allowing the implementation of ROC curves in GAMLSS needs of a preset threshold α, able to dichotomize the response variable in two exhaustive partitions. This approach could be followed both in the presence of continuous quantitative variables and as in this application, of discrete variables. For the sake of simplicity, the α threshold has been fixed to 1, this is equal to divide tweets with at least one like and those without a like, the result of this partition has been reported in Table 2.

For the selected model it is assumed the number of likes follows a Sichel distribution Sichel (1973). This distribution has been chosen among those implemented in GAMLSS for discrete variables using the GAIC (Generalized Akaike Information Criterion) Akaike (1973). Since this is a three-parameter distribution, the final result

Table 2 Dichotomization of Twitter values

Number of likes	$D = 0$	$D \geq 1$	Totale
Frequency	391	346	737

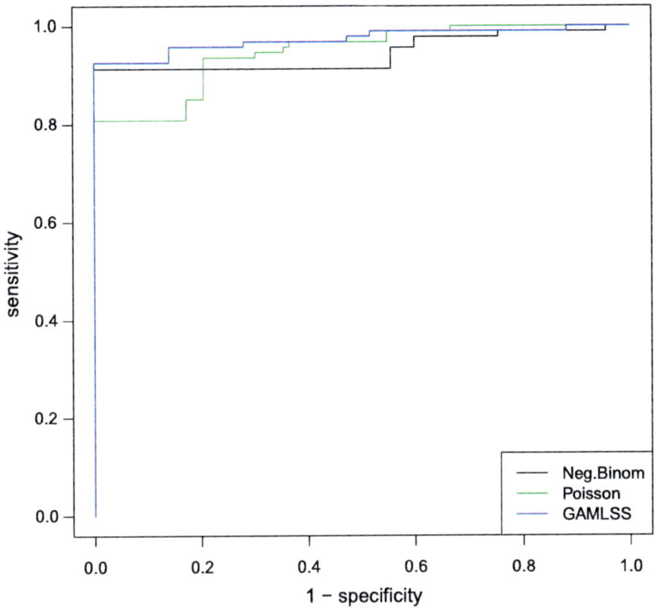

Fig. 2 ROC curves for different statistical models

will be a three-equation model. Once chosen the probability distribution, the model has been selected through a (backward and forward) stepwise procedure Chambers and Hastie (1991). Since the number of like is a counting variable, the comparison based on predictions can be realized only using other regression models with this feature like Poisson or Negative Binomial. The double comparison is represented in Fig. 2. Notably, the blue curve, the one corresponding to GAMLSS, is the highest curve and closest to the high-left angle with the best prediction levels. Only for low values of specificity, there are some intersections among curves, but all models show very high values for prediction since their curve is definitely over the bisector representing a random classification test. Moreover, since this technique provides for a sampling procedure for splitting data into training and validation set, the graph could depend on the sample.

For this reason, the second way of comparing different models tends to be more robust, in fact since it is based on the average $\overline{AUC}$, it takes into consideration different samples. The average $\overline{AUC}$ for GAMLSS is equal to 0.967 and it is slightly over the $\overline{AUC}$ value of the other two models (Table 3).

Figure 3 shows other two ways of representing the superiority of the proposed approach for this application, on the left panel, AUC value has been represented for each sample, with GAMLSS values (in blue) higher than the compared models in the majority of the cases. On the other hand, on the right panel, AUC values have been

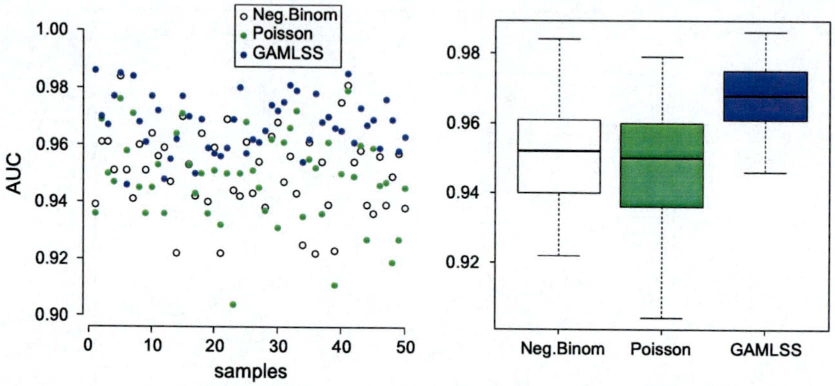

Fig. 3 AUC indices for 50 test and validation samplings

displayed using the box-plot; the blue box is higher than others and the minimum of the AUC values for GAMLSS is very close to the average of the distribution of the other two models.

6 Discussion and Final Remarks

This paper proposes the implementation of the ROC Curve in GAMLSS as a prediction tool in a Big Data context. Moreover, an application of the proposed approach has been presented on data extracted from Twitter. The counting of likes of a Twitter user could be modeled using GAMLSS assuming the Sichel distribution as probability distribution. The choice of these models has been motivated because of their flexibility as a possible alternative choice in the presence of Big Data. Notably, the implementation of the ROC Curve in GAMLSS proves to be a good prediction tool better than some existing models in literature. In the proposed approach, the predicted values p, between 0 and 1, useful for drawing the ROC curve have been obtained as complementary to one of the density function for Sichel distribution, for an established threshold α, fixed to 1 for simplicity.

Splitting up the dataset in training and validation set, 50 samples have been generated and 50 ROC curves with relative AUC indexes have been computed. In order to compare the selected GAMLSS with other statistical models, the AUC index has

Table 3 AUC indexes for compared statistical models

Model	Neg. Binom	Poisson	GAMLSS
$\overline{AUC}$	0.951	0.949	**0.967**

been obtained for Poisson and Negative-Binomial regression. The result of the possible choice has been measured in terms of AUC and these values for GAMLSS are higher compared to other models. Future works could concern the use of other values for the fixed threshold α or the use of GAMLSS with non-linear factors.

References

Akaike, H. (1973). Maximum likelihood identification of Gaussian autoregressive moving average models. *Biometrika, 60*(2), 255–265.
Alonzo, T. A., & Pepe, M. S. (2002). Distribution-free ROC analysis using binary regression techniques. *Biostatistics, 3*(3), 421–432.
Chambers, J. M., & Hastie, T. J. (1991). *1991*. Chapman and Hall, New York: Statistical Models. CRC.
Cole, T. J., & Green, P. J. (1992). Smoothing reference centile curves: the LMS method and penalized likelihood. *Statistics in medicine, 11*(10), 1305–1319.
Dunn, P. K., & Smyth, G. K. (1996). Randomized quantile residuals. *Journal of Computational and Graphical Statistics, 5*(3), 236–244.
Finger, L. (2013). *Ask*. Measure, Learn: Using Social Media Analysis to Understand and Influence Costumer Behaviour.
Hastie, T. J., Tibshirani, R. J. (1990). Generalized additive models (Vol. 43). CRC press.
Liberati C. and Mariani P. (2016). Big Data meet pharmaceutical industry: an application on social media data. Book of Abstracts, 10th Scientific Meeting of the Classification and Data Analysis Group of the Italian Statistical Society.
Nelder, J. A., Wedderburn, R. W. M. (1972). Generalized linear models. JR Statist. Soc. A 135, 370-384. Nelder370135J. R. Statist. Soc A, 1972.
Pepe, M. S. (2003). *The statistical evaluation of medical tests for classification and prediction*. USA: Oxford University Press.
Rigby, R. A., Stasinopoulos, D. M. (2001, July). The GAMLSS project: a flexible approach to statistical modelling. In New trends in statistical modelling: Proceedings of the 16th international workshop on statistical modelling (pp. 249-256).
Rigby, R. A., & Stasinopoulos, D. M. (2005). Generalized additive models for location, scale and shape. *Journal of the Royal Statistical Society: Series C (Applied Statistics), 54*(3), 507–554.
Sichel, H. S. (1973). Statistical valuation of diamondiferous deposits. *Journal of the Southern African Institute of Mining and Metallurgy, 73*(7), 235–243.

Social Media in Disasters. Big Data Issues in Public Communication Field

Francesco Marrazzo and Gabriella Punziano

Abstract With the growth and the changing nature of the (big) data, the role of social sciences researchers has been enhanced, producing an emerging assemblage of tools and techniques for managing and making sense of such data. Furthermore, a web content analysis (WCA) approach could become the basis for the use of techniques that enhance the relational context in which the production of messages and texts puts itself. In light of these premises, our contribution aims to explore the way in which new research strategies of WCA—in particular the adoption of a mixed-methoda perspective that moves back and forth qualitative and quantitative approach—could be useful in the analysis of social media use and functions in the process of disasters implementation. As disaster social media framework includes users such as communities, governments, individuals, organizations, and media outlets, the use of a broader range of techniques in scientific study of disaster social media effects could facilitate the creation of disaster social media tools in the public communication field.

1 Introduction

The article aims at highlighting the role of social big data in public communication field, toward a more sophisticated transfer of knowledge among the affected civil society and the actors devoted to manage the emergency responses, such as civil protection agencies as well as local and national administrators.

Although the paper has to be understood as a joint work of the authors, Francesco Marrazzo is the author of paragraph 1 and 3, Gabriella Punziano is the author of paragraph 2 and 4.

F. Marrazzo (✉) · G. Punziano
Social Sciences Department, University of Naples "Federico II", Naples, Italy
e-mail: francesco.marrazzo2@unina.it

G. Punziano
e-mail: gabriella.punziano@unina.it

P. Mariani and M. Zenga (eds.), *Data Science and Social Research II*,
Studies in Classification, Data Analysis, and Knowledge Organization,
https://doi.org/10.1007/978-3-030-51222-4_20

In fact, social media is rapidly emerging as a potential resource of information capable to support natural disasters management because a more and more increasingly amount of people makes use of them during the evolving of disasters generating big geosocial data of different formats and quality that must be quickly processed (Athanasis et al. 2018).

The methodological challenge dwells in the rapid and comprehensive analysis of a variety and immense universe of data so much so that, over the years, many analyses have been produced on decisional supporting tools for natural disaster management that take into account the georeferenced information. However, online content goes far beyond this component. These data are dense and loaded with a semantic and communicative component that cannot be ignored. In order to improve the management of the disaster, the recovery of this component becomes more and more essential, as well as an integrated and multidisciplinary approach becomes more and more essential to be able to really generate meaning from the huge flow of social big data.

According to Pu and Kitsuregawa (2013), the potential benefits of Big Data for disaster management stay in the five C characteristics of social media data: collectively, connectedness, completeness, clarity, and collaboration. In the reported perspective, big social media data are imbued with relational, interactional, and action values that impact and could make use of the power imbedded therein in any phase of disaster management—prevention, preparedness, and response and recovery.

In a disaster situation, social media are used by the citizens in four primary ways: family and friend's communication; situation updates; situational/supplemental awareness; services access assistance. During disaster social media help to communicate, and after the disaster they help communities come together again and enhance capabilities to build better recovery efforts and distribution of assistance (Joseph et al. 2018). Social media, as a backchannel of communications (used both to receive and post messages), increase the social capacity of information generation and dissemination (Xiao et al. 2015). Their generated data in disaster management are real-time data that bring social media to be the fourth most popular source for retrieving emergency information (Lindsay 2016). The efficient integration, aggregation, and visualization of this huge information will assist emergency managers to optimize the situational awareness and could result in a better decision-making procedure.

Methodologically reasoning, in this optic, new strategies of analysis—more hybrid, capable to go jointly in depth and in extension, able to fight with new, multi-format, and big data—impose themselves among the contingent needs for research that aims to focus in the frame of media disaster communication analysis. It is in this precise situation that, among the analytical solutions and the approaches that can be pursued in this type of research, a particular variant of the *Web Content Analysis* is presented in the next paragraphs: *Mixed-Web Content Analysis*.

2 Big Data Issues: The Role of Web Content Analysis

With the growth and changing nature of the (big) data also the role of social sciences researchers has been enhanced. At the same time, it could be at the same time possible to assist to the assemblage of tools and techniques for managing and making sense of all this data—often with no more than simple software on a standard computer (Lewis et al. 2013), and to the merging of different knowledge and sciences domains (Savage 2012). This means that the future of research on social, political, and communication fields may depend on building intellectual and technical alliances with other ways of knowing. Only by this way, the overabundance of data should end up to be a fool's gold (Karpf 2012), becoming simply much more complicated to analyze (Tinati et al. 2014), even if computational techniques for large-scale data analysis that once required supercomputers now can be deployed on a desktop computer (Manovich 2012).

As it could be evident, technical, ethical, computational question, as well as the nature, the use and the re-use of the data, and so on, are the issues always recalled when we talk about big data (Amaturo and Aragona 2016; Lauro et al. 2017). In particular, the issues generated around big data could be summarized in the Skalski et al. (2017) statement:

> Big data often take the form of information produced by human behavior and collected and archived by the programs behind social media platforms, web sites, and mobile media applications (Lewis et al. 2013). We now have the ability to search, aggregate, and cross-reference large data sets from a variety of interactive platforms, giving researchers the ability to overcome traditional sampling and coding limitations (Boyd and Crawford 2012). However, by definition, big data implies that the data are too big and complicated to handle or even be fully conceived by humans—computer power must be employed to collate, massage, and analyze. Thus, big data are removed from human experience, so only gross summarizations of the outcomes of analyses can be comprehended, making the implications of findings on big data rather abstract and not always directly applicable to human experience (204).

Therefore, hybrid and mixed solutions (Amaturo and Punziano 2016) are needed because the structural features of new media can be more fully subjected to algorithmic and quantitative analysis (because of the forms and structures) (Amaturo and Punziano 2017), while the sociocultural contexts built up around those features need the careful attention of manual methods and the deepness of qualitative approaches (Marrazzo and Punziano 2018).

But another challenge is under our look, the fact that interactive media has replaced traditional media and modes of communication such as newspapers, magazines, old-school television, and even the traditional telephone. So, not only hybrid approaches but also hybrid objects and media insist on the possibility of content analyzed the expanding sea of hybrid data characterizing our era.

Among the approaches, methods, and techniques, *web content analysis* (WCA) allows scholars to expand the horizons of the possible questions that every research can arise in relation to communication and online participation analysis by offering the ability to jointly analyze both the content and the way it is used and re-used in any context in which it is realized (Auriemma et al. 2015). Hence, WCA could become the basis for the use of techniques that boosts the relational context in which the production of messages and texts puts itself (Amaturo and Punziano 2013). WCA also perfects its way of existing, contemplating together techniques of analysis in extension (from the quantitative point of view) and techniques of analysis in depth (from the qualitative point of view) making itself a fundamental approach in the emerging stream of *Mixed Methods*.

In light of these premises, our contribution is going to explore the way in which new research strategies of *web content analysis* (Herring 2010)—in particular, the application of a mixed form that moves back and forth qualitative and quantitative approach—could be useful in the analysis of social media use and functions in the disaster implementation process.

3 Social Media in Disasters: A *Mixed Web Content Analysis* Approach

The *Mixed Web Content Analysis* (MWCA) could be used to understand the mechanisms generating mainly the influence on perception building and alternative public spheres generation when a disaster occurs. Starting from the analysis of already existing online textual big data coming from social media, by applying a mixed qualitative and quantitative perspective of *web content analysis*, every phase of disaster—prevention, preparedness, and response and recovery—could be managed in a more efficiently and timely way. In other words, thanks to MWCA, researchers could face the role and practical experiences of public (interest) communication (Rolando 2004) in disaster management, with particular reference to the linkage between top-down strategies and community's resilience strategies.

For example, the accompaniment to the disaster management for both policymakers and affected communities following this approach can be given by using a trivial example. The big data related to the increase of instantaneous communications on social platforms (the intent activity of flow data), can be considered a spy, an alarm bell, in order to understand that something is happening in a given zone (big data on platforms are also georeferenced data) which is influenced by this increase of data flow. This quantitatively marks the start of the process of a mixed analysis. The immediate analysis of the contents disseminated with this breadth and speed, from a qualitative, thematic, and content point of view, can be modeled to understand what is happening, what extent this event has, and where it is amplifying its effects. This would allow local administrators to direct aid by organizing the available forces rationally. Such a timely response would serve to manage panic for the community, as well as to increase security and a feeling of protection. The constant monitoring of flow and content data during the evolution of the disaster phases would allow an efficient and informed decision-making process and would limit the damages to the population involved, which in turn directs policies and actions through social communication, which it would also be produced without this monitoring-action purpose. The extent of the use of social media in disasters through a *web content mix analysis* is quickly rendered in these few lines, although much further can be pushed.

According to Bruns et al. (2012), the use of a broader range of techniques in the scientific study of disaster social media effects could facilitate the creation of disaster social media tools in public (interest) communication field. The implementation of research design on disaster social media aimed to facilitate the creation of social media tools should take into account some critical administrative questions (Lindsay 2016), as well as the relevance of sense-giving strategies in disaster management held by government agencies (Marx et al. 2018) and the informal trust in their relationships with citizens (Mehta et al. 2017).

3.1 Social Media in Disasters: A Literature Review

Communication has even been considered a core component of disaster planning, response, and recovery (Rodríguez et al. 2007). Since the creation of the Disaster Communication Intervention Framework (DCIF) model, social media have made their entry in the disaster management field (Houston 2012), covering different functions in the various phases characterizing a disaster (see Table 1).

Disaster social media framework includes users such as communities, government, individuals, organizations, and media outlets (Houston et al. 2015). During disasters, social media can influence traditional news media more so than the other way around (Valenzuela et al. 2017).

In the same way, social media characteristics are said to provide so many advantages over traditional media for disaster communication, that some government agencies have been interested in their use for disaster management since 2008 (Lindsay 2016).

According to Mehta et al. (2017), emergency management agencies (EMAs) can be distinguished among three social media models in disaster management, based on online trust: intelligence gathering; quasi-journalistic verification; crowdsourcing.

For Marx et al. (2018), EMAs should have the conception of themselves to be a publisher of crisis information during extreme events. As EMAs are often organized as local and nationwide branches, they can take, for the amplification mechanics, publishing strategies of media organizations as a paragon for information distribution. For the same scholars, it is possible to identify three different sense-giving strategies of media organizations (and local news media outlets): popularity arbitrage; bound amplification; open amplification (Marx et al. 2018).

Analyzing the different role played by official information—on the mainstream media or provided by government agencies—and the user-generated content in the Net—on the social media—in building perceptions and in generating alternative public spheres in communities hit by different kinds of disasters and facing high risks means conducting a systematic reflection on two sides. The former is more methodological and implies the changes that occurred in the formation of public opinion spheres and individual perceptions, influenced commonly by the media but

Table 1 Functions of disaster social media

Disaster social media use	Disaster phase
Provide and receive disaster preparendness information	Pre-event
Provide and receive disaster warnings	Pre-event
Signal and detect disasters	Pre-event → Event
Send and receive requests for help or assistance	Event
Inform others about one's own condition and location and learn about a disaster-affected individual's condition and location	Event
Document and learn what is happening in the disaster	Event → Post-event
Deliver and consume news coverage of the disaster	Event →Post-event
Provide and receive disaster response information; identify and list ways to assist in the disaster response	Event →Post-event
Raise and develop the awareness of an event; donate and receive donations; identify and list ways to help or volunteer	Event → Post-event
Provide and receive disaster mental/behavioral health support	Event → Post-event
Express emotions, concerns, well-wishes; memorialize victims	Event → Post-event
Provide and receive information about (and discuss) disaster response, recovery, and rebuilding; tell and hear stories about the disaster	Event → Post-event
Discuss socio-political and scientific causes and implications of and responsibility for events	Post-event
(Re)connect community members	Post-event
Implement traditional crisis communication activities	Pre-event →Post-event

Source: Houston (2012)

also guided by a new kind of information, that could be recognized in the online social data.

The latter goes in depth in the implications in terms of crisis communication and the relative impact on social, economical, and political structures of local post-disaster communities. Top-down processes of reconstruction and risk management (Alexander 2014), on one hand, and bottom-up strategies to resist in a broken-link community (Dufty 2012); on the other hand, come together in defining new and very complex scenarios in a post-disaster community.

So, it becomes important to consider three key factors: official information coming from mainstream media (disseminated online and on social media too); public communication (spread online and on social media too); opinion and emotion along local population involved in the post-disaster communities (expressed online and on social media too).

3.2 *Social Media Data in Disaster Management: The Role of the Mixed Web Content Analysis*

In disaster media management, social media data are increasingly being used for enhancing situational awareness and assisting disaster management. The reasons are different: social media data can characterize a (natural) disaster across space and over time, and thus are applicable to provide useful information on disaster situations. People have strong geographical awareness during (natural) disasters and are interested in communicating situational updates; news media and local authorities are opinion leaders and play a dominant role in the communication network (Wang et al. 2016).

In public risk communications, the initial focus on developing and executing best practices for outward communications is now giving way to discussions about augmenting response efforts with the inclusion of data from the public (Palen and Hughes 2017).

According to scholars, "it is almost impossible to make sense of the large amount of socially-generated data for applications to emergency management without adequate tools to filter, analyse, and visualize the data" (Palen and Anderson 2016); "information systems research can support emergency management agencies in using social media data for efficient crisis management by enhancing awareness of the benefits of social media analytics and helping to overcome organisational and technological challenges" (Stiegltiz et al. 2018).

Content analysis can help scholars and EMAs to understand how online participants communicate among and between the different crisis convergence behaviors, and which would be the best communication practices to assist crisis response efforts (Subba and Bui 2017).

As the new big data core issue in crisis and disaster management is to extract from the mass of incoming information what is important for situational awareness during emergencies, an empirical, MWCA-oriented, contribution for the implementation of public (interest) communication tools for disasters and emergencies management could be very useful.

Using this approach, it is possible to frame the disaster management not only as a question of what communicate but also of who communicate, with which intent, and in what direction, in order to assess the general sentiment of: (i) who experiences the disaster—and of course produces user-generated online content to describe and inform other people in the same situation, and also to express himself, increasingly online and on social platforms; (ii) who is storytelling the disaster—media mainstreams *in primis* for their natural function not only on the official but also on social and interactive platforms; (iii)who is called to manage with the disaster—such as local and national authorities, governors, agencies devoted to intervene, and so on, that, as well as the two previous categories, use social and minus social platforms.

All these flows of data are immersed in the general flows of data produced by, on, and for the occurred disaster, and need to be assessed, framed, and processed to become useful knowledge.

4 Conclusion and Final Remarks: Social Media in Disaster Tools

Inserting these argumentations in the framework of the MWCA, our intention was to look at employ Web 2.0 and 3.0 technologies for future collaborative decision-making (Zielinski et al. 2013)—including the use of social media—without giving up taking into account that the contents are not readily accessible and interpretable, since online texts, data, and images, are unstructured, noisy, and multilingual.

The principal result could be enclosed in a warning automated system that is able to continuously monitor the situation, the mood and the possible directions of information distortion, such as fake news, false alarms, and so on (Starbird 2017), and that could be useful for the decision-makers as well as for the inhabitants or for the official media.

In doing that, data science—that could be conceived as inclusive spaces of domino—joint to MWCA should overcome the limits experienced by the single disciplines in giving responses to all the questions arisen in media communication disasters. This is true both when social scientists are called to work on the computational side than when they are called to fight with the interpretation of the results recalling their specific expertise on the dominos side. Only the perfect mix among the two sides, the related disciplines and their main approaches, could give to the research also the function of public utility, especially in the situation in which it is needed a real-time understanding of phenomena, such as in a disaster situation.

The most solid reflection that comes from the union of big data to social media content data concerns the union of the analytics in the study of the former and the mining elaborations of the latter. The new tools to be produced in this area must take charge of the knowledge of communication experts, as well as of the ones generated by the professionals in disaster management or the ones produced from below, locally, in a social manner, capable of moving effective resolutions, in real time and closely related to opinions, moods, and contingent needs of the populations experiencing the disaster as well as of the governors who must be ready to manage the emergency situation.

References

Amaturo, E. and Aragona, B. (2016). La rivoluzione dei nuovi dati: quale metodo per il futuro, quale futuro per il metodo? In Corbisiero, F. and Ruspini, E. (a cura di). *Sociologia del futuro. Studiare la società del ventunesimo secolo.* Milano: CEDAM.

Amaturo, E., & Punziano, G. (a cura di). (2013). *Content analysis tra comunicazione e politica.* Milano: Ledizioni

Amaturo, E., & Punziano, G. (2016). *I Mixed Methods nella ricerca sociale.* Roma: Carocci.

Amaturo, E., & Punziano, G. (2017). Blurry boundaries: Internet, Big-New data and mixed-method approach. In C. Lauro, E. Amaturo, M. G. Grassia, B. Aragona, & M. Marino. (2017). *Data science and social research. Epistemology, methods, technology and applications*, New York: Springer International Publishing.

Athanasis, N., Themistocleous, M., Kalabokidis, K., Papakonstantinou, A., Soulakellis, N., & Palaiologou, P. (2018). The emergence of social media for natural disasters management: a big data perspective. *International Archives of the Photogrammetry, Remote Sensing and Spatial Information Sciences, 42*(3), W4.

Auriemma, M., Esposito, E., Iadicicco, L., Marrazzo, F., Polimene, A., Punziano, G., & Sarnelli, C. (2015). Euroscetticismo a 5 Stelle: Stili comunicativi e online text data nel caso delle elezioni europee 2014. *Sociologia Della Comunicazione, 49,* 36–54.

Boyd, D., & Crawford, K. (2012). Critical questions for big data: Provocations for a cultural, technological, and scholarly phenomenon. *Information, Communication & Society, 15*(5), 662–679.

Bruns, A., Burgess, J., Crawford, K., & Shaw, F. (2012). *#qldfloods and @QPSMedia: Crisis Communication on Twitter in the 2011 South East Queensland Floods.* Kelvin Grove: ARC Centre of Excellence for Creative Industries and Innovation.

Dufty, N. (2012). Using social media to build community disaster resilience. *The Australian journal of emergency management, 27*(1), 40–45.

Herring, S. C. (2010). Web Content Analysis: Expanding the Paradigm. In J. Hunsinger, M. Allen, & L. Klastrup (Eds.), *The international handbook of internet research* (pp. 233–249). Berlin and Netherlands: Springer Verlag.

Houston, J. B. (2012). Public disaster mental/behavioral health communication: Intervention across disaster phases. *Journal of Emergency Management, 10*(4), 283–292.

Houston, J. B., Hawthorne, J., Perreault, M. F., Park, E. H., Goldstein Hode, M., Halliwell, M. R., et al. (2015). Social media and disasters: A functional framework for social media use in disaster planning, response, and research. *Disasters, 39*(1), 1–22.

Joseph, J. K., Dev, K. A., Pradeepkumar, A. P., & Mohan, M. (2018). Big data analytics and social media in disaster management. In *Integrating disaster science and management* (pp. 287–294). Elsevier.

Karpf, D. (2012). Social science research methods in Internet time. *Information, Communication & Society, 15*(5), 639–661.

Lauro, C., Amaturo, E., Grassia, M. G., Aragona, B., & Marino, M. (2017). Data science and social research. In *Epistemology, methods, technology and applications.* New York: Springer International Publishing.

Lewis, S. C., Zamith, R., & Hermida, A. (2013). Content Analysis in an Era of Big Data: A Hybrid Approach to Computational and Manual Methods. *Journal of Broadcasting & Electronic Media, 57*(1), 34–52.

Lindsay, B. (2016). *Social media for emergencies and disasters: Overview and policy considerations.* Congressional Research Service

Manovich, L. (2012). Media visualization: Visual techniques for exploring large media collections. *The international encyclopedia of media studies.*

Marrazzo, F., & Punziano, G. (2018). Online textual data and political communication analysis. Methodological issues and research perspectives. *Sociologia Italiana – AIS Journal of Sociology,* 11, 143–158.

Marx, J., Mirbabaie, M. and Ehnis, C. (2018). Sense-Giving Strategies of Media Organisations in Social Media Disaster Communication: Findings from Hurricane Harvey. In *Australasian conference on information systems*, Sidney

Mehta, A. M., Bruns, A., & Newton, J. (2017). Trust, but verify: Social media models for disaster management. *Disasters, 41*(3), 549–565.

Palen, L., & Anderson, K. M. (2016). Crisis informatics—New data for extraordinary times. *Science, 353*(6296), 224–225.

Palen, L., & Hughes A. L. (2017). Social Media in Disaster Communication. In H. Rodríguez, W. Donner, & J. E. Trainor (Eds.), *Handbook of disaster research.* New York: Springer International Publishing

Pu, C., & Kitsuregawa, M. (2013). *Big Data and disaster management: A report from the JST/NSF Joint Workshop* (pp. 1–28). CERCS: Georgia Institute of Technology.

Rodríguez, H., Díaz, W., Santos, J. M., & Aguirre, B. E. (2007). Communicating risk and uncertainty: Science, technology, and disasters at the crossroads. In H. Rodríguez, E. L. Quarantelli, & R. R. Dynes (Eds.), *Handbook of disaster research* (pp. 476–488). New York, NY: Springer.

Rolando, S. (2004). *Comunicazione di pubblica utilità*, Voll. 1 e 2, Millano: Franco Angeli.

Savage, M. (2012). *Identities and social change in Britain since 1940: The politics of method*. Oxford: Oxford University Press.

Skalski, P. D., Neuendorf, K. A., & Cajigas, J. (2017). Content analysis in the interactive media age. In K. A. Neuendorf (Ed.), *The content analysis guidebook* (pp. 201–242).

Starbird, K. (2017). *Information wars: A window into the alternative media ecosystem*. https://www.medium.com

Stieglitz, S., Mirbabaie, M., Fromm, J., & Melzer, S. (2018). The adoption of social media analytics for crisis management-challenges and opportunities. In *Proceedings of the 26th European conference on information systems (ECIS)*, Portsmouth, UK.

Subba, R., Bui, T. (2017). Online convergence behavior, social media communications and crisis response: An empirical study of the 2015 Nepal Earthquake Police Twitter Project. In *Proceedings of the 50th Hawaii International Conference on System Sciences.*

Tinati, R., Halford, S., Carr, L., & Pope, C. (2014). Big data: Methodological challenges and approaches for sociological analysis. *Sociology, 48*(4), 663–681.

Valenzuela, S., Puente, S., & Flores, P. M. (2017). Comparing disaster news on Twitter and television: An intermedia agenda setting perspective. *Journal of Broadcasting & Electronic Media, 61*(4), 615–637.

Xiao, Y., Huang, Q., & Wu, K. (2015). Understanding social media data for disaster management. *Natural Hazards, 79*(3), 1663–1679.

Wang, Z., Ye, X., & Tsou, M. H. (2016). Spatial, temporal, and content analysis of Twitter for wildfire hazards. *Natural Hazards, 83*(1), 523–540.

Zielinski, A., Middleton, S. E., Tokarchuk, L.N., & Wang, X. (2013). Social media text mining and network analysis for decision support in natural crisis management. In *Proceedings of 10th international ISCRAM conference,* Baden-Bade, Germany.

Divorce in Italy: A Textual Analysis of Cassation Judgment

Rosanna Cataldo, Maria Gabriella Grassia, Marina Marino, Rocco Mazza, Vincenzo Pastena, and Emma Zavarrone

Abstract The dissolution of marriage is a complex social phenomenon that needs new topics of investigation, especially concerning the role of legal institutions in the conflict between partners. The research aims to identify the main issues that emerge in the institutional dimension of the phenomenon, identifying evolution and complexity of this within the sentences of the Italian Court of Cassation. Through judgments' analysis we can trace the variety of the phenomenon and identify interpretations of law in line with the evolution of contemporary institutions. The sentences are inserted in a demographic framework and are subsequently explored with topic probabilistic model (Latent Dirichlet Allocation), aimed to trace latent topic. In conclusion, the topics extracted refer to three main-dimensions, one to the procedural phases, another concerns the difficulty of leaving the separation phase and ending up in divorce, and finally the debate on the social-economic measures of divorce maintenance.

R. Cataldo · M. G. Grassia · M. Marino · R. Mazza (✉)
Department of Social Sciences, University of Naples Federico II, Naples, Italy
e-mail: rocco.mazza@unina.it

R. Cataldo
e-mail: rosanna.cataldo2@unina.it

M. G. Grassia
e-mail: mgrassia@unina.it

M. Marino
e-mail: mari@unina.it

V. Pastena
Studio Legale Pastena, Pastena, Italy
e-mail: avv.vincenzo.pastena@gmail.com

E. Zavarrone
IULM University, Milan, Italy
e-mail: emma.zavarrone@iulm.it

P. Mariani and M. Zenga (eds.), *Data Science and Social Research II*,
Studies in Classification, Data Analysis, and Knowledge Organization,
https://doi.org/10.1007/978-3-030-51222-4_21

1 Introduction

This paper aims to identify the main issues of the case law concerning divorce. These will be included in the scientific debate regarding divorce studies. In particular with regard to the role of institutions within the conflict that is established between the parterns. This objective is achieved through the application of a model for identifying latent topics to a textual corpus composed of sentences of the Italian court of cassation. The contributions to the state of the art are two: first the innovative methodological procedure, wich uses a topic model to a content analisys of italian divorce case-law mixed with textual network analisys; than the detection of main conceptual dimensions discuss by the court. Divorce in Italy is governed by the civil code art. 149 C.C. and by laws n. 898/1970 and n. 74/1987. This institute allows the dissolution of the bond (in case of civil marriage) or the cessation of the legal effects of the marriage rite (in case of the concordat marriage) if the deliberative court deems that between the parties the so-called spiritual and material communion of life has disappeared. In Italy, divorce was introduced in 1970 with the so-called Fortuna-Baslini law, promoted by the Radical Party and the Italian League for the Establishment of Divorce (LID) and presented to the Chamber by Member of Parliament Loris Fortuna; in 1971 the request for a repeal referendum by the catholic movements and the Christian Democrats was filed and the referendum was then held in 1974 with the victory of the "no" and the consecutive maintenance of the law (Fig. 1). From this moment, a procedure that would shorten the time for obtaining the dissolution began, in fact while the law 898 of 1970 provided one-year-separation period before the actual procedure starts, with the 74/1987 additions and 55/2015 this period is reduced first to twelve and then to six months. Divorce should not be confused with legal separation. This form is characterized by a temporary solution pending the reconciliation of the parties or the initiation of a divorce order. The procedure can have two alternative outcomes:

- Joint divorce, if there is an agreement between the parts on all the conditions adopted, in this case the separation period is six months
- Judicial divorce, when the parties do not agree with each other. In this case the application can also be presented by one of the two parties. By doing so, the separation period will be extended up to one year.

It is important to point out that for separation and/or divorce by mutual consent, from 2014 it is no longer necessary for married couples without minor or incapacitated children or those who are seriously disabled or economically not self-sufficient to appear in court. The dissolution can take place with a shared declaration and joint with the mayor as a registrar, with the optional assistance of a lawyer. Instead, married couples with children who have the aforementioned problems can divorce by negotiation assisted by at least one lawyer per part.

Outlined the normative bases of the institute, below is a selection of data and indicators of demographic and socio-economic nature to describe the phenomenon. According to the National Institute of Statistics, in 2017 the number of separations remains substantially stable compared to the previous year. On the other hand,

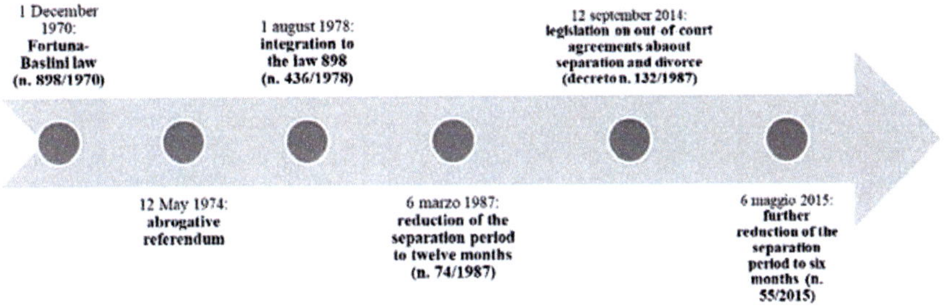

Fig. 1 Timeline of the divorce law

Fig. 2 Historical series of divorce and separation rates (values for 10.000 abitants). *Source* ISTAT, our elaboration

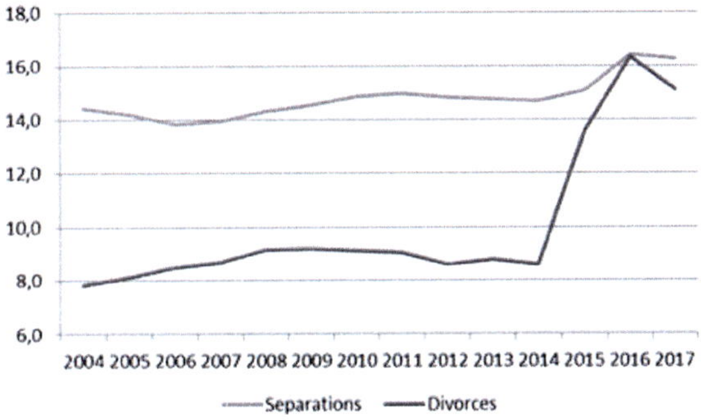

divorces showed a process of settling down after a growth in 2015–2016 (ISTAT 2018), years in which it is recorded a surge compared to the past; in this case the change in the regulatory framework considerably facilitates the start of the procedure, making it more streamlined and smooth from the bureaucratic point of view. However, there is no evidence of an increase in the propensity of spouses to dissolve unions, as it can easily be seen from the more contained tendency of the rate of separations (ISTAT 2016b) (Fig. 2).

In Italy, people between 40 and 50 years are more keen to divorce. The phenomenon can also strongly vary on a territorial level: northen regions show a much higher divorce rate than southern regions. The majority of divorced people have a job and the women are commonly housewives. These data show that people decide to deal with the divorce process regardless of their occupations. The average duration of marriage for people who decide to divorce is 18 years, 19 years for those in court and 17 years for extrajudicial proceedings divorces. Among them, 33% follow an extrajudicial procedure while 67% go to court, the average duration of the consensual divorce proceeding is equal to 102 days, 149 if the court is called and 40 if instead the extrajudicial procedure is preferred, the judicial divorce takes 585 days on average. In Italy the ratio between underaged children and foster children is 3.1 per thousand, the absolute value equal to 31,653 children; 10.8% of divorces provide the spouse's check, of which 97% is paid by the husband. These data, which offer a demographic and socio-economic description of the actors involved, help to outline a very articulated and constantly evolving phenomenon, which is difficult to define in terms of legislation and regulation. For this reason, in recent years there have been a series of attempts to dispel regulatory ambiguity through a series of interventions by the legislature rappresentative or case-law approach. In this paper there is an attempt to understand the issues that emerge from the case-law approach of the italian Court of Cassation, by referring to the latent topics that emerge from the court sentences. Outlined the demographic and legislative framework about divorce, the paper is subsequently structured as follows: in the Sect. 2 the model used is explained, in Sect. 3 there are goals and design of the research, in Sect. 4 the results and in Sect. 5 the conclusions.

2 Latent Dirichlet Allocation, A Model to Identifying Latent Topics in a Corpus

When a corpus presents large volume of information is difficult to identify its semantic structure, especially if the aim is to detect latent themes in the textual collection. To this end, there are methods that solve this problem by offering a statistically robust solution. Topic model is a model that allows us to identify a series of latent topics within a collection of texts, these topics are a collection of words that help to characterize and semantically define itself. In the large reference literature we rely on the following definition (e.g., Blei et al. 2003; Griffiths and Steyvers 2002, 2003, 2004; Hofmann 1999, 2001):

> Topic models are probabilistic latent variable models of documents that exploit the correlations among the words and latent semantic themes (Blei and Lafferty 2007).

We can summarize this basic idea in three steps (Blei and Lafferty 2009):

1. Uncover the hidden topical patterns that pervade the collection
2. Annotate the documents according to those topic
3. Use the annotations to organize, summarize, and search the texts.

This models aim to trace the latent semantic structure within a corpus and to analyse the information contained in it, they start with a fundamental assumption:

> Documents are mixtures of topics, where a topic is a probability distribution over words (Steyvers and Griffiths 2007).

From this assumption, we can define these models as a generative model for documents: to generate a new text a topic is extracted, and subsequently a term from the distribution on the corresponding vocabulary; the process must be iterated along the entire length of the document. Obviously the process can be reversed through statistical techniques, in order to make inference on the set of topics that generated the document. A variety of probabilistic topic models have been used to analyze the content of documents and the meaning of words (Blei et al. 2003; Griffiths and

Steyvers 2002, 2003, 2004; Hofmann 1999, 2001), one of these is the Latent Dirichlet Allocation (from now LDA). The most important reference to the model is the paper of Blei et al. (2003), in addition the model has been extensively studied in Griffiths and Steyvers (2004), Heinrich (2005), Blei and Lafferty (2009), Berry and Kogan (2010) and others; in this section there is an LDA topic model overview, largely based on the original authors' work. At the base of the LDA we find the assumption previously defined, it is a generative and inferential model; we observe only documents and words, no topics, these are partly hidden, latent precisely, within the structure of the document. The goal is to infer the latent structure of topic, consisting of documents and words. The model does this by recreating the documents in the corpus considering the relative weight of the topic in the document and the word in the topic, in an iterative way. The Latent Didrichlet Allocation is a Bayesian model and assumes that document and topic distributions can be described by a Dirichlet distribution (Blei et al. 2003). As it is the conjugate distributions of the Multinomial, it is very convenient to use it as a priori, and it follows that it can be an excellent tool as regards the problems of model inference (Neapolitan 2003). By not delving further into Dirichlet distribution, we can consider it as the multivariate oversimplification of Beta distribution which itself has been used in Bayesian statistics for modeling belief (Ponweiser 2012). Taking a function a posteriori from the Dirichlet distribution, parameterized by higher weights on a single topic, we can derive a distribution that goes to make up every topic mix. To define the process schematically: from a $Dir\ (\alpha)$ dirichlet distribution we carry out a random sampling which represents the distribution of the topics of a particular document, this topic distribution is θ; from θ, we select a particular *topic* Z based on the distribution; then from another Dirichlet distribution $Dir\ (\beta)$, we select a random sample representing the word distribution of the topic Z, this distribution is ψ, from ψ we choose the word w.

LDA generally works better than other models because it can easily generalize new documents. In this model, If we have not seen a document, we can easily sample it from the Dirichlet distribution and move on from there.

2.1 Number of Topics

In topic modeling usually the number of topics to extract from the corpus is a parameter to be defined a priori, this can be a problem if an exploratory approach is adopted and an automatic extraction of the topics is carried out. There are a variety of methods and algorithms for evaluating the model with the optimal number of topics (Blei and Lafferty 2009, Wallach et al. 2009; Buntine 2009; Chang and Blei 2009). To find the optimal topics numberwe ran the model a defined number of times by entering various parameters and choose the best performing one among the various models, to measure the performance of the model we could rely on held-out data: "Estimating the probability of held-out documents provides a clear, interpretable metric for evaluating the performance of topic-related models related to other topic-based models as well as other non-topic-based generative models" (Wallach et al. 2009). In this paper the best model was selected through the harmonic mean method (Griffiths and Steyvers 2005; Wallach et al. 2009; Buntine 2009). It is possible to formalize the method in this way (Griffiths and Steyvers 2004):

> In our case, the data are the words in the corpus, w, and the model is specified by the number of topics, K, so we wish to compute the likelihood p(w|K). The complication is that this requires summing over all possible assignments of words to topics z [, i.e., p(w|K) =Rp(w|z,K)p(z)dz]. However, we can approximate p(w|K) by taking the harmonic mean of a set of values of p(w|z,K) when z is sampled from the posterior p(z|w,K).

3 Our Proposal

The appeal in Cassation may be lodged against the measures issued by the ordinary courts at the appellate level or in degree only: the reasons given to support the use may be, in civil matters, the violation of the material (errore in iudicando) or procedural (errore in procedendo) right, the vices of motivation (lack, insufficiency or contradiction of motivation) of the judgment under appeal; or, again, the grounds for jurisdiction (Basic Law on the Judiciary of 30 January 1941 no. 12 (art. 65)). The Supreme Court of Cassation represents the last resort's judge of legitimacy of the judgments emitted by the ordinary magistracy. Its sentences constitute a guiding criteria of the national jurisprudence, which before making its decisions can take into account the judgments issued by the Court. Therefore, analysing sentences assumes an interest in the study in order to monitor the evolution of specific phenomena, for which the current legislation presents a certain degree of ambiguity. The rulings of the Supreme Court are normally followed by the judges of lower courts (in particular the sentences of the United Sections). The institutional dimension of the phenomenon is explored through the study of jurisprudential sources. This is to outline the social phenomenon's evolution from the point of view of the sources examined. There are two research questions that animate this work:

- RQ1: What are the issues around which the attention of giuridic authority is concentrated?
- RQ2: Is it possible identify a semantic structure that links the issues together and defines the phenomenon within the reference jurisprudence?

From a methodological point of view, it is possible to answer the questions through two phases. First identifying latent topics within the sentences issued by the Court; Than understanding the semantic relationships that exist between the extracted topics and their lexical forms. The work is divided into six steps that constitute a methodological procedure to identify a set of topics. The first two steps involve the extraction and automatic pre-treatment of the textual corpus (Bolasco and De Mauro 2013; Lembart 1994), followed by the creation of a document—term matrix. With the application of the Latent Dirichlet Allocation topic modeling (Blei and Lafferty 2007), it has been possible to extract a group of latent topics (RQ1). Subsequently, we created a word-topic affiliation matrix and outline the network. In conclusion, through

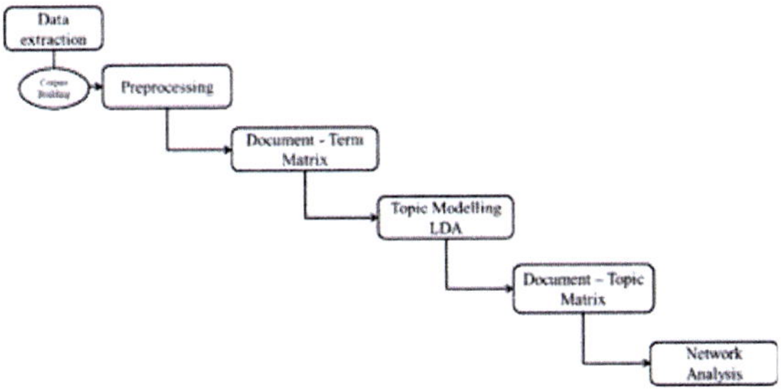

Fig. 3 Methodological procedure of work

the calculation of centrality measures, it was possible to understand semantic text's structure (RQ2) (Fig. 3).

4 Results

The analyzed corpus has 193 texts, extracted from the official court database. [1] The texts have an average length of 12,667 characters, and each text consists of a final sentence issued by the court. The period to which most judgments refer is from 2013 to 2014 (71%), while for the period 2015–2016 is 24% and 2017–2018 6%, it is possible to see that before the 2015 regulatory intervention there was a greater tendency to resort to Cassation. As the demographic framework shows, in the corpus the majority of judgments comes from the regions where the number of divorces is higher, in fact 47% refers to courts in the North and 41% to courts in the Center, while only 15 and 10% is shown to the South and Islands. A further preliminary analysis was carried out with the exploration of the lexical matrix through the creation of a specific vocabulary, based on Lembart's statistical measure *valor test* (Lebart et al. 1995). The peculiar dictionary of lexical forms was constructed regarding the reference area of the lower court that issued the sentence appealed in Cassation. The following semantic groups can be outlined (Fig. 4):

- North: the vocabulary refers to the economic aspects, with reference both to court costs and to the maintenance of the parties.
- Center: there are references to marriage and family from a legal standpoint and regarding *rights*.
- South: the relationship to the ecclesiastical and family sphere appears, which in this context becomes an expression of crisis.
- Islands: here there are mostly references to mainly judicial and procedural aspects.

Before showing the analysis results, the output in Fig. 5 shows the calculation of the best model. The figure shows the optimal number of topics; this is equal to the highest point of the curve, rappresented by points 7 and 11, drawn by the interpolation line.

[1] http://www.italgiure.giustizia.it/sncass/.

Fig. 4 Specific vocabulary

Fig. 5 Best model graphix

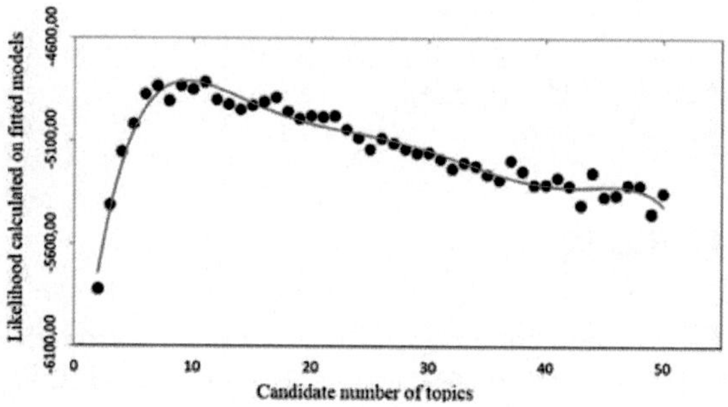

Table 1 Extracted topics

Topic I	Topic II	Topic III	Topic IV	Topic V	Topic VI	Topic VII
Cassazione	Spese	Mantenimento	Separazione	Matrimonio	Assegno	Ricorso
Giudizio	Sentenza	Minorenni	Immobile	Sentenza	Economica	Ricorrente
Decisione	Tribunale	Genitori	Sentenza	Legge	Divorzile	Ex
Violazione	Condanna	Giudice	Diritti	Diritti	Assegno.divorzile	Relazione
Ricorso	Ricorso	Diritti	Cassazione	Convivenza	Reddito	Merito

For reasons related to a correct and comprehensive semantic interpretation of the individual topics and a non-significant difference in model quality, it was preferred to adopt the K = 7 model. In Table 1 the exctracted topics are shown, with lemmas associated with them with greater probability:

1. This topic includes judgments that refer to some violation to the lower courts' sentences
2. The second topic refers to the costs for the process
3. The third topic refers to child economic support and parenting duties
4. In this topic there are references to the phase of separation of the couple
5. This topic refers to marriage
6. In the sixth topic the maintenance check is the main topic
7. In the seventh topic there are references to the causes of divorce and to the personal relationship between the spouses.

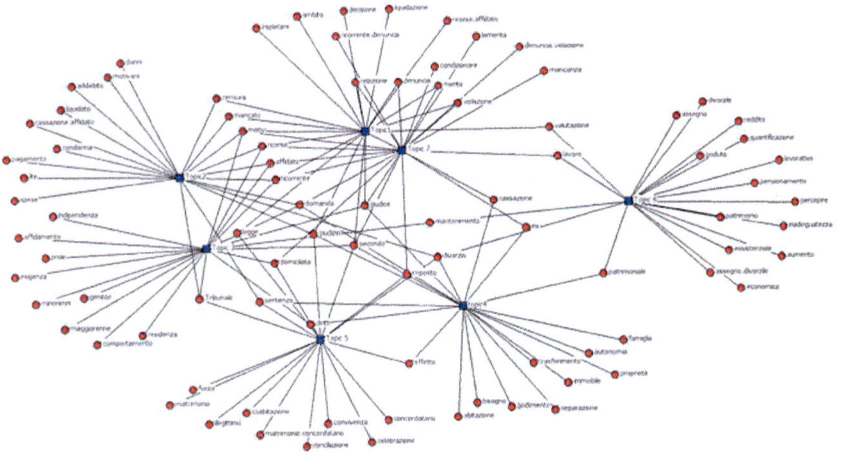

Fig. 6 Topics' network

The topics have been named by the word associated with the highest probability. If several words had the same probability, the one that best expressed the topic semantically was assigned. Once the topics have been identified within the corpus, the second question reseach involves the study of the relationships between them. After the creation of an adjacency matrix *lemmas by topics*, a network was generated. Through the description of the network and the study of centrality measure (Faust 1997), it is possible to understand the links that exist between the elements taken under consideration. A first description of the network can be made by undestanding its structure. To study the framework of the relationships between lemmas and topics it is necessary to analyze the core-periphery structure that allows to identify strongly connected words (those in the core) and to evaluate their role in establishing relations with the peripheral partners. The network structure is outlined in Fig. 6. This is divided into a periphery, in which the lemmas that exclusively characterize each single topic are present, and in a central core, more dense, in which the lexical forms link the various topics. Regarding the topics, we can read the relation beetween themes extracted with the model thanks to the core-periphery structure, specially for the core, that helps us to understand the semantic relationships between the various topics. The indicators used for the study of centrality are degree centrality and eigenvector degree, these are specifically calculated for two-mode networks. The core density is 0.508 and is composed of topics *separazione (separation), spese (expenses) and ricorso (recourse)* and from the lexical forms *cassazione (cassation), imposto (imposed), domanda (demand), affidato (entrusted), diritti (rights), ricorrente (applicant), divorzio (divorce), cassazione.affidato (cassation.entrusted), condanna (condemnation), motivi (reasons), sentenza (sentence), secondo (second), giudice (judge), giudizio (judgment), censura (censorship), denuncia.violazione (denunciation.violation), ricorso (appeal), indipendenza (indepen-*

dence), mancanza (lack), motivare (motivate). Within the core the words with the highest degree centrality are *affidato (entrusted), cassazione (cassation), censura (censorship), diritti (rights), divorzio (divorce), domanda (demand), giudice (judge), giudizio (judgment), imposto (imposed), motivi (reasons), ricorrente (recurrent), sentenza (sentence)*. The network also contributes to represent the words positioned in the periphery of the network, which, although with low probabilities, help to better define the various topics. The topics *mantenimento (maintenance), ricorso (recourse), cassazione (cassation), spese (expenses), separazione (separation)* have higher degree (0.227); while *assegno (check) e matrimonio (wedding)* have a slightly lower degree (0.216); the most central nodes represent semantically richer topics, even if the difference is small, we can affirm that these topics have a greater semantic articulation, within the study of a corpus of sentences this can be translated into a greater attention to the specific problematic node represented by the topic. The eigencentrality index was calculated to understand which topics are most influential within the network, the themes with the highest value are *mantenimento (maintenance), ricorso (recourse), cassazione (cassation)* and the words are *affidato (entrusted), cassazione (cassation), censura (censorship), denuncia (complaint), diritti (rights), divorzio (divorce), domanda (demand), domiciliata (domiciled), ex, giudice (judge), giudizio (judgment), imposto (imposed), merito (merit), motivi (reasons), relazione (relationship), ricorrente (recurrent), ricorso (appeal), sentenza (sentence), tribunale (court), violazione (violation)*. These are the nodes that have more connections with the more central nodes.

5 Conclusions

The model has been an efficient analytical tool for our exploratory objectives and it has allowed us to extract from the judgments the main legal issues around which the Court of Cassation has been deliberating in the period under examination. By the reading of the analysis' results it is possible to identify three main dimensions, referable to the following macro-argoments:

1. The first referring to the procedural phase, here the topics recall the phases and aspects related to the legal procedure in progress.
2. Another referred to transition from the separation to the divorce phase, which begins with separation and takes form in the couples' assets evaluation.
3. The last argoment recalls the economic-social aspects related to the existence of one of the two parts.

The dimensions defined above are conceptually linked to each other, this is due to the particular shape assumed by the analyzed network and by the presence of a core that has linked several themes. From the texts examined, the Court of Cassation appears to be an institution that acts with the aim to re-establish order within a conflict between social partners. The sentence, in this perspective, is a decisive act from which a family with a new balance is shaped. This is possible by the intervention of the juridical body. In the network core, it is clear that the aspects that concern the economic sphere meet the needs of foster care, especially as regards the transition from being separated to being divorced. Moreover, the headwords that recall a more technical and process-centered register play a central role. This could refer precisely to the role of conflict solver operated by the sentence, which is going to be included in a phase of redefining the role of the individual within the family institution.

References

Berry, M. W., & Kogan, J. (2010). *Text mining: applications and theory*. Wiley.

Blei, D. M., & Lafferty, J. D. (2007). A correlated topic model of science. *The Annals of Applied Statistics*, *1*(1), 17–35.

Blei, D. M., Lafferty, J. D. (2009). Topic models. In *Text mining* (pp. 101–124). Chapman and Hall/CRC.

Blei, D. M., Ng, A. Y., & Jordan, M. I. (2003). Latent dirichlet allocation. *Journal of Machine Learning Research*, *3*, 993–1022.

Bolasco, S., & De Mauro, T. (2013). L'analisi automatica dei testi: fare ricerca con il text mining, Carocci Editore.

Buntine, W. (2009). Estimating likelihoods for topic models. In: *Asian Conference on Machine Learning* (pp. 51-64). Berlin, Heidelberg: Springer.

Chang, J., & Blei, D. (2009). Relational topic models for document networks. In *Artificial Intelligence and Statistics* (pp. 81–88).

Faust, K. (1997). Centrality in affiliation networks. *Social Networks*, *19*(2), 157–191.

Griffiths, T. L., & Steyvers, M. (2002). A probabilistic approach to semantic representation. In *Proceedings of the annual meeting of the cognitive science society*, vol. 24, no. 24.

Griffiths, T. L., & Steyvers, M. (2003). Prediction and semantic association. In *Advances in Neural Information Processing Systems* (pp. 11–18).

Griffiths, T. L., Steyvers, M., Blei, D. M., & Tenenbaum, J. B. (2005). Integrating topics and syntax. In *Advances in Neural Information Processing Systems* (pp. 537–544).

Griffiths, T. L., & Steyvers, M. (2004). Finding scientific topics. *Proceedings of the National Academy of Sciences*, *101*, 5228–5235.

Heinrich, G. (2005). Parameter estimation for text analysis. Technical report.

Hofmann, T. (1999). Probabilistic latent semantic analysis. In *Proceedings of the Fifteenth Conference on Uncertainty in Artificial Intelligence* (pp. 289–296). Morgan Kaufmann Publishers Inc.

Hofmann, T. (2001). Unsupervised learning by probabilistic latent semantic analysis. *Machine Learning*, *42*(1–2), 177–196.

ISTAT (2016b). Le trasformazioni demografiche e sociali: una lettura per generazione.

ISTAT (2018). Annuario statistico italiano.

Lebart, L., Morineau, A., & Piron, M. (1995). *Statistique exploratoire multidimensionnelle* (Vol. 3). Paris: Dunod.

Neapolitan, R. E. (2003). *Learning Bayesian networks*, vol. 58, no. 4 (pp. 1064–1082). Prentice-Hall.

Ponweiser, M. (2012). Latent Dirichlet allocation in R.

Steyvers M., & Griffiths T. (2007). Probabilistic topic models. In: T. Landauer, D. Mcnamara, S. Dennis & W. Kintsch (Eds.), *Latent semantic analysis: a road to meaning* (p. 427). Lawrence Erlbaum.

Wallach, H. M., Murray, I., Salakhutdinov, R., & Mimno, D. (2009). Evaluation methods for topic models. In *Proceedings of the 26th Annual International Conference on Machine Learning* (pp. 1105–1112). ACM.

A Bayesian Mixture Model for Ecotoxicological Risk Assessment

Sonia Migliorati and Gianna Serafina Monti

Abstract In ecotoxicological risk assessment, the estimation of a Species Sensitivity Distribution (SSD) is a routine method used to derive hazardous levels of concentrations for chemical substances. Here, we propose a Bayesian hierarchical approach leading to the definition of a new SSD. Our approach allows to use all information available at chemical-class-species levels to make inferential decisions. We estimate parameters via computer-intensive methods based on Markov Chain Monte Carlo methods, and we propose a way to derive the estimates of concern levels of toxicants that could be easily adopted in ecotoxicological risk management.

1 Introduction

A Species Sensitivity Distribution (SSD) is a distribution of species tolerance to a chemical stressor in a defined ecosystem. In the context of ecotoxicological risk assessment, several regulatory, and governmental programs routinely adopt the SSD concept to derive concern levels of concentrations for chemical substances potentially harmful for a given proportion of the species in the ecosystem (Posthuma et al. 2002; Stephan 2002; ECHA 2008). A standard SSD is commonly represented by a cumulative probability distribution function which relates the potential affected proportion (PAF) of the community with (usually log-transformed) concentrations of a chemical. In particular, the sensitivities of the tested species are ordered and represented in a cumulative fashion with the aim of estimating the concentration of the chemical stressor that is hazardous to p% of the corresponding biological assemblage (Hickey and Craig 2012). The use of SSDs has been criticized essentially due to the strong assumptions it is based on, and in particular, because of the hypothesis of

S. Migliorati (✉) · G. S. Monti
Department of Economics, Management and Statistics, University of Milano Bicocca, Milan, Italy
e-mail: sonia.migliorati@unimib.it

G. S. Monti
e-mail: gianna.monti@unimib.it

P. Mariani and M. Zenga (eds.), *Data Science and Social Research II*, Studies in Classification, Data Analysis, and Knowledge Organization,
https://doi.org/10.1007/978-3-030-51222-4_22

exchangeability of the tested species (Newman et al. 2000; Forbes and Calow 2002). A simple and promising way of overcoming the problem is to enrich the simple model defining standard SSD by means of random effects, that account for different sources of variability. This leads to the definition of a new SSD based on hierarchical (random-effects) models able to fully exploit the information provided by data. Estimation issues for this kind of model can be dealt with by a Bayesian approach, which allows to overcome some flaws typical of frequentist estimation (Aldenberg and Jaworska 2000; Grist et al. 2006) such as the difficulty of expressing the marginal likelihood in closed form, as well as issues connected with small sample sizes. Indeed, the use of (computer-intensive) Bayesian statistics for the derivation of SSDs is becoming a standard tool in the environmental sciences (O'Hagan et al. 2005; Grist et al. 2006; Hickey et al. 2008). The new SSD we define allows to derive the estimate of concern levels of toxicants that could be easily adopted in ecotoxicological risk management.

2 Materials and Methods

2.1 Risk Measures

Let y_j $(j = 1, \ldots, n)$ denote the log(base 10)-transformed toxicity data for n species exposed to a given chemical stressor, and suppose that y_j values are ordered in an increasing fashion. Note that the typical toxicological measure of a species sensitivity is the median effect concentration (denoted by EC50). Moreover, let $p_j = j/(n+1)$ be the corresponding plotting positions, as given by the Weibull formula (Shabri 2002), that is the relative rankings of y_j $(j = 1, \ldots, n)$. The standard SSD approach, used for European risk assessment, assumes that the y_j's represent an independent and identically distributed sample from a normal distribution with mean μ and variance σ^2. Therefore, the plug-in estimate of the SSD is simply obtained by evaluating the normal cumulative distribution function (CDF) at the sample mean and variance of log-concentrations. In other words:

$$\text{SSD}_i = \Phi_i\left(\frac{y-\mu_i}{\sigma_i}\right), \tag{1}$$

where Φ is the cumulative distribution function of a standard normal distribution, and μ_i and σ_i are the mean and standard deviation which are evaluated at the sample mean $\bar{y}_i$ and the sample standard deviation s_i of the log-transformed distinct species toxicity values, respectively.

Given the estimated SSD, the hazardous concentration HCp corresponds to its pth percentile, e.g., HC5 is the hazardous concentration affecting 5% of species, and thus HC5 would protect 95% of the species in a specific community (Aldenberg and Luttik 2002). Clearly, the quantity HC5 plays an important rule in ecotoxicological risk assessment as it is considered the predicted no-effect concentration (PNEC) by ECHA

(2008). PNEC represents an estimation of the concentration of the toxicant which is below the lowest species tolerance value in the generic ecological community (Posthuma et al. 2002).

Several methods are present in the literature to estimate HCp's, depending on the distributional assumptions on the SSD, the choice between parametric or non-parametric approaches, and the adopted inferential paradigm, i.e., Bayesian versus frequentist. The most widespread estimates are the following.

The plug-in estimate $\widehat{\text{HCp}}_{PI}$ of HCp ($0 < p < 1$) is based on the assumption that the SSD follows a log-normal distribution, and it is simply given by

$$\widehat{\text{HCp}}_{PI} = \bar{y} - z_{1-p/100}\, s\,, \tag{2}$$

where $z_{1-p/100}$ is the $(100 - p)$th percentile of a standard normal distribution (Posthuma et al. 2002).

Aldenberg and Jaworska (2000) proposed to incorporate the uncertainty of this estimate considering the γ% lower confidence limit of the estimation of HCp. Under the normal approach to SSD, this leads to the form

$$\widehat{\text{HCp}}_{L,\gamma} = \bar{y} - \frac{1}{\sqrt{n}}\, t_{n-1,ncp;\gamma/100}\, s \tag{3}$$

where $t_{n-1,ncp;\gamma/100}$ is the γth percentile of a noncentral t-distribution with $n-1$ degrees of freedom and noncentrality parameter $ncp = z_{1-p/100}\sqrt{n}$. A common value for γ in a conservative and protective perspective is 5 (Newman et al. 2000; Wagner and Lokke 1991). Note that such interval can also be viewed as a Bayesian credible interval using the conventional "vague" prior for sampling from a normal distribution.

Hickey and Craig (2012) proposed a backwards regression estimation, applying a probit transformation to the set of relative rankings p_j, i.e. ,$x_j = \Phi^{-1}(p_j)$, and estimating a simple linear regression model on the data pairs (y_j, x_j):

$$\widehat{\text{HCp}}_R = \bar{y} - z_{1-p/100}\frac{(n-1)s^2}{\sum_{j=1}^{n} y_j x_j}. \tag{4}$$

2.2 *A Mixture Model for the Definition of SSDs*

The probabilistic ecotoxicological risk assessment is essentially based on the derivation of thresholds of risk from SSDs. The calculation of a single SSD assumes the availability of data derived from toxicity tests for selected species on a specific compound. Indeed, the selected species are generally tested also on other compounds. These compounds can be typically classified into chemical classes according to their mode of action. Moreover, the selected species can be pooled according to the taxonomic groups they belong to. Therefore, a typical ecotoxicological database, used to calculate SSDs, contains data on different levels: data on various compounds together with information on the classes of compounds and data related to different species as well as information on their taxonomic groups. We propose to incorporate all the available information on data into a hierarchical model defining a new SSD.

Such a hierarchical model for a given ecotoxicity database can be expressed as follows:

$$\begin{aligned} & y_{isjg}|\mu, \delta_i, \eta_s, \gamma_j, \lambda_g, \sigma^2 \sim \mathrm{N}(\mu + \delta_i + \eta_s + \gamma_j + \lambda_g, \sigma^2), \\ & \mu|\sigma_\mu \sim \mathrm{N}(0, \sigma_\mu^2), \\ & \delta_i|\sigma_\delta \sim \mathrm{N}(0, \sigma_\delta^2), \quad \text{(random substance effect)}, \\ & \eta_s|\sigma_\eta \sim \mathrm{N}(0, \sigma_\eta^2), \quad \text{(random chemical class effect)}, \\ & \gamma_j|\sigma_\gamma \sim \mathrm{N}(0, \sigma_\gamma^2), \quad \text{(random species effect)}, \\ & \lambda_g|\sigma_\lambda^2 \sim \mathrm{N}(0, \sigma_\lambda^2), \quad \text{(random taxonomic group effect)}, \end{aligned} \tag{5}$$

where y_{isjg} is the log-(base 10) toxicity value for compound i, $(i = 1, \dots, k)$ in the chemical class s, $(s = 1, \dots, S)$, tested on species j $(j = 1, \dots, n)$ belonging to taxonomic group g $(g = 1, \dots, G)$, and all variance components are supposed to be strictly positive. In (5), all random effects are supposed to be (conditionally) independent. Therefore, for each chemical class–species–group combination (i, s, j, g) the residuals are a random sample from a normal distribution centered about zero with variance σ^2 that accounts for measurement error in data. Note that σ^2 is supposed to be homogeneous with respect to chemical class, species, and group. Moreover, given that substances, chemical classes, species, and taxonomic groups are accounted for by the model by means of random effects, it is possible to enrich the model variability structure, usually represented by σ^2, by four further components, namely σ_δ^2, σ_η^2, σ_γ^2, and σ_λ^2 (Gelman and Hill 2007; Gelman 2015). This aspect is expected to lead to a better model interpretation as well as more accurate inferential conclusions with respect to the usual SSD approach.

Within a Bayesian approach, the model specification can be completed by assigning suitable hyper-prior distributions to all hyper-parameters present in (5). A widespread choice, which represents "vague" priors, is the following:

$$\begin{aligned} & \sigma_\delta \sim \mathrm{Uniform}(0, 100), \quad \sigma_\eta \sim \mathrm{Uniform}(0, 100) \\ & \sigma_\gamma \sim \mathrm{Uniform}(0, 100), \quad \sigma_\lambda \sim \mathrm{Uniform}(0, 100), \end{aligned} \tag{6}$$

which can be completed with the same prior for the measument error.

The posterior distributions of the parameters of interest can be obtained via Markov Chain Monte Carlo (MCMC). In particular, the Gibbs algorithm can be used since the model specification is fulfilled by means of conditional distributions, which makes the full conditionals readily available.

Thus, the proposed SSD for substance i of chemical class s, namely $\mathrm{MSSD}_{is}(y)$, can be expressed in terms of the hierarchical model (5) as a finite mixture of normal

CDFs:

$$\mathrm{MSSD}_{is}(y) = \frac{1}{n}\sum_{j=1}^{n}\frac{1}{G}\sum_{g=1}^{G}\Phi_{isjg}\left(\frac{y-(\mu+\delta_i+\eta_s+\gamma_j+\lambda_g)}{\sigma}\right). \tag{7}$$

Furthermore, the posterior CDF of the hazardous concentration HCp_M can be estimated from the posterior of $\mathrm{MSSD}_{is}(y)$ by observing that, given a value y, the following equality holds

$$P(\mathrm{MSSD}_{is}(y) \leq p) = P(\mathrm{HCp}_M \geq y) \ . \tag{8}$$

Therefore, the CDF of HCp_M at point y can be estimated with the proportion of MCMC sample values of $\mathrm{MSSD}_{is}(y)$ that are greater than or equal to p.

3 Results

3.1 Data Description

The motivation of our work is derived from the analysis of a subset of the aquatic ecotoxicity database described in De Zwart (2002), Hickey et al. (2012) integrated with the database US EPA ECOTOX (https://cfpub.epa.gov/ecotox/ last upload June 15, 2016).

In particular, we selected 31 insecticide compounds, belonging to 4 chemical classes (Organophosphorus insecticides, Carbamate insecticides, Pyrethroids insecticides, and Organochlorine insecticides), on 17 species (11 fishes, 2 insects, 2 crustaceans, 1 mollusk, and 1 alga), belonging to 5 taxonomic groups. The endpoint selected was the median effect concentration EC50, except for algae (EC50 growth inhibition) and *Daphnia magna* (EC50 immobilization). The exposure times selected were: 3–4 days for algae, 2 days for *Daphnia magna*, and insect larvae, 4 days for all other taxonomic groups.

3.2 Model Estimation and Risk Measures

We fit model (5) running three MCMC chains in parallel. For each chain, after a burn-in period of length 5,000 to ensure reached stationarity of the chain, we generated samples of size 10,000 for each random variable. Moreover, we chose a thinning interval of 50, i.e., we discarded all but every 50th observation, with the goal of reducing autocorrelation. All analyses were performed using the software R (Core Team 2019).

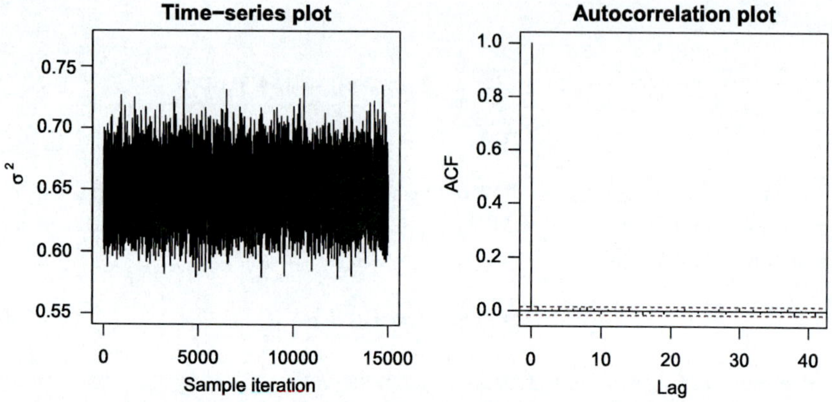

Fig. 1 Diagnostic tools to assess convergence of MCMC sample with respect to the parameter σ^2. Left panel: time-series plot (after burn-in period and thinning regime). Right panel: autocorrelation function

In Fig. 1, we have reported a time-series plot and autocorrelation plot referred to σ^2. These diagnostic tools allow to assess convergence properties. For all the remaining variance components we obtained similar patterns.

Figures 2, 3, 4 and 5 report the caterpillar plots of the random effects, i.e., they show their Bayesian estimates (posterior median) in rank order together with their 90% credible intervals.

These plots highlight that the random effects referred to substances and species (Figs. 2 and 4, respectively) can be considered highly different from one another. Indeed, the intersection among all intervals is empty in both plots, thus confirming the appropriateness of including a random effect into the model. Similar, although less strong, conclusions can be drawn for taxonomic groups (Fig. 5).

To appreciate the good fitting of the $\text{MSSD}_{is}(y)$ given by (7), let us consider Fig. 6, which shows the estimates of the MSSD and of the standard SSD, together with their 90% pointwise confidence bands for Organophosphorus pesticide Chlorpyrifos. The new model MSSD clearly outperforms the standard one, indeed it fits observed data much better and shows smaller confidence bands, thus leading to more precise inference. This can be ascribed to its capacity of better capturing the variability of the sample data due to the presence of many sources of variability, each one linked to a random effect.

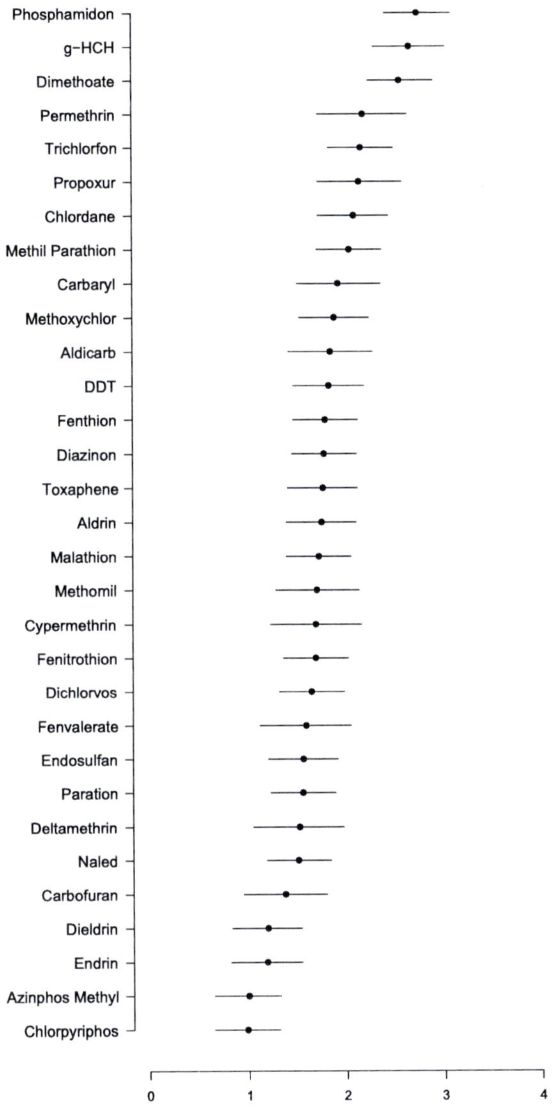

Fig. 2 Caterpillar plot of the substances random-effects δ_i (the dots correspond to posterior median) along with their 90% credible intervals (solid line segments)

Finally, as already underlined, the proposed method can be used by risk assessors to derive risk measures such as HC5. Table 1 reports the estimated hazardous concentrations for 5% of species $\widehat{HC5}_M$ derived from the MSSD curve, together with the alternative estimates computed according to the different approaches described in Sect. 2.1 (see (2), (3) and (4)).

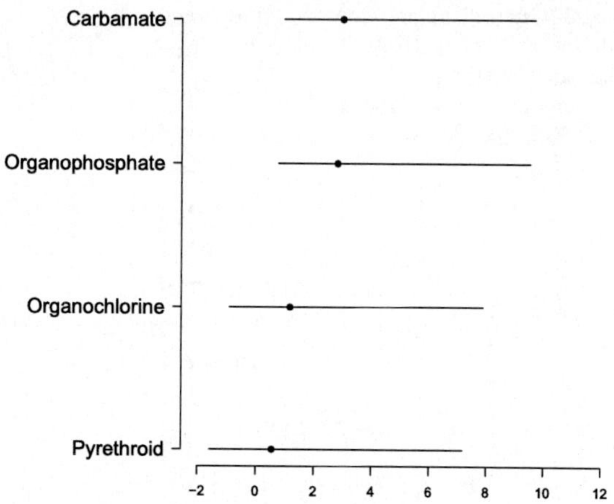

Fig. 3 Caterpillar plot of the chemical class random-effects η_s (the dots correspond to posterior median) along with their 90% credible intervals (solid line segments)

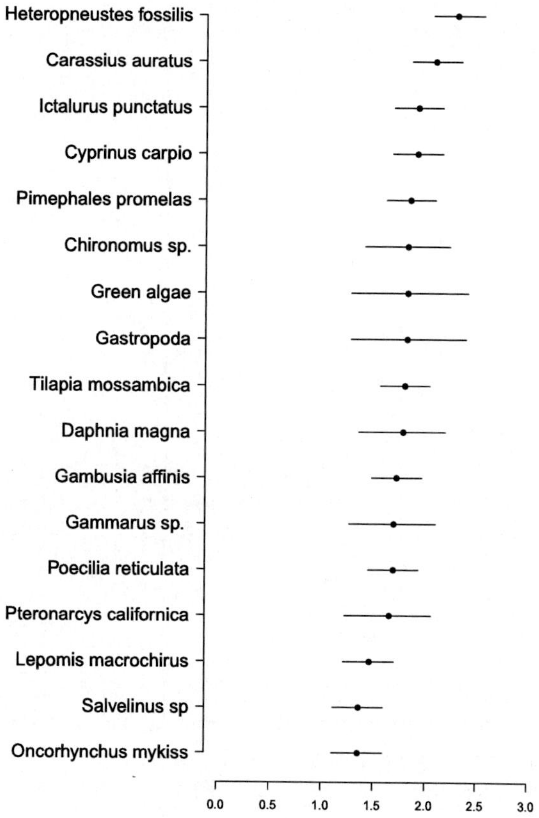

Fig. 4 Caterpillar plot of the species random-effects γ_j (the dots correspond to posterior median) along with their 90% credible intervals (solid line segments)

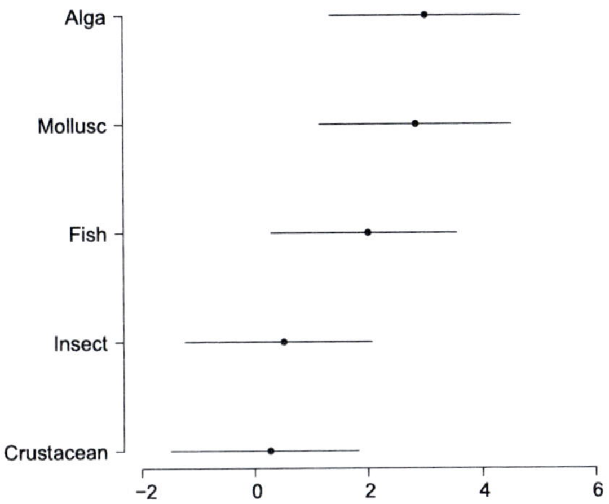

Fig. 5 Caterpillar plot of the taxonomic group random-effects λ_g (the dots correspond to posterior median) along with their 90% credible intervals (solid line segments)

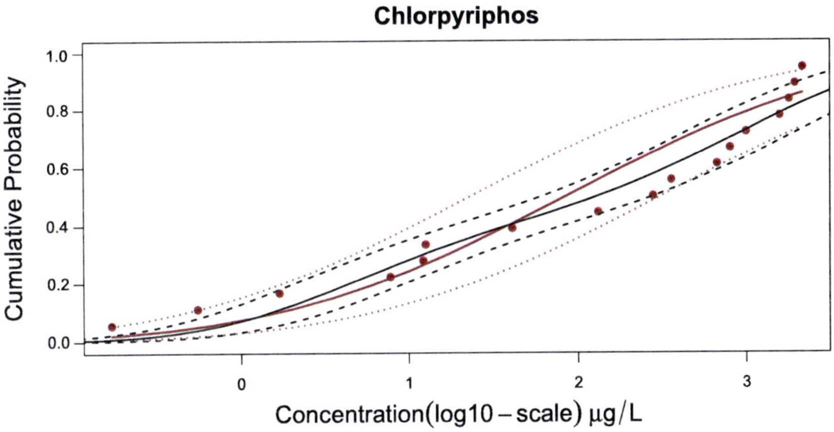

Fig. 6 Fitted SSD curves for the Organophosphorus pesticide Chlorpyrifos. Standard SSD with 90% bands (dark red solid and dashed lines), MSSD with 90% pointwise confidence bands (solid and dashed black lines)

It emerges that in most cases $\widehat{\mathrm{HC5}}_M$ takes values lower than the plug-in or backward regression estimates, thus leading to conservatism. Moreover, often its values are similar to $\widehat{\mathrm{HC5}}_L$, which is expressly designed to lead to conservative estimates. This aspect is particularly valuable for risk assessors, and it is likely due to the very good fit shown by the random-effects model.

Table 1 HC5 estimates on original scale for the selected compounds: sample quantile, ($\widehat{HC5}_q$), plug-in estimate ($\widehat{HC5}_{PI}$), backward regression estimate ($\widehat{HC5}_R$), Aldenberg and Jaworska estimates ($\widehat{HC5}_L$ with $\gamma = 5$), and estimates derived from MSSD curve ($\widehat{HC5}_M$)

Compound	$\widehat{HC5}_q$	$\widehat{HC5}_{PI}$	$\widehat{HC5}_R$	$\widehat{HC5}_L$	$\widehat{HC5}_M$
Deltamethrin	0.07	0.06	0.04	0.01	0.01
Dieldrin	0.07	0.07	0.04	0.01	0.03
Fenvalerate	0.10	0.07	0.04	0.01	0.02
Dichlorvos	0.14	0.49	0.08	0.01	3.24
Endrin	0.24	0.11	0.06	0.02	0.03
Azinphos Methyl	0.30	0.22	0.09	0.01	0.72
Cypermethrin	0.39	0.19	0.10	0.05	0.02
Chlorpyriphos	0.43	0.55	0.21	0.04	0.71
Permethrin	0.48	0.26	0.14	0.04	0.06
Methoxychlor	0.56	0.67	0.29	0.12	0.13
Aldrin	1.18	1.66	1.11	0.56	0.09
Endosulfan	1.26	0.26	0.10	0.04	0.06
Paration	1.39	3.04	1.08	0.26	2.63
Malathion	1.93	3.60	1.28	0.27	3.80
Diazinon	2.52	6.26	2.46	0.59	4.37
DDT	2.56	0.20	0.03	0.02	0.11
Methil Parathion	2.64	6.38	0.75	0.43	7.94
Fenitrothion	3.19	4.41	1.26	0.39	3.55
Toxaphene	3.36	1.10	0.54	0.30	0.10
Naled	5.40	5.29	2.17	0.64	2.34
Fenthion	7.47	17.37	4.90	2.69	4.47
Carbaryl	7.56	13.72	3.32	1.25	9.33
Chlordane	9.23	9.89	8.11	5.44	0.20
Phosphamidon	10.50	23.59	8.42	1.24	38.90
Carbofuran	11.57	10.74	6.11	1.76	2.69
g-HCH	13.62	5.13	3.02	0.96	0.76
Trichlorfon	18.84	9.50	2.64	0.67	10.47
Aldicarb	21.59	29.62	15.40	4.45	7.76
Propoxur	28.20	32.98	11.55	3.55	15.49
Methomil	39.38	36.30	18.72	7.26	5.62
Dimethoate	121.79	119.57	57.08	18.23	25.70

References

Aldenberg, T., & Jaworska, J. S. (2000). Uncertainty of the hazardous concentration and fraction affected for normal species sensitivity distributions. *Ecotoxicology and Environmental Safety, 46*, 1–18.

Aldenberg, T., & Luttik, R. (2002). Extrapolation factors for tiny toxicity data sets from species sensitivity distributions with known standard deviation. In L. Posthuma, G. W. Suter, & T. P. Traas (Eds.), *Species sensitivity distributions in ecotoxicology* (pp. 103–118). Boca Raton: Lewis Publishers.

Core Team, R. (2019). *R: a language and environment for statistical computing*. Vienna, Austria: R Foundation for Statistical Computing.

ECHA (2008). Guidance for the implementation of REACH: Guidance on information requirements and chemical safety assessment. *Chapter R.10: Characterisation of dose [Concentration]-response for environment*, May 2008. http://guidance.echa.europa.eu/docs/guidancedocument/informationrequirementsr10en.pdf. Accessed June 2014.

Forbes, V. E., & Calow, P. (2002). Species sensitivity distributions revisited: a critical appraisal. *Human and Ecological Risk Assessment: An International Journal, 8*(3), 473–492.

Gelman, A., & Hill, J. (2007). *Data analysis using regression and multilevel/hierarchical models*. Cambridge, New York.

Gelman, A. (2015). Analysis of variance—why it is more important than ever. *The Annals of Statistics, 33*(1), 1–33.

Grist, E. P. M., O'Hagan, A., Crane, M., Sorokin, N., Sims, I., & Whitehouse, P. (2006). Bayesian and time-independent species sensitivity distributions for risk assessment of chemicals. *Environmental Science & Technology, 40*(1), 395–401.

Hickey, G., & Craig, P. (2012). Competing statistical methods for the fitting of normal species sensitivity distributions: recommendations for practitioners. *Risk Analysis, 32*(7), 1232–1243.

Hickey, G. L., Craig, P. S., Luttik, R., & de Zwart, D. (2012). On the quantification of intertest variability in ecotoxicity data with application to species sensitivity distributions. *Environmental Toxicology and Chemistry, 31*(8), 1903–1910.

Hickey, G. L., Kefford, B. J., Dunlop, J. E., & Craig, P. S. (2008). Making species salinity sensitivity distributions reflective of naturally occurring communities: using rapid testing and bayesian statistics. *Environmental Toxicology and Chemistry, 27*(11), 2403–2411.

Newman, M. C., Ownby, D. R., Mézin, L. C. A., Powell, D. C., Christensen, T. R. L., Lerberg, S. B., et al. (2000). Applying species-sensitivity distributions in ecological risk assessment: assumptions of distribution type and sufficient numbers of species. *Environmental Toxicology and Chemistry, 19*(2), 508–515.

O'Hagan, A., Crane, M., Grist, E. P. M., & Whitehouse, P. (2005). Estimating species sensitivity distributions with the aid of expert judgements. *Research Report*, No. 556/05.

Posthuma, L., Suter, G. W., & Traas, T. P. (2002). *Species sensitivity distribution in ecotoxicology*. Boca Raton, FL: Lewis Publishers.

Shabri, A. (2002). A comparison of plotting formulas for the Pearson type III distribution. *Journal Teknologi, 36*(C), 61–74.

Stephan, C. E. (2002). Use of species sensitivity distributions in the derivation of water quality criteria for aquatic life by the U.S. Environmental Protection Agency. In L. Posthuma, G. W. Suter, & T. P. Traas (Eds.), *Species sensitivity distributions in ecotoxicology* (pp. 211–220). Boca Raton: Lewis Publishers.

Wagner, C., & Lokke, H. (1991). Estimation of ecotoxicological protection levels from NOEC toxicity data. *Water Research, 25*(10), 1237–1242.

Zwart, D. D. (2002). Observed regularities in SSDs for aquatic species. In L. Posthuma, G. W. Suter, & T. P. Traas (Eds.), *Species sensitivity distributions in ecotoxicology* (pp. 133–154). Boca Raton: Lewis Publishers.

Virtual Encounter Simulations: A New Methodology for Generating Conflict Data

Georg P. Mueller

Abstract This article presents a new methodology for generating knowledge about group conflicts by the use of social surveys, which mainly inform about characteristics and values of the interviewees. The method is based on the idea of simulating virtual encounters between pairs of persons from the same and different groups in order to determine the value-conflicts between the related individuals. For a more subjective assessment of the situation, inter-group conflicts are compared with intra-group conflicts. This results in a new typology, which allows to conceptualize asymmetrical conflict. The proposed method is applied to the analysis of the national identities in the French- and German-speaking parts of Switzerland. It turns out, that the two groups have less conflict about "Swissness" than the traditional methods of analysis suggest.

1 Introduction and Overview

Representative standardized interviews have for a long time been the usual method for collecting data about societies. For reuse by other researchers, many of these social surveys are stored as computer-readable data files in national and international data archives like, e.g. UniData (2019), GESIS (2019), or ICPSR (2019). Such archives are among the earliest institutions holding *open data,* freely accessible to interested researchers (Kitchin 2014: Chap. 3). As the meta-data of these archives suggest, most of their holdings inform about the attitudes, opinions or socio-demographic characteristics of interviewed persons. To the contrary, data about interpersonal relations like *conflicts* are rather rare in these archives. Among others, this may have to do with the technical format of the widespread SPSS-files as well as the analytical technics for the statistical analyses of these files (MacInnes 2017).

This article proposes to fill the gap of missing conflict data by virtual encounter simulations, which in earlier works of the author (Mueller 2011, 2017) proved to be

G. P. Mueller (✉)
Faculty of Economics and Social Sciences, University of Fribourg, Fribourg, Switzerland
e-mail: georg.mueller_unifr@bluewin.ch

P. Mariani and M. Zenga (eds.), *Data Science and Social Research II*,
Studies in Classification, Data Analysis, and Knowledge Organization,
https://doi.org/10.1007/978-3-030-51222-4_23

useful for conflict analysis and which relate the current article to artificial societies (JASSS 2019) and computational social science (Templ 2016). It proposes random-matching of conventional monadic data-records from survey research, which yields a new type of data-records that combines information about pairs of randomly selected persons. These new dyadic records can be used for the calculation of value- or opinion-conflicts between pairs of persons belonging to different specific groups. After aggregating the pairs' conflict-scores it is possible to determine the amount of conflict within and between the two mentioned groups. Inter-group conflict can this way be compared with two group-specific benchmarks of intra-group conflicts. The result is a new conflict classification, which distinguishes between nine different categories of inter-group conflict.

In order to demonstrate the usability of the virtual encounter method for practical purposes, the author analyzes the differences between French- and the German-speaking Switzerland with regard to national identities. Political literature (Schmid 2001: Chap. 7; Büchi 2003: 243–287; Hermann and Leuthold 2003: 48–51) and conventional data analyses suggest important differences between the two linguistic groups, which live in different regions of Switzerland and have different political and cultural traditions. However, the present virtual encounter analyses show *only for two* out of eight facets of Swiss national identity ("Swissness") differences, which may lead to conflicts. This has to do with the "noise" produced by the internal benchmarks of conflict, which hides for many facets of Swissness the inter-group conflict.

2 Conflict Simulations

This section describes how inter-group conflict can be simulated on the basis of conventional "monadic" data, in which each analysed individual has a particular data-record with his/her opinions, attitudes and socio-demographic characteristics. The procedure corresponds to Monte Carlo simulation (Mooney 1997) and depicts modern urban societies, where spontaneous contacts with "strangers" prevail (Toennies 1979). Consequently, it makes for these societies sense to analyze dissent between randomly matched persons. In practice, this leads to the following three-step procedure.

Step 1 is the construction of a "dyadic" dataset, which unifies the monadic data of two groups A and B, as described in Fig. 1. In order to ensure statistical independence of the pairs of persons in the resulting virtual encounter file, the monadic files have to be trimmed to the same length by randomly removing some individuals of the longer of the two files: this guarantees that each person from group A or B appears at most once in the new dyadic encounter file. In Fig. 1, the result of this operation is represented by the hatched area of the file of group B. To ensure the randomness of the record matching it is recommended to reorder the sequence of persons in the monadic files by a random process. This is especially important if virtual encounters

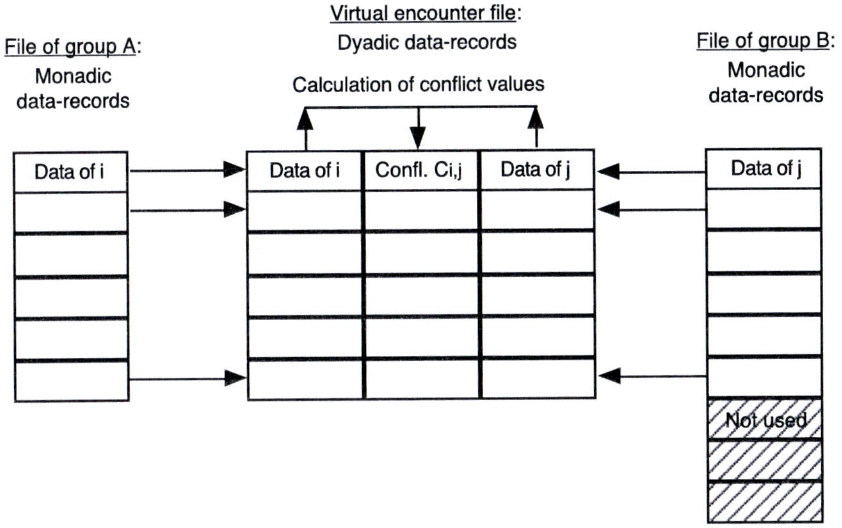

Fig. 1 The construction of a virtual encounter file

within the *same* group are studied, such that in Fig. 1 group A is identical to group B.

Step 2 is the calculation of the conflict $C_{i,j}$ between the randomly matched persons i and j (see Fig. 1) with regard to a characteristic X. For *interval-* or *ratio-scaled* values X_i and X_j of two persons $i \in A$ and $j \in B$, $C_{i,j}$ can either be defined as

$$C_{i,j} = (X_i - X_j)^2 \tag{1a}$$

or alternatively as

$$C_{i,j} = |X_i - X_j| \tag{1b}$$

Since (1b) is also sensitive to small differences between i and j, we prefer (1b) over (1a). For *nominal-scaled* X_i and X_j, we suggest to define

$$C_{i,j} = 0, \text{ if } X_i = X_j \tag{2a}$$

and

$$C_{i,j} = 1, \text{ if } X_i \neq X_j \tag{2b}$$

Step 3 is the aggregation of the inter-individual conflicts $C_{i,j}$ into a collective inter-group conflict, which has the "natural" definition[1]

$$C_{A,B} = \mathrm{MEAN}(C_{i,j} | \, i \in A, \; j \in B) \tag{3}$$

$C_{A,B}$ can be calculated for different conflict items like party preferences, attitudes towards the EU, etc. Comparisons of such items with regard to the amount of conflict between the groups A and B are easily possible, e.g. by ordering the items according to increasing values of $C_{A,B}$. By focussing on the minimum, the maximum and the median of the $C_{A,B}$-values it is possible to identify the least serious, the most serious and the "mean" conflict item.

However, with regard to the behavioural consequences of the concerned individuals, $C_{A,B}$ is only of limited interest: whether interactions with the other group are perceived as conflictive or harmonic depends very much on the respective benchmark of the intra-group conflicts, which is for members of A

$$C_{A,A} = \mathrm{MEAN}(C_{i,j} | \, i \in A, \; j \in A) \tag{4a}$$

and for members of B[2]

$$C_{B,B} = \mathrm{MEAN}(C_{i,j} | \, i \in B, \; j \in B) \tag{4b}$$

Thus, there are *relative* conflicts of group A with group B

$$K_{A,B} = C_{A,B} - C_{A,A} \tag{5a}$$

and of group B with group A[3]

$$K_{B,A} = C_{B,A} - C_{B,B} = C_{A,B} - C_{B,B} \tag{5b}$$

The relative conflicts $K_{A,B}$ and $K_{B,A}$ are by definition independent each of the other. They can be positive, negative or zero. If $K_{A,B}$ or $K_{B,A}$ are *negative* the respective relations between the two groups are *harmonic*, since the external conflicts are lower than the internal ones. Similarly, if $K_{A,B}$ or $K_{B,A}$ are *zero* the respective relations between the two groups are *neutral*. Thus, there are $3 \times 3 = 9$ possible scenarios of inter-group relations, which are *defined* by the signs of $K_{A,B}$ and $K_{B,A}$. Figure 2 gives an overview of these nine types of inter-group relations, which are in many cases *asymmetrical*.

[1]This definition can also be applied to *nominal-scaled* data with conflict definitions described by (2a) and (2b). In this case, it corresponds to the share of virtual encounters of persons with different opinions.

[2]For nominal-scaled data, $C_{A,A}$ and $C_{B,B}$ are the shares of virtual encounters of persons within group A or group B, who differ with regard to their opinions: thus, the definitions (4a) and (4b) can also be applied to this type of data.

[3]From the Eqs. (3) and (1a), or (1b), or (2a, b) follows that $C_{B,A} = C_{A,B}$.

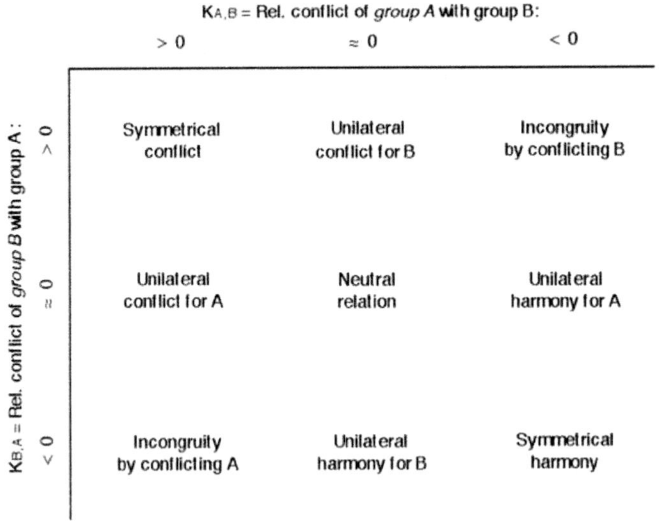

Fig. 2 The different types of inter-group conflict

As a matter of course, Fig. 2 contains also three classical symmetrical situations, which are represented by its main diagonal: *symmetrical conflict, neutral relations* and *symmetrical harmony,* depending on whether the relative conflicts $K_{A,B}$ and $K_{B,A}$ are both positive, zero or negative. Moreover, there are two types of asymmetrical *unilateral conflicts* if one of the parties perceives more external than internal conflict, whereas for the other party the situation between internal and external conflict is balanced and the relative conflict is zero. An analogous reasoning justifies the existence of two cases of *unilateral harmony:* one of the groups has a positive perception of the other, which in turn has a neutral perception of the first. Finally, there are also two *incrongruities,* if one of the parties perceives conflict with the other, whereas the second party has so much internal conflict, that the relations with the first appear as rather harmonic.

3 An Exemplary Analysis: Swiss National Identities

Switzerland is a federal state, built in the middle of the nineteenth century by the unification of formerly independent small republics with very heterogeneous political, religious and linguistic traditions. Consequently, Switzerland is supposed to still have a rather fragmented national identity, with cleavage lines reflecting the mentioned traditions. One of these cleavages, which has persisted over time is the conflict between French-speaking Western Switzerland and the German-speaking East of the country. As compared to the East, the French-speaking West of Switzerland is more Europe- and welfare-oriented and stresses the importance of the local government as a counterbalance to the central state in Berne (Hermann and Leuthold 2003: 48–49). Consequently, we will in this section analyze whether the known cleavages between the French- and the German-speaking citizens lead to conflicts about *Swissness,* i.e. the Swiss national identity.

One of the data sources for this purpose are the interviews of the International Social Survey Programme 2003 on national identities, which contain the following variables:

V56: Nationality: Swiss, or other. This variable was used in order to *exclude* from the present analysis of those *foreigners* with Swiss residence, who took part in the Swiss EVS-interviews.
V64: Primary language that is spoken at home: French or German. This allowed to identify the two linguistic groups of the present analysis. Since this language-question was dropped in the more recent ISSP surveys about national identities, the present analysis was done with relatively old data, collected in 2003.
V11: Importance of having been *born in Switzerland* for being "truly Swiss": 1 = very important, 2 = fairly important, 3 = not very important, 4 = not important at all.
V12: Importance of *having the Swiss citizenship* for being "truly Swiss": response categories as for V11.
V13: Importance of having *lived most of life in Switzerland* for being "truly Swiss": response categories as for V11.
V14: Importance of *speaking a Swiss national language* for being "truly Swiss": response categories as for V11.
V15: Importance of *Christian faith* for being "truly Swiss": response categories as for V11.
V16: Importance of *respecting Swiss laws and political institutions* for being "truly Swiss": response categories as for V11.
V17: Importance of *feeling Swiss* for being "truly Swiss": response categories as for V11.
V18: Importance of *having Swiss ancestry* for being "truly Swiss": response categories as for V11.

Following the methodology of Sect. 2, we calculated for the previously listed criteria of Swissness (V11 to V18) the inter-group conflict between the French- and German-speakers as well as the intra-group conflicts of the two language groups. The results of these virtual encounter simulations are presented in Table 1, ordered by increasing inter-group conflict between the French- and the German-speakers. The series of criteria of Swissness starts not so surprisingly with the least conflictive *Respecting Swiss laws/political institutions* (V16) and ends with the most conflictive item *Christian faith* (V15). The two columns of the intra-group conflicts of the French- and the German-speakers are rather similar with regard to the order of the items. As a consequence, the relative conflict between the two linguistic groups is rather small, entailing for most variables neutral relations (see Fig. 2). There are only two exceptions to this lack of relative conflict: One is *Having the Swiss citizenship* (V12), which triggers for the German-speakers a unilateral conflict, which does, however, not exist for the French-speakers. The other exception is the variable *Speaking a Swiss nat. language* (V14), which leads to an incongruity (see Fig. 2) between the German- and the French-speakers. In the German-speaking East, this is a consensual must of Swissness, whereas in the French West it is on the average less important but highly debated. Consequently, the mutual perceptions are incongruent, i.e. negative from the perspective of the East and positive for the West of Switzerland.

4 Comparisons with Traditional Analyses

The standard method for comparing groups with regard to values and attitudes is obviously different from the previously presented virtual encounter simulations: it is usually based on t-tests of *mean values* of group characteristics (Kanji 1993: 23) instead of comparisons of external and internal *group-conflicts.* Not only the method but also the results and their interpretation can be very different, as Table 2 demonstrates. It shows straightforward the differences between the German- and the French-speakers with regard to different criteria of Swissness, which are in most cases statistically *significant.* This is in sharp contrast to Table 1, where most inter-group conflicts were not too different from the related intra-group conflicts. Moreover, the ordered sequence of *inter-group conflict* is obviously not the same as the ordered sequence of the *difference of group-means.*

The disparities between Tables 1 and 2 suggest the question, which of the two tables is closer to reality. The answer depends on the type of "noise", which is considered. The traditional analysis of Table 2, based on classical inferential statistics, refers only to sampling errors. This is mainly the perspective of the *external* observers, not directly involved in the analysed processes. Virtual encounter simulation considers in addition also the "noise-level" of the intra-group conflicts. This reflects the perspective of the simulated *insiders,* directly involved in the analysed processes. Their threshold of perception is generally higher than the purely statistical benchmark of the external observers. Consequently Table 1, which is based on virtual encounter simulations reports less conflict than Table 2 that refers to conventional statistical analysis.

For *human conflicts* the perspective of the internal spectator is probably more relevant than the external perspective. This assumption is further supported by the fact that the virtual encounter approach with two different group-specific benchmarks is able to explain the often observed phenomena of unilateral conflict and unilateral harmony, which the traditional approach even fails to conceptualize. Moreover, the virtual encounter approach helps to avoid ecological fallacies (Crow 2006): the equality of the mean values of two groups with regard to a criterion of Swissness is blind to inter-group conflicts at the individual level, which are correctly identified by the virtual encounter approach.

5 Summary and Outlook

This article proposes a new methodology for identifying hidden political and social conflicts in survey data, which are abundantly available in open data archives for secondary analyses (Stewart and Kamins 1998). It is based on the idea of simulating virtual encounters between the participants of the mentioned surveys. This way it becomes possible to define benchmarks based on intra-group conflicts, which can be used in order to assess the level of inter-group conflict. Since each of the analysed groups has its own internal conflict benchmark, the proposed analysis leads to a new classification of conflict, which allows to identify asymmetrical behaviour like incongruities and unilateral conflict or harmony.

In this article, the method of virtual encounters is applied to the national identities of the French- and German-speaking regions of Switzerland. It turns out that conflicts between these groups with regard to criteria of Swissness are much smaller than the theoretical literature (Schmid 2001: Chap. 7; Hermann and Leuthold 2003: 48–51) and conventional statistical analyses (see Table 2) suggest. The conflicts identified by the proposed virtual encounter method refer only to the importance of *citizenship* and the *linguistic performance.* The question of citizenship causes a unilateral conflict, where the German-speakers perceive a difference to the French-speakers, for whom the resulting conflict is not really salient. The criterion of linguistic performance creates an incongruity between the French- and the German-speakers, where the first group seems to have a rather positive perception of the second, which in turn has a negative attitude towards the first. Since the relations between the two linguistic groups have changed over time (Büchi 2003: 243–287), the observed incongruity may also be the result of the statistical aggregation of different *generations* with different political attitudes. Hence the disaggregation of the simulated virtual encounters according to the age of the matched pairs is a project for the future, which may contribute to a better understanding of the somewhat paradoxical situation of the conflict incongruity between the French- and the German-speakers.

Table 1 Comparison of inter- and intra-group conflicts of the German- and French-speakers

Criterion of Swissness	Intra-German conflict	German–French conflict				Intra-French conflict	N
Respecting Swiss laws/pol. institutions (V16)	0.491	≈ (0.064)	*0.585*	≈ (0.351)		*0.561*	171-171
Speaking a Swiss nat. language (V14)	*0.485*	< (0.016)	0.629	< (0.006)		0.794	165-169
Feeling Swiss (V17)	<u>0.788</u>	≈ (0.116)	0.892	≈ (0.154)		0.799	165-169
Having the Swiss citizenship (V12)	0.707	< (0.008)	<u>0.933</u>	≈ (0.441)		<u>0.926</u>	163-167
Lived most of life in Switzerland (V13)	<u>0.901</u>	≈ (0.095)	<u>1.00</u>	≈ (0.223)		<u>0.940</u>	167-171
Born in Switzerland (V11)	1.09	≈ (0.242)	1.03	≈ (0.234)		0.939	163-169
Having Swiss ancestry (V18)	1.17	≈ (0.500)	1.17	≈ (0.120)		**1.08**	153-161
Christian faith (V15)	**1.29**	≈ (0.305)	**1.23**	≈ (0.085)		**1.08**	161-163

Legend (): *p*-values of one-tailed one-sample t-tests, comparing intra- with inter-group conflicts. <, >: Difference statistically significant at error-level 5 % (see p-value). ≈: Difference statistically not significant (see p-value). V11, …, V18: Variables referring to the ISSP (2003) dataset. Criteria of Swissness ordered by increasing German–French inter-group conflict. Bold: Maximum of col. Underlined: Adjacent to median of col. Italics: Minimum of col.
Note Median of Intra-German conflict = 0.845; median of German–French conflict = 0.967; median of Intra-French conflict = 0.933

Table 2 Differences between the French- and the German-speakers with regard to group-means of Swissness

Criterion of Swissness	G-speakers: mean values	Diff. of mean values	F-speakers: mean values	N of cases
Respecting Swiss laws/pol. institutions (V16)	1.591	0.163**	1.754	171-171
Speaking a Swiss nat. language (V14)	1.400	0.344***	1.744	168-170
Feeling Swiss (V17)	1.845	0.284***	2.129	168-170
Having the Swiss citizenship (V12)	1.621	0.505***	2.126	167-169
Lived most of life in Switzerland (V13)	2.135	0.072	2.207	169-171
Born in Switzerland (V11)	2.288	0.203*	2.491	167-170
Having Swiss ancestry (V18)	2.554	0.353***	2.907	162-166
Christian faith (V15)	2.645	0.421***	3.066	166-167

Legend Criteria of Swissness ordered by increasing inter-group conflict in Table 1. Significances of 1-tailed two-sample t-tests: *: 5%, **: 1%, ***: 0.1%. G-speakers: German-speakers; F-speakers: French-speakers

Note The higher the mean value of a criterion of Swissness, the *less* important it is for the concerned group

References

Büchi, C. (2003). *"Röstigraben" (The "roasted potatoes" cleavage)* (3rd ed.). Zürich: Verlag Neue Zürcher Zeitung.

Crow, I. (2006). Ecological Fallacy. In V. Jupp (Ed.), *The Sage dictionary of social research methods* (pp. 82–83). London: Sage Publications.

GESIS. (2019). *Datenbestandskatalog (Data catalogue).* https://www.gesis.org/angebot/archivieren-und-registrieren/datenarchivierung/datenzugang. Accessed 14 Sep 2019.

Hermann, M., & Leuthold, H. (2003). *Atlas der politischen Landschaften (Atlas of the political landscapes).* Zürich: vdf Hochschulverlag.

ICPSR. (2019). *Find & Analyze Data.* https://www.icpsr.umich.edu/icpsrweb/ICPSR/. Accessed 14 Sep 2019.

ISSP. (2003). ZA Study 3910: International Social Survey Programme: National Identity II, Codebook. In GESIS. https://dbk.gesis.org/dbksearch/sdesc2.asp?no=3910&tab=4&db=e. Accessed 15 Sep 2019.

JASSS. (2019). *Journal of Artificial Societies and Social Simulation.* http://jasss.soc.surrey.ac.uk/JASSS.html. Accessed 14 Sep 2019.

Kanji, G. (1993). *100 statistical tests.* London: Sage Publications.

Kitchin, R. (2014). *The data revolution.* Los Angeles: Sage Publications.

MacInnes, J. (2017). *An introduction to secondary data analysis with IBM SPSS statistics.* London: Sage Publications.

Mooney, C. (1997). *Monte Carlo simulation.* Thousand Oaks: Sage Publications.

Mueller, G. (2011). Microsimulation of virtual encounters: a new methodology for the analysis of socio-cultural cleavages. *International Journal of Microsimulation, 4*(1), 21–34.

Mueller, G. (2017). On the use of microsimulation for investigating ideological dissent: exemplary analyses of the values of the European political left. *ASK Research & Methods, 26*(1), 61–80.

Schmid, C. (2001). *The politics of language: conflict, identity, and cultural pluralism in comparative perspective.* Oxford: Oxford University Press.

Stewart, D., & Kamins, M. (1998). *Secondary research: information sources and methods.* Newbury Park: Sage Publications.

Templ, M. (2016). *Simulation for data science with R.* Birmingham: Packt Publishing.

Toennies, F. (1979). *Gemeinschaft und Gesellschaft (Community and Society).* Darmstadt: Wissenschaftliche Buchgesellschaft.

UniData (2019). *Bicocca Data Archive.* https://www.unidata.unimib.it/?lang=en. Accessed 14 Sep 2019.

Is Public Service Motivation–Performance Relationship Mediated by Other Factors?

Raffaela Palma, Anna Crisci, and Luigi D'Ambra

Abstract Although the association between public service motivation (PSM) of public employees and Performance has received increased attention, there are yet inconsistencies in the literature regarding how the PSM–performance relationship may be mediated by other factors. This study is based on a sample of 618 Italian public teachers and considers a set of hypotheses in public education in which the relationship is mediated by person–organization fit (P-O fit) and organizational commitment (OC). This mediated relationship varies depending on how performance is considered.

1 Introduction

Public service motivation (PSM) is described as the motivation to contribute to society (Perry and Wise 1990), can drive employees to perform better in public sector work (Perry and Wise 1990), when working on meaningful public services.

Empirical research has found that PSM is positively related to organizational commitment (OC), to job satisfaction and to several types of performance such as job performance (Vandenabeele 2009), extra-role behaviour (Gould-Williams et al. 2015), supervisor ratings (Bright 2007) and even student grades as a measure of teacher performance (Andersen et al. 2014). This has made PSM a promising concept in improving public services (Andersen et al. 2014). However, there is a need to better clarify the effect of PSM on individual performance (IP) (Ritz et al. 2016; Vandenabeele, 2009). To better clarify the PSM–performance link, scholars have proposed two interesting theoretical approaches: PSM could have both a direct and an indirect influence on employee performance (Bright 2007; Vandenabeele 2009), as the

R. Palma · A. Crisci (✉) · L. D'Ambra
Department of Economics, Management and Institutions, University of Naples "Federico II", Naples, Italy
e-mail: anna.crisci@unina.it

R. Palma
e-mail: raffaela.palma@unina.it

P. Mariani and M. Zenga (eds.), *Data Science and Social Research II*,
Studies in Classification, Data Analysis, and Knowledge Organization,
https://doi.org/10.1007/978-3-030-51222-4_24

environment plays a role in determining behaviour. The relationship between PSM and performance thus seems much more nuanced and complex than the original proposition that PSM is positively related to performance (Perry and Wise 1990). Using two types of self-reported performance, specifically organizational citizenship behaviours (OCBs) as a form of contextual performance and IP, as a form of employee performance, this research suggests that the relationship between PSM and performance may be mediated by both P-O fit and OC, and that P-O improves employee performance through its influence on OC (Kristof-Brown et al. 2005; Kim 2012; Vandenabeele 2009). To do so, based on an existing study on higher education (Jin et al. 2018), we explore the direct and indirect effects of PSM on different types of performance in public education.

Understanding how PSM influences both contextual and individual performance will contribute to both theory and practice.

In Sect. 2, the literature review, regarding both the direct and indirect effects of PSM on individual performance, the research hypotheses are shown. In Sect. 3, we describe the data, the method used and the considered variables. Finally, we proposed final discussion in Sect. 4

2 Public Service Motivation and Performance

Many empirical studies have linked PSM to specific employee outcomes. One of the most studied is individual employee performance (Ritz et al. 2016; Bright 2007; Vandenabeele 2009). Perry and Wise (1990) have claimed that public sector individuals with high PSM perform better because such individuals would embrace work that has a positive impact on the well-being of other people (Grant 2007). Despite the aforementioned empirical findings, however, causality between PSM and performance remains unclear.

As regards performance in public sector, there is no a systematization of the conceptual space of performance in public organizations (Andersen et al. 2016).

Two important aspects of the multifaced concept of performance in public sector are a form of contextual performance which is linked to the specific context where employees work and the individual performance linked to the specific employees, but easily comparable if used in other sectors (Vandenabeele 2009).

This study includes both aspects, contextual and individual, in the form of OCBs, and IP.

2.1 *Organizational Citizenship Behaviours*

OCBs are a form of contextual performance (Borman and Motowidlo 1997) that are defined as individual behaviours that are beneficial to the organization and are discretionary, not directly or explicitly recognized by the formal reward system (Organ

1988). This concept can be extended to classroom settings and directed at students (Oplatka 2006), as teachers interact and spend more time with students than with their co-workers. Teachers spend extra time meeting parents, preparing lessons, correcting homework, helping colleagues in their work. PSM addicted employees are prepared to sacrifice personal interests to help the community even without tangible personal rewards (Perry and Wise 1990). These dynamics, therefore, comprise the theoretical underpinnings for the argument that PSM can be expected to correlate positively with OCBs in public education settings. In fact OCBs could lead to a significant, direct influence on the quality of classroom relations with students and on student achievement (Oplatka 2006), that is on helping society, the noble aim of PSM. We, therefore, consider the following hypothesis:

Hypothesis 1: PSM is positively associated with the OCBs
Hypothesis 2: PSM is positively associated with IP.

2.2 *The Mediating Role of Pearson-Organization Fit*

P-O fit theory suggests that "fit" between individuals and organizations influences individuals' attitudes and behaviours (Kim 2012). Scholars argue that, as the similarity between individuals and organizations increases, employees become more committed to and thus more productive in their jobs (Farooqui and Nagendra 2014; Bright 2007). It is reasonable to argue that PSM addicted individuals are more likely than others to have values and goals similar to those of the public service organization. Other scholars found that P-O fit is related to both OCBs and task performance (Hoffman and Woehr 2006) arguing that the alignment of organizational culture and personal values leads individual to exhibit extra-role behaviours in the workplace (Jin et al. 2018). It is, therefore, reasonable to assume that those who are a good fit with their organizations will be motivated to achieve better results both in individual and contextual terms. Thus, we explore the following hypothesis:

Hypothesis 3: PSM has an indirect, positive effect on Performance through its influence on P-O fit.

2.3 *The Mediating Role of Organizational Commitment*

OC is the strength of an individual's identification with and involvement in a particular organization (Porter et al. 1974). Several researchers found a significant positive correlation between PSM and OC (Crewson 1997), OC and individual performance (Park and Rainey 2007), PSM and individual performance through its influence on OC (Vandenabeele 2009). Individuals with PSM have motives and public values which lead to greater job commitment and a more positive attitude of employees toward their job (Park and Rainey 2007). Then, we have the following research hypothesis:

Hypothesis 4: PSM has an indirect, positive effect on Performance through its influence on OC.

Does P-O Fit influence OC? The answer to this question could be positive. That is, teachers with high levels of PSM will more likely seek a high congruence between their characteristics and those of their school. Moreover, they will be more committed to their work, which commitment has, in turn, a positive impact on their performance behaviour and individual productivity.

In other terms, P-O fit and OC are both involved in the PSM–individual performance relationship. Not unexpectedly, research has found that P-O fit and OC factors are inter-correlated. Therefore, PSM has an indirect, positive effect on OC through its influence on P-O fit (Kim 2012). It means that public employees with high levels of PSM develop a sense of belonging to their organization and are willing to make sacrifices to contribute to its well-being insofar they feel that their values meet those of their organization (Pandey et al. 2008). On the basis of an existing study on higher education (Jin et al. 2018), we explore a three-path mediation as follows:

Hypothesis 5: PSM has an indirect, positive effect on OCBs through the influence of P-O Fit on OC.

3 Method

3.1 *Sample and Procedure*

The survey took place from September to December 2015 through a hand-delivered structured questionnaire. The sample consisted of 618 public teachers working in all public school grades, yielding an overall response rate of 95.8 %. It included more females (78.3 %) than males (21.7 %). Given that we used a convenience sample, we acknowledge possible response biases might limit the generalizability of the current findings. For this reason, we adopted some remedies to reduce the risk of method bias. For instance, all participants were assured anonymity in completing the survey; the researchers accurately explained to respondents why the questions were important and the necessity of accurate answers and separated motivation and performance in the survey. Furthermore, before testing hypotheses, in order to verify the statistical detection of common method bias (CMB) for the dataset, we carried out the Harman single factor (Harman 1960; Podsakoff et al. 2003. The sample of schools consisted of 6 high schools, and 3 Comprehensive Schools including 9 nursery schools, 8 elementary schools, 5 lower secondary schools. A Comprehensive School has a single school principal but includes different school grades, generally Nursery, Primary and Lower Secondary. The median age was 55 years, ranging from 51–60 years and respondents had been in service in school (job tenure) for just over 20 years at the time of the survey. The predominance of females and their age (aged 50 or older) reflects the profile of the European teacher published by Eurostat (Report

Eurostat 2015). Schools in the Campania Region of Italy were selected including contexts regarding both city centre and suburban location.

3.2 Measures

Responses to all questionnaire items associated with each construct in the study took the form of a choice on a 5-point Likert-type scale ranging from 1 (*strongly disagree*) to 5 (*strongly agree*).

The variables are self-reported performance measures. A better way to collect performance (contextual or not) data is from the physical records of organizations or from supervisors, we chose the present approach for several reasons. The first is that teachers were granted anonymity; the second, the participants themselves may be able to assess their own contextual and individual performance behaviour better than their supervisors, as they have much more information (Organ and Konovsky 1989). The third these statistics were not available at the time of data collection. Therefore, a four-item measurement scale of perceived IP was used in the survey (Vandenabeele 2009). OCBs were measured using four items modified to fit the context of public education (Benkhoff 1997).

The independent variable, PSM, was measured using an aggregate of four items from Perry's (1996) original scale, which has been validated as a multi-item, unidimensional measure of PSM. These items capture the three dimensions—commitment to public interest, compassion and self-sacrifice—as representative of the affective or normative motives most closely associated with the altruistic appeal of public sector values (Perry and Wise 1990). The fourth dimension, attraction to policymaking, is removed, as in other studies (Perry and Wise 1990) as it has consistently proven difficult to interpret and validate (Bozeman and Su 2015). Limiting the number of items in the analysis increases clarity as well as limits the possible amount of Type I errors. P-O fit was measured using the four items originally developed by Bright (2007) based on a review of existing research (Kristof-Brown et al. 2005). OC was measured using the six items from Vandenabeele (2009).

3.3 Data Analysis

Before testing hypotheses, in order to verify the statistical detection of common method bias (CMB) for the dataset, we carried out the Harman single factor (Harman 1960), whose value is equal to 41.3 %, which does not exceed the commonly accepted threshold of 50 % (Podsakoff et al. 2003).

Figure 1 shows the research framework. To test the hypotheses from an explorative viewpoint, we adopted the PLS-Path Modelling algorithm (Sarstedt et al. 2017), by using the module R-package. In addition, we used bootstrapping techniques that can more accurately show the significance of mediation processes (Zhao et al. 2010).

Fig. 1 Research framework

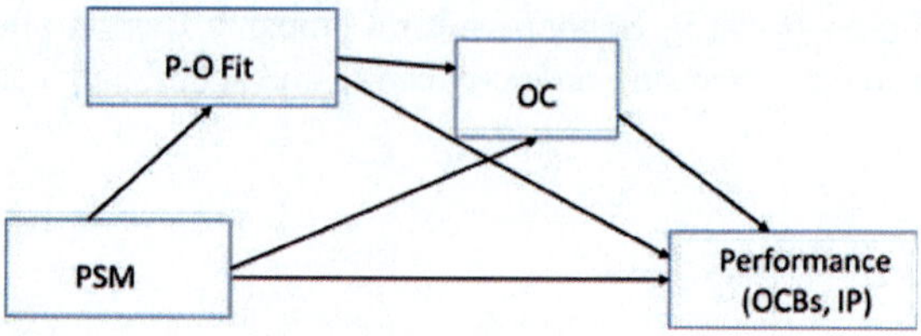

Table 1 Bootstrap validation for path coefficients

	Original	Mean.Boot	Std.Error	perc.025	perc.975
PSM - > P-O fit	0.3044	0.3125	0.0441	0.2328	0.399
PSM - > OC	0.2790	0.2821	0.0333	0.2123	0.338
PSM - > OCBs	0.2301	0.2376	0.0482	0.1462	0.315
PSM - > IP	0.3568	0.3597	0.0482	0.2624	0.447
P-O fit - > OC	0.5675	0.5655	0.0354	0.5004	0.644
P-O fit - > OCBs	0.0781	0.0812	0.0619	-0.0356	0.197
P-O fit - > IP	0.1267	0.1342	0.0591	0.0226	0.236
OC - > OCBs	0.2014	0.194	0.0631	0.087	0.308
OC - > IP	0.1376	0.1358	0.0543	0.0457	0.240

The indirect effects of PSM - > IP and PSM - > OCBs are 0.1007 and 0.1147, respectively

Before proceeding to the estimation of parameters, we have verified the unidimensionality of the manifest variables (VMs) blocks by means of Dillon-Goldstein's rho (Dillon and Goldstein 1984). A block is unidimensionality if this index is greater than 0.7. The value of the index is >0.7 for all the observed VMs Blocks. As regards the measurement model, all loading coefficients are significant and positive.

Table 1 shows the bootstrap results for path coefficients. All path coefficients are significant at 5 %, except the link between P-O Fit and OCBs. The highest path-coefficient value is that for the link between P-O Fit and OC. It means that public service motivated teachers feel that their values match those of their organization, they develop a sense of belonging to their organization and are willing to give something of themselves to contribute to its well-being, success, and to benefit individuals within (students, co-workers) and outside (parents, and thus community) their work context.

4 Discussion and Conclusion

Given that the PSM–Performance relationship is more complex than what is expected (Perry 1996), some scholars tried to verify this link through a mediation analysis recognizing that PSM affects performance indirectly (e.g. Bright 2007; Gould-Williams et al. 2015; Vandenabeele 2009). Other scholars explored the context dependency of the PSM–performance relationship (Lynggaard et al. 2018).

Our study includes multiple mediators between PSM and performance in the model in a specific context that is the public school.

Here, two important mediators come into play: OC and P-O Fit. Our study, however, also shows that PSM–performance relationships may vary depending on how the performance is measured and in which specific context the performance itself is measured (van Loon et al. 2017).

The present study offers important insights. For example, it appears successful in a selection of teachers who are highly motivated by public service. A teacher highly motivated by public service who understands and shares the missions and goals of the school has greater potential for commitment to the organization, which in turn increases his/her performance.

Finally, our data are based on a convenience sample collected from teachers in some public school in the South of Italy. More research is needed to generalize our findings to include universities, sectors and other countries.

References

Andersen, L. B., Boesen, A., & Pedersen, L. H. (2016). Performance in public organizations: clarifying the conceptual space. *Public Administration Review, 76*(6), 852–862.

Andersen, L. B., Heinesen, E., & Pedersen, L. H. (2014). How does public service motivation among teachers affect student performance in schools? *Journal of Public Administration Research and Theory, 24,* 651–671.

Benkhoff, B. (1997). Disentangling organizational commitment: the dangers of the OCQ for research and policy. *Personnel Review, 26*(1/2), 114–131.

Borman, W. C., & Motowidlo, S. J. (1997). Task performance and contextual performance: the meaning for personnel selection research. *Human Performance, 10*(2), 99–109.

Bozeman, B., & Xuhong, S. (2015). Public service motivation concepts and theory: a critique. *Public Administration Review, 75*(5), 700–710.

Bright, L. (2007). Does person-organization fit mediate the relationship between public service motivation and the job performance of public employees? *Review of Public Personnel Administration, 27*(4), 361–379.

Crewson, P. E. (1997). Public-service motivation: building empirical evidence of incidence and effect. *Journal of Public Administration Research and Theory, 7,* 499–518.

Dillon, W. R., & Goldestein, M. (1984). *Multivariate analysis: method and applications.* New York: Wiley.

Farooqui, M. S., & Nagendra, A. (2014). The impact of person organization fit on job satisfaction and performance of the employees. *Procedia Economics and Finance, 11,* 122–129.

Gould-Williams, J. S., Mostafa, A. M. S., & Bottomley, P. (2015). Public service motivation and employee outcomes in the Egyptian public sector: testing the mediating effect of person-organization fit. *Journal of Public Administration Research and Theory, 25,* 597–622.

Grant, A. M. (2007). Relational job design and the motivation to make a prosocial difference. *Academy of Management Review, 32,* 393–417.

Harman, H. H. (1960). *Modern factor analysis.* Chicago, IL: University of Chicago Press.

Hoffman, B. J., & Woehr, D. J. (2006). A quantitative review of the relationship between person-organization fit and behavioral outcomes. *Journal of Vocational Behavior, 68,* 389–399.

Jin, M. H., McDonald, B., & Park, J. (2018). Does public service motivation matter in public higher education? Testing the theories of person–organization fit and organizational commitment through a serial multiple mediation model. *the American Review of Public Administration, 48*(1), 82–97.

Kim, S. (2012). Does person-organization fit matter in the public sector? Testing the mediating effect of person-organization fit in the relationship between public service motivation and work attitudes. *Public Administration Review, 72,* 830–840.

Kristof-Brown, A. L., Zimmerman, R. D., & Johnson, E. C. (2005). Consequences of individuals' fit at work: a meta-analysis of person-job, person-organization, person-group, and person-supervisor fit. *Personnel Psychology, 58,* 281–342.

Lynggaard, M., Pedersen, M. J., & Andersen, L. B. (2018). Exploring the context dependency of the PSM–performance relationship. *Review of Public Personnel Administration, 38*(3), 332–354.

Oplatka, I. (2006). Going beyond role expectations: toward an understanding of the determinants and components of teacher organizational citizenship behavior. *Educational Administration Quarterly, 42,* 385–423.

Organ, D. W. (1988). *OCB: the good soldier syndrome.* Lexington, MA: Lexington Books.

Organ, D. W., & Konovsky, M. (1989). Cognitive versus affective determinants of organizational citizenship behavior. *Journal of Applied Psychology, 74,* 157–164.

Pandey, S. K., Wright, B. E., & Moynihan, D. P. (2008). Public service motivation and interpersonal citizenship behavior: testing a preliminary model. *International Public Management Journal, 11,* 89–108.

Park, S. M., & Rainey, H. G. (2007). Antecedents, mediators, and consequences of affective, normative, and continuance commitment: empirical tests of commitment effects in federal agencies. *Review of Public Personnel Administration, 27,* 197–226.

Perry, J. L., & Wise, L. R. (1990). The motivational bases of public service. *Public Administration Review,* 367–373.

Podsakoff, P. M., MacKenzie, S. B., Lee, J. Y., & Podsakoff, N. P. (2003). Common method biases in behavioral research: a critical review of the literature and recommended remedies. *Journal of Applied Psychology, 88*(5), 879–903.

Porter, L. W., Steers, R. M., Mowday, R. T., & Boulian, P. V. (1974). Organizational commitment, job satisfaction, and turnover among psychiatric technicians. *Journal of Applied Psychology, 59,* 603–609.

Report Eurostat. (2015). https://ec.europa.eu/eurostat/documents/2995521/7017572/302102015-BP-EN.pdf/5a7b5406-4a0d-445b-8fa3-3558a8495020.

Ritz, A., Gene, A. B., & Oliver, N. (2016). Public service motivation: a systematic literature review and outlook. *Public Administration Review, 73*(6), 414–426.

Sarstedt, M., Christian, M. R., & Joseph, F. H. (2017). Partial least squares structural equation modeling. In *Handbook of Market Research* (pp 1–40). Springer International Publishing.

Van Loon, N. M., Vandenabeele, W., & Leisink, P. L. M. (2017). Clarifying the relationship between public service motivation and in-role and extra-role behaviors: the relative contributions of person-job and person-organization fit. *American Review of Public Administration, 47,* 699–713.

Vandenabeele, W. (2009). The mediating effect of job satisfaction and organizational commitment on self-reported performance: more robust evidence of the PSM—performance relationship. *International Review of Administrative Sciences, 75*(1), 11–34.

Zhao, X., Lynch, J. G., Jr., & Chen, Q. (2010). Reconsidering Baron and Kenny: myths and truths about mediation analysis. *Journal of Consumer Research, 37*(2), 197–206.

A Classification Algorithm to Recognize Fake News Websites

Giuseppe Pernagallo, Benedetto Torrisi, and Davide Bennato

Abstract "Fake news" is information that generally spreads on the web, which mimics the form of reliable news media content. In this paper, we use a classifier to distinguish a reliable source from a fake news website based on information potentially available on websites, such as the presence of a "contact us" section or a secured connection. This framework offers a concrete solution to attribute a "reliability score" to news websites, defined as the probability that a source is reliable or not, and based on this probability a user can decide if the news is worth sharing.

1 Introduction

The internet age has redefined the idea of information in all its forms: the idea that information can spread almost everywhere in less than seconds is exciting and alarming at the same time. Social networks are a chaotic environment, in the words of Piedrahita et al. (2018) "*Digital technologies have turned interpersonal networks into massive, pervasive structures that constantly pulsate with information*". Although social media platforms are recognized as useful sources of knowledge sharing (Leonardi 2017), they can also be channels for misinformation (Kumar and Geethakumari 2014).

The phenomenon of "fake news" has gained relevant interest and the need for truth has grown exponentially not only for internet users, but also for authorities and companies. To answer to this necessity, several tools (such as *Hoaxy* or *Botometer*)

G. Pernagallo (✉)
Collegio Carlo Alberto, Piazza Vincenzo Arbarello 8, 10122 Turin, Italy
e-mail: giuseppepernagallo@yahoo.it; giuseppe.pernagallo@carloalberto.org

B. Torrisi
Department of Economics and Business, University of Catania, Catania, Italy
e-mail: btorrisi@unict.it

D. Bennato
Department of Humanities, University of Catania, Catania, Italy
e-mail: dbennato@unict.it

P. Mariani and M. Zenga (eds.), *Data Science and Social Research II*,
Studies in Classification, Data Analysis, and Knowledge Organization,
https://doi.org/10.1007/978-3-030-51222-4_25

have been developed to detect the veracity of a news. Indeed, false information can cause relevant economic consequences; for example, "*false beliefs about a bank's allegedly limited solvency can lead to a bank run and induce the destabilization of the bank*" (Mäs and Opp 2016, p. 116). Nonetheless, limiting or prohibiting the diffusion of information cannot represent the solution, because it would damage the freedom of speech and of information. As we can see the problem of "anomalies" detection in social networks is a challenging task, given the existence of many interconnections among users and among different social networks (Savage et al. 2014).

In this paper, we aim to tackle the problem using a different perspective: instead of classifying a news, we tried to classify its source producing a score on how much reliable the originator of the news is. In this way, the user will have a numerical datum to decide if the source is trustworthy or not. The algorithm in this paper is based on logistic regression and assigns a probability that a website is fake based on few predictors obtainable via the website of the originator. This algorithm can offer a concrete solution not only for users but also for owners of platforms (damaged by the diffusion of erroneous information) and for authorities (to reestablish trust in institutions).

The paper is structured as follows. Section 2 shows how fake news can affect society and economy and why the issue should be adequately handled by the policy-maker. Section 3 discusses the mechanisms of propagation of fake news. Section 4 presents the statistical and conceptual framework used in this work and Sect. 5 provides the results of the model. The final section concludes the paper.

2 The Socioeconomic Impact of Fake News and the Role of the Policy-Maker

Fake news can be defined as unfounded information that mimics the form of reliable news media content (Lazer et al. 2018). In most cases, originators of fake news lack the structure that characterizes reputable editorial companies, an aspect that is easily recognizable from the website of the source. The phenomenon of fake news can be considered as a form of misinformation or disinformation (Lazer et al. 2018): misinformation occurs when false information is shared, but no harm is meant, whereas disinformation is false information knowingly shared to mislead people (Wardle and Derakhshan 2017). Although fake articles are better known, recently, for their political content, they can potentially convey any type of information with equal detrimental effects.

A famous case that illustrates the deep impact of fake news on markets was the ImmunoCellular Therapeutics case.[1] ImmunoCellular Therapeutics is a clinical-stage biotechnology company specialized in immune-based therapies for the treatment of cancer. On January 18, 2012, an article published on Seeking Alpha reported that the company discovered an important cancer treatment, cheaper than the existing

[1]https://www.ft.com/content/a37e4874-2c2a-11e7-bc4b-5528796fe35c.

Fig. 1 ImmunoCellular therapeutics adjusted stock price (in dollars) from 2010-12-31 to 2014-04-30. *Source* our elaboration on yahoo! finance data

products. This (false) news, architected by the company and the author of the article, pushed up strongly the stock price of the company. Figure 1 shows the evolution of ImmunoCellular Therapeutics stock price before and after the diffusion of the fake news. The fake news date is indicated by the first dashed line. It is evident that after the spread of this false information the market overreacted. The second dashed line, December 2013, indicates the moment of truth: a discouraging clinical update on the new product of the company caused the price to fall drastically. During 2018, ImmunoCellular Therapeutics stock has been traded at less than $0.50. This is just an example of how information shared via internet can influence financial markets (for another application, see Liu and Ye 2016).

Fake news affects also the reputation of social networks. Social networks that promote deceiving news lose credibility and rise the hostility of users. For this reason, established social networks invest huge amounts of money to prevent the spread of hoaxes among users. Even public authorities are concerned about the problem because it undermines the foundations of the democratic information system, furthermore, the diffusion of hoaxes is dangerous on many levels. From a political perspective, fake news may be used to drive public opinion toward certain voting choices. From an economic perspective, we have seen in the case of ImmunoCellular Therapeutics how fake news can affect investment or consumption decisions. Finally, from an ethical perspective, governments should always pursue and promote transparency of information for all citizens.

The case of Italy is emblematic; as reported by Tambini (2017, p. 13), "*In February 2017 a draft law was introduced to the Italian Parliament in response to the issue of 'Fake News'. This attempted to criminalise the posting or sharing of 'false, exaggerated or tendentious news', imposing fines of up to 5000 Euros on those responsible. In addition, the law proposed imprisonment for the most serious forms of fake news such as those that might incite crime or violence, and also imposed an obligation on social media platforms to monitor their services for such news*". Of course, these measures should find a balance between the need for truthful information and freedom of expression, but this is not the place for such discussions.

3 The Diffusion Process of a Fake News

The causes of propagation of fake news are not different from those that characterize common news. Lee and Ma (2012) designed and administered a survey to 203 students in a large local university and, via Structural Equation Modeling, they found out that important determinants of news sharing are

- gratification of information seeking;
- socializing and status seeking;
- prior experience with social media.

Vosoughi et al. (2018) described the process of a rumor cascade on Twitter. The process starts when a user tweets an argument in the form of text, images, videos, or links to online articles. Via retweeting, the rumor is propagated to other users based on the dimension of the network in which the originator is placed. Because nowadays social networks are highly interconnected, the process is accelerated by the propagation of the rumor on other platforms. This diffusion process can be characterized by several cascades, each for every user that, independently from the others, originates a tweet regarding the same claim. The dimension of each cascade depends on how many times these tweets are retweeted. The process of "social contagion" is comparable to epidemic diseases (there is a vast literature that investigated the functioning of this process: Granovetter 1978; Strogatz 2001; Valente 2012; Berry et al. 2019). Figure 2 shows an example of a diffusion network of a news obtained via *Hoaxy* (Shao et al. 2016, 2018a, 2018b).[2] The title of the news is "Louis Farrakhan Chants 'Death to America' in Iran"; the plot represents the number of tweets containing this news as an object. We can see that the two principal cascades are the profiles of Brigitte Gabriel and Donald Trump Jr., and from their accounts a series of retweets expands densely. We omitted from the plot isolated cascades, i.e., users that independently shared the news that was retweeted only a few times. Bigger nodes represent bigger users in term of connections, and the color of nodes indicates the similarity of accounts to humans, from light blue (human like) to red (bot like). This network shows how the news spread in only seven days: a fake news can spread quickly and

[2]Hoaxy is a platform promoted by researchers at the Indiana University Network Science Institute (IUNI) and the School of Informatics and Computing's Center for Complex Networks and Systems Research (CNetS). The platform is available online, and it can be used for the automatic tracking of both online fake news and fact-checking on social media. The goal of the platform is to trace how hoaxes are diffused online and how many users are involved in the process. The interested reader can find further information at https://hoaxy.iuni.iu.edu/faq.php#faq-q1

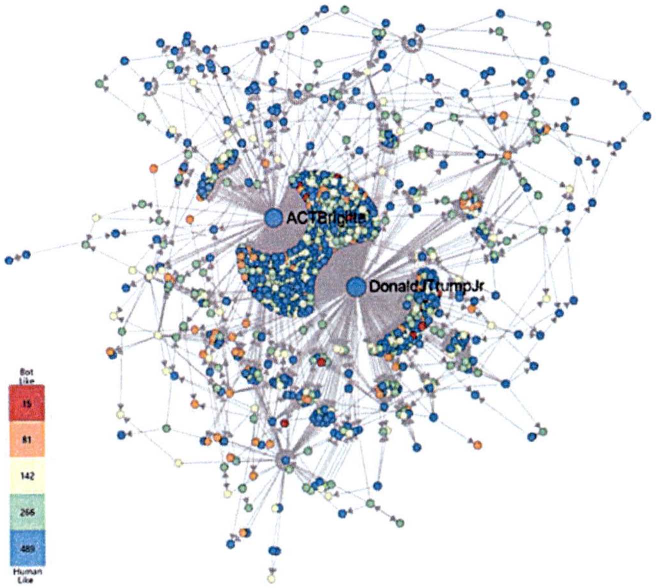

Fig. 2 Diffusion network of the news "Louis Farrakhan Chants 'Death to America' in Iran" from November 05, 2018 to November 17, 2018. *Source* adapted from Hoaxy

widely, and once entered in the system it becomes difficult to stop it. The best way to arrest the propagation of a fake news is to avoid its sharing. Because very often it is difficult to verify the goodness of the content of the news (think of scientific fake news with highly sophisticated terms) a good alternative is to verify the goodness of the original source of the article. Adopting this strategy, users can have a glimpse on how reliable the news is and may choose to arrest its propagation. The problem is that users may not have the skills, the time, or the will to distinguish a reliable source from a deceiving one, for this reason we propose in this paper a simple tool to assess the probability that a website is not reliable.

4 Methodology and Data

As pointed out by Figueira and Oliveira (2017, p. 820), there are two main approaches to face the problem: "*human intervention to verify information veracity*", such as the International Fact Checking Network (IFCN) that allows users to signal fake articles, or "*using algorithms to fight algorithms*". Our attempt falls within the second approach, but differently from available tools such as *Hoaxy* or *Botometer*, we focused on the source of the news. Focusing on the source partly overcomes the problem of anomalies detection in social networks, which is very difficult given the presence of interactions between individuals (Savage et al. 2014). A possible solution, in our thoughts, is to provide the users with adequate knowledge and tools to detect false information and to stop its diffusion.

The dataset used in this study has been prepared choosing 200 confirmed fake news websites[3] and 200 established news websites from different countries. We mainly gathered websites from United States, Britain, India, Italy, Philippines, France, Germany, Mexico, Spain, and some other Asian sources to work with an heterogenous dataset. The choice of 400 websites is not derived from a specific sample design. This choice was obligated by the lack of data and the high cost involved with dealing with a longer data gathering process. Anyway, researchers have pointed out that in some situations (especially for studies regarding public opinion) non-representative surveys can be fast, cheap, and (mostly) accurate tools (Goel et al. 2015). One may argue that considering the same number of fake news websites and reliable sources may lead to an over-representation of fake news websites. This is a serious concern, but without more data we are not able to assess the issue. Consequently, future works should indeed focus on how to enlarge the dataset.

We chose logistic regression to quantify the reliability of a website because it gives us the probability that the dependent variable assumes one of two possible outcomes: the website produces fake news (Fake news website = 1) or the website is reliable (Fake news website = 0). The choice of a logit makes the interpretation of the results easier than other models. For example, albeit the probit model holds similar results (Gujarati 2011), it is more complex. However, we estimated also probit models to show that the logit framework is the better choice in this case. Furthermore, Liu et al. (2015), in a study with a very similar topic, adopted the logistic framework because of the presence of numerous dichotomous variables, as it is in our work. Mathematically the model is

$$\begin{aligned} &\mathrm{P}(\mathit{Fake\,news\,websites} = 1 | X_1 = x_1, \cdots, X_5 = x_5) = \\ &\mathit{exp}(\mathrm{b}_0 + \mathrm{b}_1 x_1 + \cdots + \mathrm{b}_5 x_5)/[1 + \mathit{exp}(\mathrm{b}_0 + \mathrm{b}_1 x_1 + \cdots + \mathrm{b}_5 x_5)] \end{aligned} \tag{1}$$

We used 5 dummy variables to predict the dependent:

X_1 = *Padlock*, a dummy variable equal to 1 if the website uses the SSL protocol (a data transfer security standard) or the TLS protocol, 0 otherwise;

X_2 = *Contact*, a dummy variable equal to 1 if the website has a "contact us" section or something similar ("connect with us", "gives us a tip", etc.), 0 otherwise;

X_3 = *Telephone*, a dummy variable equal to 1 if the website makes available a telephone and/or a fax number, 0 otherwise;

X_4 = *About*, a dummy variable equal to 1 if the website has an "about us" section or something similar ("information", "who we are", etc.), 0 otherwise;

---

[3]There are several articles online that report this information such as the page on Wikipedia, "List of fake news websites", blog.feedspot.com/fake_news_blogs/, usnews.com or the Italian website bufale.net. There are also websites that mimic existing reputable website domain or deliberately share absurd and ironic news.

$X_5 = Terms$, a dummy variable equal to 1 if the website has a "terms and conditions" section or something similar ("terms", "legal notes", "terms of use", etc.), 0 otherwise.

The inclusion of these variables is justified by the fact that established news websites clearly expose these elements, which are manifestation of an organized structure compliant with editorial norms and processes. The advantage of this model is that it uses only the website to take the needed inputs and compute a probability that the source is reliable. Gonzalez-Bailon (2009) emphasized the importance of links in the centrality of sites and their visibility, therefore, a variable such *Padlock* consents immediately to see if the website link refers to a secured source or not. Reputable websites are, in most cases, endowed by a secure connection easily verifiable via their URL. In general, the variables used in the model serve to measure the credibility of the source (Kumar and Geethakumari 2014).

Figure 3 shows the entire process of recognition. The representation of an algorithm via a flow chart is common in literature (for an application to information diffusion, see Kumar and Geethakumari 2014). The process starts when a user doubts on the veracity of a news. Inside the circle are reported the operations of the machine. Hypothesize that a software based on our model is freely available, at this point the user can insert the URL of the website and the algorithm operates a first screening recurring to an internal database: if the domain or the name of the website explicitly mimics or copies an older and established source of news, the algorithm attributes to that website probability of 1, which means that it is a fake news website. If it is not the case, the algorithm computes the probability using our logit model producing a certain probability x. In the last stage of the procedure, the user receives this probability and decides, based on the tolerance threshold T, to share or not to share the news.

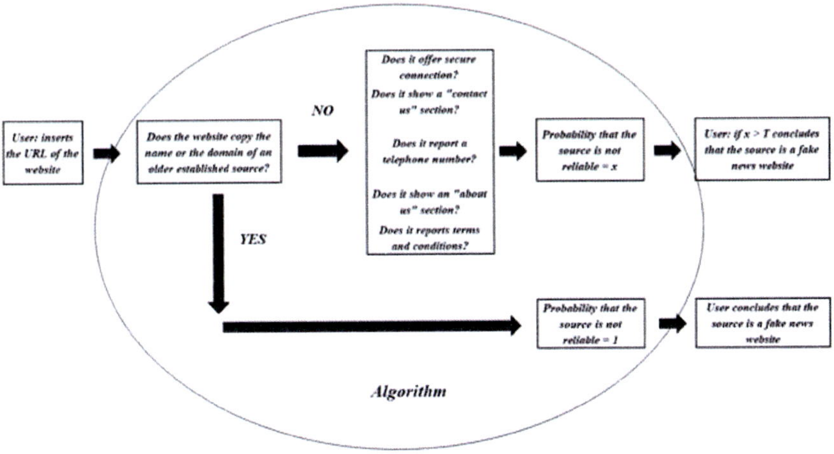

Fig. 3 The process of recognition of a fake news website using our model

The reader should not confuse the threshold used by the logit model with *T*. The logit model classifies 1s and 0s based on a threshold (generally 0.5). If the score attributed to an observation is higher than the threshold, then it is classified as a 1, otherwise as a 0. In this stage, the individual's belief is irrelevant, the entire process is automatized. When the result of the model, e.g., 0.7, is produced, based on this quantity the user will decide whether to share or not the news given the tolerance *T*. This means that in a society where users have high tolerance levels to false information, fake news will spread anyways. Given that we cannot limit the freedom to share information, the only plausible solution in our thoughts is to better educate users and provide them with tools that raise awareness against fake news.

5 Results

Table 1 shows the frequency distributions of the explanatory variables. We computed a tetrachoric correlation matrix to test the level of agreement between our dichotomous variables as showed in Table 2. The first column of the matrix shows the agreement between our dependent variable and all the independent variables. There is a strong and negative agreement between the variable *Fake news websites* and all

Table 1 Frequency distributions of the explanatory variables

	Yes		No	
	Absolute frequency	Relative frequency	Absolute frequency	Relative frequency
Padlock	251	0.6275	149	0.3725
Contact	310	0.775	90	0.225
Telephone	147	0.3675	253	0.6325
About	239	0.5975	161	0.4025
Terms	200	0.5	200	0.5

Table 2 Tetrachoric correlation matrix for dichotomous variables. *indicates 0.01 % significance

	Fake news website	Padlock	Contact	Telephone	About	Terms & Conditions
Fake news website	1					
Padlock	−0.8398*	1				
Contact	−0.7293*	0.5321*	1			
Telephone	−0.7673*	0.6689*	0.8673*	1		
About	−0.3263*	0.316*	0.7523*	0.3285*	1	
Terms	−0.729*	0.6532*	0.6899*	0.503*	0.4608*	1

Table 3 Chi-square test of independence. Expected frequency assumption is respected for all the variables

Associated variables	Empirical test statistic	p-value
Fake-Padlock	146.4103	< 0.0001
Fake-Contact	74.3226	< 0.0001
Fake-Telephone	114.1029	< 0.0001
Fake-About	17.4745	< 0.0001
Fake-Terms	108.1600	< 0.0001

the considered explanatory variables, whereas this agreement is only moderate for the variable *About*. This means that when one of this section is present, the website tends to be more reliable, i.e., the dummy *Fake news website* tends to assume value 0. The correlation between all the explanatory variables is tendentially moderate but highly significant and positive. A positive value means positive agreement, therefore, websites that present one of the considered sections tend to present also the other ones.

At this point, we questioned whether the association between the independent variables and the outcome variable is due to chance or is statistically significant. Therefore, we used the chi-square test of independence to test whether the association is significant or not. Table 3 reports the results of the test: it clearly emerges that there is association between all the independent variables and the dependent variable, and this association is highly significant.

Table 4 shows the results of the logit model tested on our dataset. Model I includes all the variables: the coefficients are all significant at 1 % except for the coefficient of the variable *About*. In model II, we excluded the non-significant variable *About*; this model yields similar results, however, based on the Akaike Information Criterion (AIC) it should be preferred to model I. The signs of the coefficients are coherent with our expectations: if a website uses a security protocol, provides a "contact us" section, reports terms and conditions and a phone number, the probability that is a fake news website diminishes. The level of VIFs for the variables is nearly 1, showing absence of multicollinearity, and the likelihood ratio test consents us to reject the null hypothesis that the model is not statistically significant. The level of the McFadden R-squared (0.4660) is moderate, however, this model can correctly predict 337 cases over 400 (Table 5), showing a prediction accuracy of 84.2 %. This result is satisfactory because an algorithm based on this model, with few predictors, is computationally fast.

We compared the result of the logit model with the estimates yielded by two probit models with the same covariates. The results of models III and IV are similar to the results of model I and II. We have the same signs on coefficients and almost the same levels of p-values. Even in the case of the probit, model IV is preferable to model III based on AIC, however, the AIC of model IV is higher than the AIC of model II, hence we preferred the logit model. Moreover, probit models are more complex computationally and mathematically, so there is no substantial reason in this case to prefer the probit framework to the logit.

Table 4 Results of the logit model based on (1) using the whole dataset (models I–II) and comparison with probit estimates (III–IV). * indicates 1 % significance

	I				II			
	Coefficient	*P*-value	Slope	VIF	Coefficient	*P*-value	Slope	VIF
Constant	3.7723*	< 0.0001			3.8405*	<0.0001		
Padlock	−2.3133*	< 0.0001	−0.5058	1.392	−2.3141*	<0.0001	−0.5053	1.392
Contact	−1.3385*	0.0049	−0.3064	1.606	−1.1682*	0.0089	−0.2714	1.347
Telephone	−1.7285*	<0.0001	−0.4060	1.357	−1.7179*	<0.0001	−0.4040	1.356
About	0.3744	0.2789	0.0932	1.331				
Terms	−1.5144*	<0.0001	−0.3605	1.402	−1.4569*	<0.0001	−0.3478	1.381
McFadden R-squared	0.4681				0.4660			
Adjusted R-squared	0.4465				0.4479			
Akaike criterion	306.9277				306.1254			
Percent correctly predicted	84.5%				84.2%			
Likelihood ratio test chi-square (p-value in brackets)	259.59 [0.0000]				258.392 [0.0000]			
Observations	400				400			
Type of model	Logit				Logit			
	III				IV			
	Coefficient	*P*-value	Slope	VIF	Coefficient	*P*-value	Slope	VIF
Constant	2.1628*	<0.0001			2.2055*	<0.0001		
Padlock	−1.3471*	<0.0001	−0.4864	1.392	−1.3474*	<0.0001	−0.4858	1.392
Contact	−0.7686*	0.0039	−0.2877	1.606	−0.6613*	0.0084	−0.2505	1.347
Telephone	−0.9691*	<0.0001	−0.3715	1.357	−0.9570*	<0.0001	−0.3674	1.356
About	0.2347	0.2288	0.0932	1.331				
Terms	−0.8602*	<0.0001	−0.3320	1.402	−0.8265*	<0.0001	−0.3195	1.381
McFadden R-squared	0.4683				0.4656			
Adjusted R-squared	0.4466				0.4476			
Akaike criterion	306.8644				306.3372			
Percent correctly predicted	84.5%				84.2%			
Likelihood ratio test chi-square (p-value in brackets)	259.65 [0.0000]				258.181 [0.0000]			
Observations	400				400			
Type of model	Probit				Probit			

Table 5 Confusion matrix of model II

	Predicted: reliable	Predicted: fake
Actual: reliable	166 (41.5%)	34 (8.5%)
Actual: fake	29 (7.3%)	171 (42.7%)

We also checked for potential bias of the coefficient estimates of model I and model II. Using bootstrap resampling (10,000 replications), we found out that the level of bias for each coefficient estimate is very low and the bootstrap standard error estimates are very close to the standard error of the actual model (see Table 6). Consequently, the original confidence intervals for model I and model II are very

Table 6 Bias and bootstrap standard errors of logit models I and II using 10,000 replications

	Model I					
	Original estimate	Bias	Original standard error	Bootstrap standard error	Original confidence interval	Bootstrap confidence interval
Constant	3.7723	0.1265	0.4770	0.5450	[2.8373, 4.7072]	[2.5775, 4.7139]
Padlock	−2.3133	−0.0685	0.3390	0.3418	[−2.9777, −1.6490]	[−2.9148, −1.5749]
Contact	−1.3385	−0.0566	0.4754	0.5591	[−2.2704, −0.4066]	[−2.3778, −0.1861]
Telephone	−1.7285	−0.0361	0.3395	0.3758	[2.3939, −1.0631]	[−2.4290, −0.9558]
About	0.3744	0.0116	0.3458	0.3478	[−0.3034, 1.0522]	[−0.3189, 1.0445]
terms	−1.5144	−0.0349	0.3101	0.3266	[−2.1221, −0.9067]	[−2.1197, −0.8393]
	Model II					
	Original estimate	Bias	Original standard error	Bootstrap standard error	Original confidence interval	Bootstrap confidence interval
Constant	3.8405	0.1109	0.4748	0.4908	[2.9099, 4.7712]	[2.7677, 4.6916]
Padlock	−2.3141	−0.0567	0.3375	0.3389	[−2.9758, −1.6525]	[−2.9218, −1.5932]
Contact	−1.1682	−0.0466	0.4463	0.4623	[−2.0430, −0.2934]	[−2.0276, −0.2155]
Telephone	−1.7179	−0.0287	0.3382	0.3775	[−2.3807, −1.0551]	[−2.4290, −0.9494]
About						
terms	−1.4569	−0.0281	0.3033	0.3142	[−2.0513, −0.8625]	[−2.0447, −0.8130]

close to the bootstrap confidence intervals. The histograms of the bootstrap statistics signal a problem of non-normality for model I for the intercept estimate and the coefficient of *Contact* (Fig. 4), whereas the histograms for model II (the best model) appear to be approximately normal (Fig. 5).

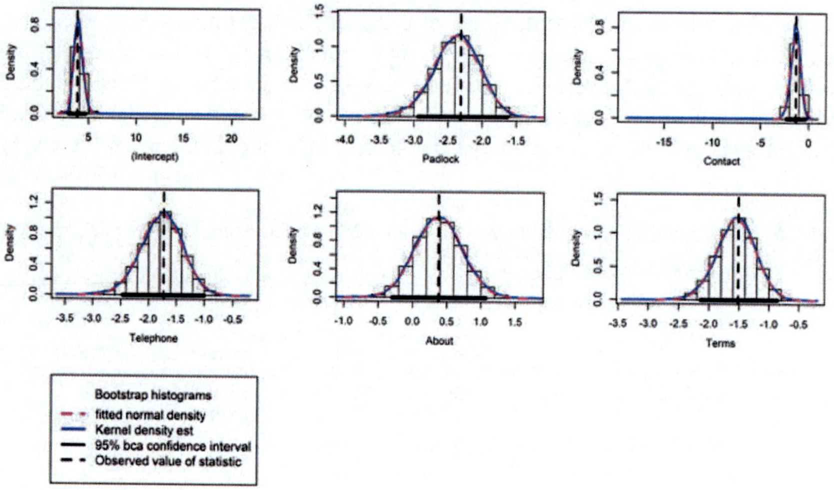

Fig. 4 Histograms of bootstrap statistics (regression coefficients) for model I using 10,000 replications

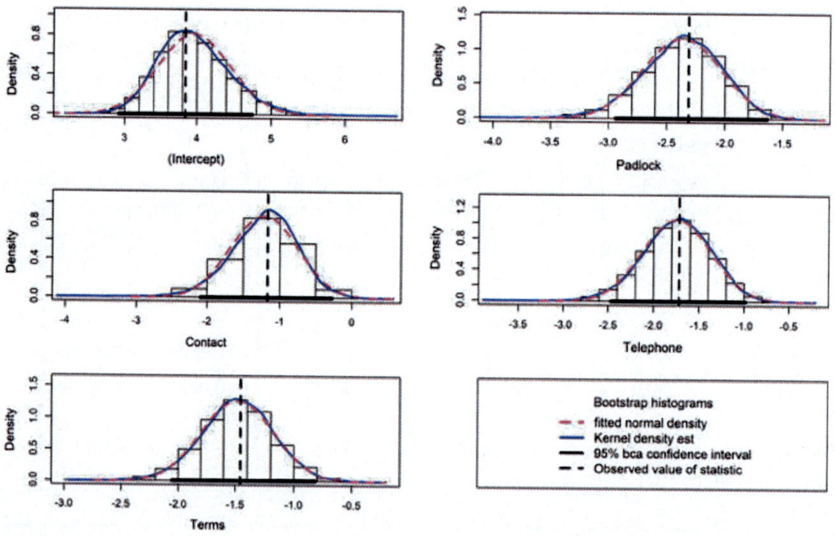

Fig. 5 Histograms of bootstrap statistics (regression coefficients) for model II using 10,000 replications

However, training the algorithm using the whole dataset generally is not a good idea (for example, it can generate overfitting). Hence, we split our dataset into a training set and a test set using the common 80–20 rule, i.e., we used 80 % of the data (320 observations) to train the algorithm and then we used it to evaluate its performance on the 20 % of observations set aside (80 observations). We maintained a balanced proportion between fake news websites and reputable websites, so in both the training and test sets half of the observations are fake news websites and the other half reliable websites. The resulting model for the training set is showed in Table 7. In this case, the variables *About* and *Contact* were not significant at 1 % but excluding them from the model (see model VI) increases the AIC and lowers the other goodness of fit measures like the McFadden R-squared. For this reason, model V, with all the variables, should be preferred in this case. From the confusion matrix in Table 8, we have that the accuracy of the algorithm for the training set is 81.9 %. Finally, we tested the performance of the algorithm on the test set. The results (Table 9) indicate an accuracy of 93.8 %, showing that the results obtained on the whole dataset and on the training set are not due by overfitting.

Table 7 Results of the logit model based on (1) using the training dataset. * indicates 1 % significance

	V				VI			
	Coefficient	*P*-value	Slope	VIF	Coefficient	*P*-value	Slope	VIF
Constant	3.2151*	<0.0001			2.7813*	<0.0001		
Padlock	−2.0965*	<0.0001	−0.4704	1.335	−2.1212*	<0.0001		1.324
Contact	−1.1844	0.0174	−0.2773	1.602				
Telephone	−1.6192*	<0.0001	−0.3820	1.313	−1.8052*	<0.0001		1.208
About	0.6665	0.0860	0.1651	1.365				
terms	−1.4366*	<0.0001	−0.3440	1.349	−1.4813*	<0.0001		1.231
McFadden R-squared	0.4105				0.3949			
Adjusted R-squared	0.3834				0.3769			
Akaike criterion	273.5004				276.4229			
Percent correctly predicted	81.9%				82.2%			
Likelihood ratio test chi-square (*p*-value in brackets)	182.114 [0.0000]				175.191 [0.0000]			
Observations	320				320			

Table 8 Confusion matrix of model V (training dataset)

	Predicted: reliable	Predicted: fake
Actual: reliable	136 (42.5%)	24 (7.5%)
Actual: fake	34 (10.6%)	126 (39.4%)

Table 9 Confusion matrix of model V (test dataset)

	Predicted: reliable	Predicted: fake
Actual: reliable	39 (48.8%)	1(1.2%)
Actual: fake	4(5.0%)	36 (45.0%)

Nonetheless, the results obtained on the test set may depend on what observations were allocated in the training and test sets. This can affect the prediction accuracy. To solve this problem, we used bootstrap resampling to assess the discrepancy between the actual statistic and the bootstrapped statistic using 10,000 replications. Via the 80–20 rule, we assigned randomly with replacement 320 observations to the training set and 80 observations to the test set. In each set, half of the websites were fake and the other half reliable; in this way, we maintained the composition of the original dataset. For each replication, we trained model V (the best model) using the training set and then we used this model to compute the percent of cases correctly predicted using the test set. Note that in each replication the training and the test sets changed thanks to random resampling so that the final bootstrap result is not affected by how the observations were placed in the two sets. The average percent correctly predicted in the 10,000 replications was 84.3 % with a standard error of only 0.0364. As we can see the bias is negligible: for model II is only of −0.1 and for model V is 2.4. The bias is obtained as the difference between the bootstrap average percent correctly predicted and the percent correctly predicted of the model (Fox and Weisberg 2017). The prediction accuracy obtained for the first test set (Table 9) probably was too high; the bootstrap statistic shows that 84 % can be considered a realistic result. The histogram of the bootstrap percent correctly predicted signals normality of the distribution (Fig. 6).

6 Conclusions

In this work, we have proposed a possible tool to distinguish a reliable source of news from a deceiving one. Our model is parsimonious, in the sense that it is made by few variables (it only uses information potentially available on every website), and it can produce a reliability score rapidly. Nevertheless, it should be noted that the results in this paper apply to the case study proposed. The dimension of the dataset and the data gathering process represent the main limitations of this study.

Promoting a more reliable informative system is essential to

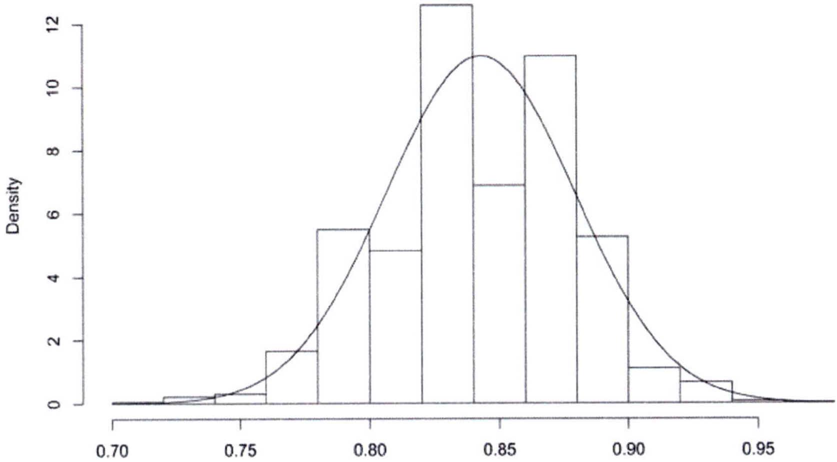

Fig. 6 Histogram (with normal density) of bootstrap percent correctly predicted for the test set using 10,000 replications

1. reduce the social impact of fake news such as the sense of mistrust toward classical and innovative means of communication;
2. reduce the economic impact of fake news such as inflated asset prices, or the economic damage suffered by social networks and information companies;
3. reestablish trust in institutions;
4. reward adequately (in terms of shares and notoriety) reputable news websites.

Furthermore, our empirical model offers an immediate rule of thumb for non-expert users of social networks: be skeptic about website sources without a secured connection, without a telephone number or a "contact us" section and that do not clearly expose terms and conditions of their services. In addition to this heuristic, our algorithm can quantify how much reliable is a source. The results of the model may seem obvious to specialists, but they are not trivial for non-educated users. Social networks are used by individuals of every age and cultural background, how many users know what is a secured connection or how to find a contact section on a website? The aim of the paper is to propose an easy and fast way to fill up this lack of knowledge in less skilled users.

As we have already mentioned, we covered the algorithm solution, but to solve the problem human intervention is also needed. Indeed, individuals face nowadays a problem of information overload, a relevant problem for modern "knowledge-intensive organizations" because it negatively affects productivity and decision-making of agents (Whelan and Teigland 2013; Pernagallo and Torrisi 2020). Even though algorithms can effectively separate good information from bad information, it is impossible for the common user to elaborate all the available information. Policy-makers have the difficult responsibility to balance the control of this relentless flow of information and at the same time ensuring the undeniable rights of expression and information. This is the most relevant future topic on which scholars should focus.

On the other hand, future quantitative topics that we have not assessed in this paper are: how to enlarge the dataset in order to increase the number of predictors and the accuracy of the model; what sampling method could be adopted to maximize the power of algorithms; what are the consequences of different levels of T, the threshold on which a user decides if a news should be shared or not. This last point is particularly interesting also from a theoretical economic perspective. For example, lower levels of T means that users are willing to accept low-quality information. This may have profound implications triggering the undesired effect of "bad" information that replaces "good" information. Threshold models are largely diffused to simulate these kinds of problems (see, for example, Granovetter 1978, or recently, Piedrahita et al. 2018, or Mäs and Opp 2016), so we think that this paper opens new research questions for future works.

We conclude with a question: why should managers of good information sources invest time and money in better and reliable articles? The question arises because "bad" information could be shared by consumers without distinction. The answer should be assessed both economically and statistically and it will be content of future research.

Acknowledgements The authors are grateful for the precious comments made by the anonymous reviewer. They also would like to thank the editors for their work.

Funding This research did not receive any specific grant from funding agencies in the public, commercial, or not-for-profit sectors.

References

Berry, G., Cameron, C. J., Park, P., & Macy, M. (2019). The opacity problem in social contagion. *Social Networks, 56,* 93–101.

Figueira, A., & Oliveira, L. (2017). The current state of fake news: challenges and opportunities. *Procedia Computer Science, 121,* 817–825.

Fox, J., & Weisberg, S. (2017). Bootstrapping regression models in R. An appendix to an R companion to applied regression, second edition. Retrieved from https://socialsciences.mcmaster.ca/jfox/Books/Companion-2E/appendix/Appendix-Bootstrapping.pdf.

Goel, S., Obeng, A., & Rothschild, D. (2015). Non-representative surveys: Fast, cheap, and mostly accurate. Working paper. Retrieved from http://researchdmr.com/FastCheapAccurate.

Gonzalez-Bailon, S. (2009). Opening the black box of link formation: social factors underlying the structure of the web. *Social Networks, 31,* 271–280.

Granovetter, M. (1978). Threshold models of collective behavior. *American Journal of Sociology, 83*(6), 1420–1443.

Gujarati, D. (2011). Econometrics by example. Palgrave Macmillan.

Kumar, K. P. K., & Geethakumari, G. (2014). Detecting misinformation in online social networks using cognitive psychology. *Human-centric Computing and Information Sciences, 4* (14).

Lazer, D. M. J., Baum, M. A., Benkler, Y., Berninsky, A. J., Greenhill, K. M., Menczer, F., et al. (2018). The science of fake news. *Science, 359*(6380), 1094–1096.

Lee, C. S., & Ma, L. (2012). News sharing in social media: the effect of gratifications and prior experience. *Computers in Human Behavior, 28*(2), 331–339.
Leonardi, P. M. (2017). The social media revolution: sharing and learning in the age of leaky knowledge. *Information and Organization, 27*(1), 47–59.
Liu, Q., Zhou, M., & Zhao, X. (2015). Understanding News 2.0: a framework for explaining the number of comments from readers on online news. *Information & Management, 52*(7), 764–776.
Liu, X., & Ye, Q. (2016). The different impacts of news-driven and self-initiated search volume on stock prices. *Information & Management, 53*(8), 997–1005.
Mäs, M., & Opp, K. (2016). When is ignorance bliss? Disclosing true information and cascades ofnorm violation in networks. *Social Networks, 47,* 116–129.
Pernagallo, G., & Torrisi, B. (2020). A theory of Information overload applied to perfectly efficient financial markets, *Review of Behavioral Finance.* https://www.emerald.com/insight/content/doi/10.1108/RBF-07-2019-0088/full/html.
Piedrahita, P., Borge-Holthoefer, J., Moreno, Y., & Gonzalez-Bailon, S. (2018). The contagion effects of repeated activation in social networks. *Social Networks, 54,* 326–335.
Savage, D., Zhang, X., Yu, X., Chou, P., & Wang, Q. (2014). Anomaly detection in online social networks. *Social Networks, 39,* 62–70.
Shao, C., Ciampaglia, G. L., Flammini, A., & Menczer, F. (2016). Hoaxy: a platform for tracking online misinformation. In *Proceedings of the 25th International Conference Companion on World Wide Web (WWW '16 Companion)* (pp. 745–750). http://doi.org/10.1145/2872518.2890098.
Shao, C., Ciampaglia, G. L., Varol, O., Yang, K., Flammini, A., & Menczer, F. (2018). The spread of low-credibility content by social bots. *Nature Communications, 9,* 4787. https://doi.org/10.1038/s41467-018-06930-7
Shao, C., Hui, P. M., Wang, L., Jiang, X., Flammini, A., Menczer, F., & Ciampaglia, G. L. (2018b). Anatomy of an online misinformation network. PLOS ONE, e0196087. https://doi.org/10.1371/journal.pone.0196087.
Strogatz, S. H. (2001). Exploring complex networks. *Nature, 410*(6825), 268–276.
Tambini, D. (2017). *"Fake News: Public Policy Responses", Media Policy Brief 20.* London: Media Policy Project, London School of Economics and Political Science.
Valente, T. W. (2012). Network interventions. *Science, 337*(6090), 49–53.
Vosoughi, S., Roy, D., & Aral, S. (2018). The spread of true and false news online. *Science, 359*(6380), 1146–1151.
Wardle, C., & Derekhshan, H. (2017). Information disorder: toward an interdisciplinary framework for research and policy making. Council of Europe report, DGI(2017)09.
Whelan, E., & Teigland, R. (2013). Transactive memory systems as a collective filter for mitigating information overload in digitally enabled organizational groups. *Information and Organization, 23*(3), 177–197.

A Comparative Analysis of the University Student Mobility Flows Among European Countries

Marialuisa Restaino, Ilaria Primerano, and Maria Prosperina Vitale

Abstract Higher education institutions' policies aimed at increasing the number of credits gained by university students abroad. Thus, the analysis of the internationalization process and the factors pulling and pushing students in a foreign country to complete their higher education are important features for academic institutions. In line with some previous studies, the present contribution aims to analyse the trend of the Erasmus student mobility flows and to capture the role played by each country by using the social network analysis approach. Data on Erasmus student exchanges among countries are gathered from the European Union Open Data Portal and used to build the network between countries. The main findings suggest that some countries in Europe are more attractive in terms of the number of incoming and outgoing students.

1 Introduction

The European Region Action Scheme for the Mobility of University Students (Erasmus) Programme, established in 1987, represents one of the initiatives introduced by the European Commission to enrich opportunities for exchanging cultural, professional and personal experiences between students, teachers and academic staff within the European Union. Its main goal is to encourage and support academic mobility within the European countries, increasing the quality of the education system and stimulating individual creativity in educational institutions. By the end of the academic year 2013–14, the Erasmus programme had supported 3.3 million students and 470,000 staff members since its launch. The new Erasmus programme for

M. Restaino (✉) · I. Primerano · M. Prosperina Vitale
University of Salerno, Via Giovanni Paolo II, 132, Fisciano Salerno, Italy
e-mail: mlrestaino@unisa.it

I. Primerano
e-mail: iprimerano@unisa.it

M. Prosperina Vitale
e-mail: mvitale@unisa.it

P. Mariani and M. Zenga (eds.), *Data Science and Social Research II*,
Studies in Classification, Data Analysis, and Knowledge Organization,
https://doi.org/10.1007/978-3-030-51222-4_26

the period 2014–2020, so-called Erasmus+, aims at doubling the current number of participants in the programme.

The programme and its impact on the university internationalization process are studied by experts from different fields (sociologists, psychologists, economists, educational researchers and so on), offering a complementary point of view on the phenomenon. Moreover, the benefits of participating in the university study programme abroad are reported in numerous works. In fact, it is pointed out that studying abroad considerably contributes to students' personal development, understanding of and interest in global affairs, language competence and inter-cultural skills (King and Ruiz-Gelices (2003); Norris and Gillespie (2009)). Students with this experience seem to work in higher status employment sectors, they are more likely to have an international job or work, and they are also less likely to remain unemployed after their graduation (King and Ruiz-Gelices (2003); Norris and Gillespie (2009); Parey and Waldinger (2011)).

Furthermore, the international flow of students is an important research topic because of the increased numbers of foreign students. In particular, in order to capture the structural features and patterns of student mobility, and to better understand the complexity of these flows between countries, some authors have suggested to analyse these flows by means of network analysis approach. Thus, it is possible to identify feeder and storer actors, i.e. good importers and good exporters (Barnett et al. (2016); Chadee and Naidoo (2009); Chen and Barnett (2000); Jiang (2014); Kondakci et al. (2018); Macrander (2017); Restaino et al. (2020)); to analyse the directions and the intensity of this phenomenon (Breznik and Ragozini (2015); Barnett et al. (2016); Doreian et al. (2005)); to capture the relationship between countries and universities involved in the Erasmus (Barnett et al. (2016); Breznik et al. (2013)).

Within this scenario, our aim is to identify the main characteristics of the Erasmus student mobility flows in European countries in six academic years, from 2008–2009 to 2013–2014. The analysis of this phenomenon is done by using some tools of social network analysis (SNA). The data under study are gathered from the European Union Open Data Portal and network data structures are defined in order to analyse and describe relationships among countries.

In SNA framework, the empirical data consists of a set of actors linked by one or more kinds of relationships. In this study, the actors are the countries and the links represent the number of students involved in the mobility exchange. Then, links among them can be defined according to the direction and/or the presence or absence of link weights. In particular, a directed and weighted network is defined when the direction of the connections is taken into account. Hence, temporal directed weighted networks are built for Erasmus data considering the students' flows (outgoing and incoming) among countries.

The paper is organized as follows. Section 2 briefly describes the methodological approach for exploring international student mobility data. In Section 3, details on the data collected from the European Union Open Data Portal are reported. Section 4 illustrates the main findings. Finally, Section 5 highlights the conclusions and suggestions for future lines of research.

2 Methodological Approach

Social network analysis (SNA) tools are here used to capture and analyse the structural characteristics and patterns of student mobility flow in the Erasmus programme. The Erasmus data can be described as a network, where countries represent the actors ($\mathcal{N}$) and student exchanges between countries define the links ($\mathcal{L}$) between them. The number of students involved in this exchange represents the weight ($\mathcal{V}$) of each link. The corresponding adjacency matrix **A** is both directed, with a link from the origin country to the destination country, and weighted, with elements a_{ij} equal to 0 if country $i \in \mathcal{N}$ does not send a link to country $j \in \mathcal{N}$, and a_{ij} greater than 0 otherwise.

By means of SNA methods, we are able to achieve a comprehensive understanding of the relationships and structure of the emerging student exchanges among countries and to identify countries who play a central role, discovering the presence of a particular configuration of the whole network structure.

In order to get this information, among the network measures used to recognize countries with central position into the network (Freeman, (1979)) the hub (HUB) and authority (AUT) centrality scores (Kleinberg, (1999)) are adopted to identify which countries are good exporters and/or good importers. By definition, a country with a high authority score is linked by many different hubs, i.e. *good importer*. A country with a good hub points to many other countries, i.e. *good exporter*. A country can play both roles being a good authority and a good hub.

3 Data Collection

The data on Erasmus student mobility flows are downloaded by the official European Commission website on Erasmus-Statistics.[1] The period under analysis consists of six academic years from 2008–2009 to 2013–2014. Two types of Erasmus mobility of students enrolled at higher education institutions are available: the *Student Mobility for Studies* (SMS), that enables students to spend a study period in another country; and the *Student Mobility for Placement* (SMP), that enables students to spend a placement period (traineeship or internship) in an enterprise or in an organization in another country. In both cases, the period abroad can vary from 3 months to 12 months.

The information available in the datasets are:

- ID of sending and hosting partner Erasmus;
- Sending and hosting countries;
- Students' gender;
- Subject area code;
- Type of mobility (SMS or SMP);

[1] For details see https://data.europa.eu/euodp/en/data/publisher/eac.

- Level of study (first cycle, second cycle, third cycle and short cycle);
- Duration of mobility in months.

Moreover, in order to better explain the international mobility flows, the aggregation of countries in five macro-areas (Northern Europe, Western Europe, Eastern Europe, Central Europe and Southern Europe) is considered in the analysis.

In this paper, we focus on the SMS dataset.

4 Trend of Erasmus Student Mobility Flows

The distribution of Erasmus students for SMS is shown in Table 1. An increasing trend in the total number of exchanges and in the mobility for studies is observed, except for the last year 2013–2014 where it registered a moderate decrease of -0.14% on the previous year. The average length of stay and the average age of students are quite stable during the period considered. Moreover, the percentage of women who participate in the Erasmus programme is always higher than that of men, and it is also quite steady over the years. The number of higher education institutions sending students registers a constant increment throughout all years.

In line with results discussed in related literature (De Benedictis and Leoni, (2020)), to study the temporal changes and the networks' characteristics that occurred in the academic years under analysis, six separate weighted directed adjacency matrices are defined. Each matrix describes the students' flows among countries involved in the Erasmus programme for each academic year.

The structure of the temporal networks shows a small increase in terms of involved countries and links among them. Specifically, the number of countries increases from 31 in 2008–2009 to 34 in 2013–2014, and the number of links goes up from 807 in 2008–2009 to 928 in 2013–2014 (Table 2).

Table 1 Distribution of Erasmus students mobility between 2008–2009 and 2013–2014

Year	Total number of exchanges	Number of SMS exchanges	Average duration (in month)	% of women	Total number of HEI	Average age
2008–2009	198,523	168,193	6.1	60.7	2,658	23.5
2009–2010	213,266	177,705	6.0	61.1	2,853	22.6
2010–2011	231,408	190,495	6.0	61.0	3,040	22.5
2011–2012	252,827	204,744	5.9	60.7	3,189	22.5
2012–2013	268,143	212,522	5.8	60.9	3,267	22.5
2013–2014	272,497	212,208	5.8	60.5	3,456	23.5

Table 2 Number of countries involved in the Erasmus student mobility for studies (SMS), from 2008–2009 to 2013–2014

Academic Year	Number of countries	Number of links	Number of students
2008–2009	31	807	168,193
2009–2010	32	823	177,705
2010–2011	33	839	190,495
2011–2012	33	919	204,744
2012–2013	33	910	211,995
2013–2014	34	928	212,208

The evolution of the Erasmus network in terms of incoming and outgoing students in all European countries is explored through the variation (%) between each year and the previous one (Tables 3 and 4).

Looking at the Table 3, it is possible to observe a positive trend in terms of incoming students for all years under analysis, even if a negative variation is revealed for some countries in 2013–2014. Moreover, the countries where a relevant positive variation is observable are Switzerland, Estonia, Hungary, Poland, Slovenia, Slovakia and Turkey. In Italy and Spain, we observe a negative variation in the last 2 years.

Looking at the variation of outgoing students (Table 4), we note that the trend is unstable over the period considered. A constant positive variation is registered in Germany, Denmark, Finland, France, Greece, Croatia, Ireland, Italy, The Netherlands, Sweden, Slovakia, Turkey and the United Kingdom. In other countries, positive and negative variations are observed along with the years.

Figure 1 displays the network visualization (digraph) of the six weighted adjacency matrices showing the increase of the relationships among countries for the SMS network over the time. The graph visualizations are enriched by colouring the countries according to European macro-areas' aggregation.

By considering the results above described in terms of variation of incoming and outgoing students, countries are classified as *good exporters* or *good importers* by means of the *hub* and *authority* network centrality indexes. Figures 2 and 3 show the maps of European countries in terms of the hubs and authorities scores in the SMS network defined for all the years under analysis. Thus, it is possible to highlight the good importers and good exporters and evaluate the attractiveness of European countries.

For all years the good importers are Spain, France, the United Kingdom, Germany and Italy (Figure 2). As good exporters, the top positions are occupied by Germany, France, Italy and Spain (Figure 3).

Table 3 Variation (%) of number of incoming students in the Erasmus student mobility for studies for all countries and for each year with respect to the previous year

	2009–2010	2010–2011	2011–2012	2012–2013	2013–2014
AT	4.13	6.32	7.94	3.71	−1.44
BE	4.77	7.64	5.76	1.78	−0.17
BG	2.04	23.94	11.07	20.47	−7.97
CH	0.00	0.00	0.00	4.66	−3.90
CY	26.92	34.34	16.04	17.71	−3.30
CZ	9.91	11.09	9.73	9.70	5.71
DE	1.16	6.65	10.97	7.10	0.37
DK	8.63	7.80	−3.42	−5.62	−12.76
EE	11.34	10.64	27.75	15.27	−1.21
ES	4.09	4.27	2.99	−0.42	−3.46
FI	−0.43	3.50	−0.03	4.10	0.91
FR	5.14	5.17	4.73	−1.23	0.36
GR	5.81	−3.79	−8.28	−19.37	20.34
HR	0.00	0.00	0.00	84.43	39.96
HU	12.20	14.75	14.27	12.70	8.51
IE	−2.54	3.66	4.19	8.63	3.81
IS	16.43	9.73	1.11	8.11	−1.22
IT	2.28	5.37	4.33	−3.34	−0.04
LI	−5.88	12.50	19.44	−9.30	−7.69
LT	7.07	11.04	20.71	23.52	5.15
LU	7.55	52.63	−1.15	8.14	17.20
LV	4.24	28.23	35.63	26.41	6.20
MK	0.00	0.00	0.00	0.00	0.00
MT	26.20	1.12	−5.52	14.95	−2.44
NL	5.00	5.26	7.90	2.71	−0.91
NO	12.10	10.79	1.75	5.39	4.35
PL	22.22	25.26	17.97	20.27	7.70
PT	15.42	14.60	6.66	7.51	3.75
RO	8.99	15.38	11.00	19.32	3.34
SE	7.09	4.62	2.76	2.74	−8.58
SI	14.83	13.09	17.79	10.88	−0.24
SK	14.87	10.51	10.41	10.06	2.72
TR	22.84	30.53	20.43	15.30	11.21
UK	4.72	4.05	2.97	0.32	2.37

AT=Austria; BE=Belgium; BG=Bulgaria; CH=Switzerland; CY=Cyprus; CZ=Czech Republic; DE=Germany; DK=Denmark; EE=Estonia; ES=Spain; FI=Finland; FR=France; GR=Greece; HR=Croatia; HU=Hungary; IE=Ireland; IS=Iceland; IT=Italy; LI=Liechtenstein; LT=Lithuania; LU=Luxembourg; LV=Latvia; MK=Macedonia; MT=Malta; NL=Netherlands; NO=Norway; PL=Poland; PT=Portugal; RO=Romania; SE=Sweden; SI=Slovenia; SK=Slovakia; TR=Turkey; UK=United Kingdom

Table 4 Variation (%) of number of outgoing students in the Erasmus student mobility for studies for all countries and for each year with respect to the previous year

	2009–2010	2010–2011	2011–2012	2012–2013	2013–2014
AT	4.47	0.17	7.26	1.17	−1.00
BE	4.52	7.29	2.72	8.99	−1.30
BG	13.09	6.75	−1.03	−0.78	−14.20
CH	0.00	0.00	0.00	2.98	4.36
CY	38.19	25.13	−14.06	29.44	2.89
CZ	−1.88	4.70	8.41	2.08	0.13
DE	2.66	4.78	9.59	4.69	3.79
DK	8.86	6.63	20.23	11.52	0.62
EE	31.58	8.55	−2.16	2.47	−9.25
ES	12.50	14.50	8.51	−1.63	−8.72
FI	2.71	11.28	4.10	4.16	1.90
FR	3.68	5.58	0.52	3.15	0.68
GR	1.94	3.91	3.07	11.28	3.94
HR	0.00	96.60	47.62	29.33	21.32
HU	−2.76	−2.16	3.73	−3.49	−8.71
IE	12.60	16.12	5.65	0.66	7.34
IS	15.59	14.88	−6.07	−1.29	−15.28
IT	7.68	3.43	3.19	4.94	2.23
LI	−5.00	84.21	−5.71	−30.30	8.70
LT	−6.10	13.31	2.21	−6.33	−5.79
LU	4.46	−1.35	1.82	−10.51	7.75
LV	14.95	9.06	4.48	−3.25	−2.29
MK	0.00	0.00	0.00	0.00	0.00
MT	−14.08	−100.00	0.00	50.00	7.09
NL	9.30	10.97	8.46	6.26	5.52
NO	−4.18	14.90	6.34	4.02	−2.87
PL	−1.45	−0.35	4.61	−1.21	−4.82
PT	−3.25	7.57	4.73	3.42	−2.28
RO	2.12	11.95	−3.51	−4.97	14.66
SE	13.05	4.33	12.58	2.22	1.50
SI	−1.24	7.25	17.68	−6.73	−2.96
SK	5.58	14.13	5.70	12.63	5.12
TR	15.84	12.19	14.18	20.22	4.89
UK	8.40	6.51	6.03	6.03	6.64

AT=Austria; BE=Belgium; BG=Bulgaria; CH=Switzerland; CY=Cyprus; CZ=Czech Republic; DE=Germany; DK=Denmark; *EE*=Estonia; ES=Spain; FI=Finland; FR=France; GR=Greece; HR=Croatia; HU=Hungary; IE=Ireland; IS=Iceland; IT=Italy; LI=Liechtenstein; LT=Lithuania; LU=Luxembourg; LV=Latvia; MK=Macedonia; MT=Malta; NL=Netherlands; NO=Norway; PL=Poland; PT=Portugal; RO=Romania; SE=Sweden; SI=Slovenia; SK=Slovakia; TR=Turkey; UK=United Kingdom

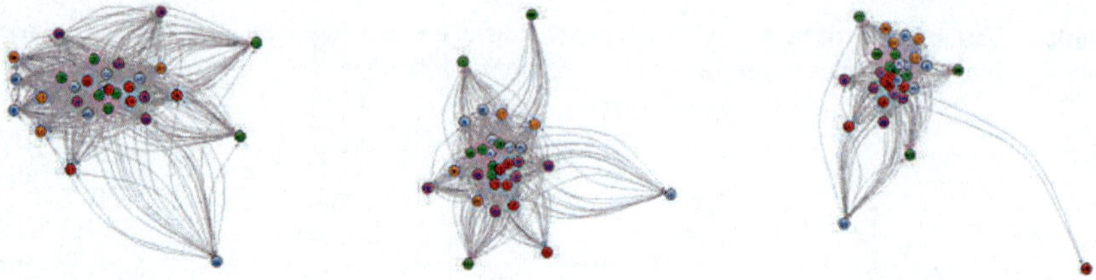

(a) SMS 2008–2009 (b) SMS 2009–2010 (c) SMS 2010–2011

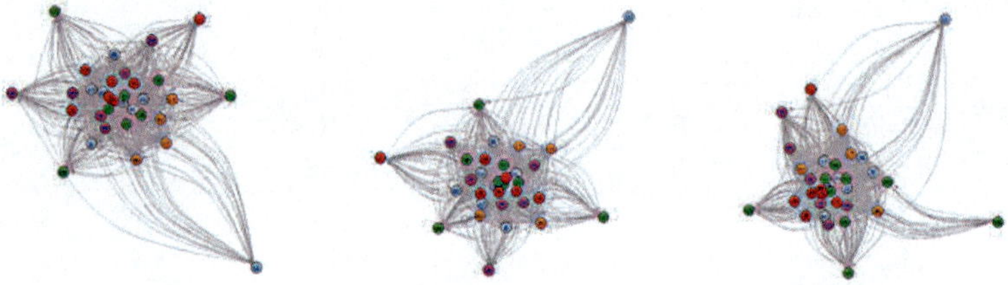

(d) SMS 2011–2012 (e) SMS 2012–2013 (f) SMS 2013–2014

Fig. 1 Graphs visualization of Erasmus student mobility for studies (SMS) according to European macro-areas' aggregation of countries, from 2008–2009 to 2013–2014. Node colour: red = Western Europe, orange = Eastern Europe, light blue = Central Europe, green = Southern Europe, and violet = Northern Europe

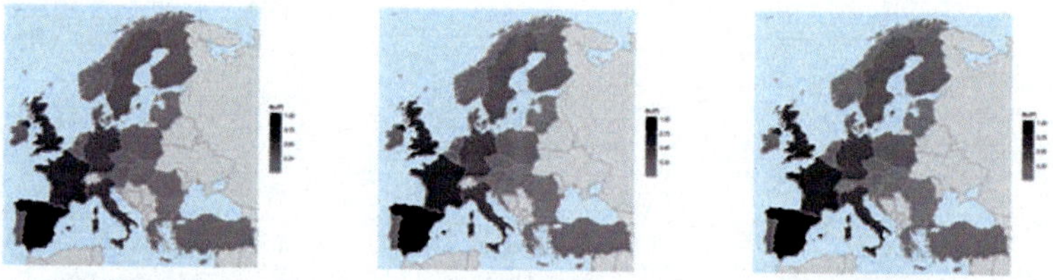

(a) 2008–2009 (b) 2009–2010 (c) 2010–2011

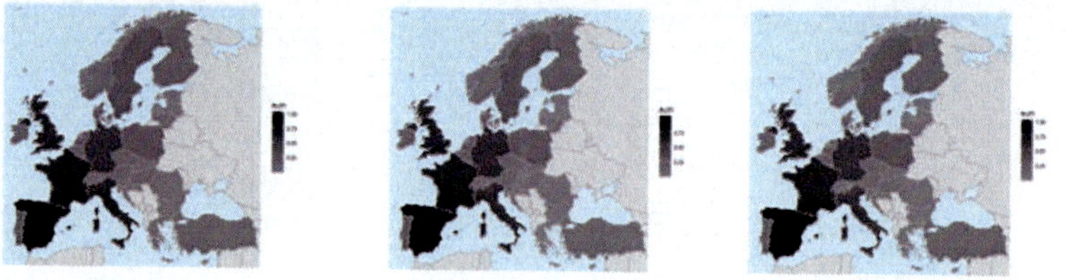

(d) 2011–2012 (e) 2012–2013 (f) 2013–2014

Fig. 2 Maps of authority score (i.e. good importers) for the SMS network for all years

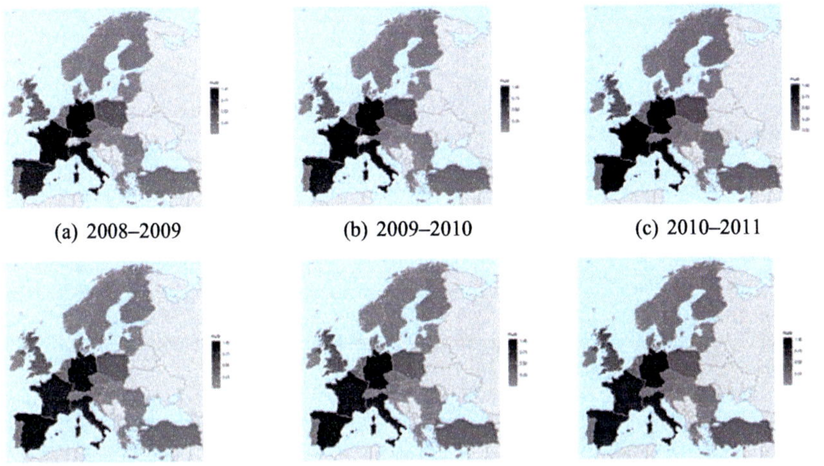

(a) 2008–2009 (b) 2009–2010 (c) 2010–2011

(d) 2011–2012 (e) 2012–2013 (f) 2013–2014

Fig. 3 Maps of hub score (i.e. good exporters) for the SMS network for all years

5 Conclusion

The present contribution focused on the analysis of the Erasmus students' mobility in order to catch the role played by countries in the internationalization process that characterized European universities. Thanks to social network analysis (SNA) approach, it is possible to sketch some conclusions that can be synthesized as follows:

- The number of students exchanges between EU countries is growing faster over the time;
- Few countries, especially those in the Mediterranean area (Spain, Italy, France, Germany), are classified as good exporters and/or good importers;
- A higher propensity to share students from different European macro-areas is shown. In more detail, students from a European country are willing to go in countries belonging to European macro-areas different from where they live.

As a future line of research proposal, the use of clustering procedure for network data (blockmodeling analysis, Derszi et al. (2011)) could be adopted to further investigate the network structure by setting up ideal network configurations on the basis of theoretical hypotheses. A deeper investigation of the relationship between network's dynamics and country's attractiveness is also required, by focusing on some indicators related to specific features of the Tertiary Education System. Thus, it could be of interest to describe the geographical distribution of hosting and sending countries, to examine the country's position in the Erasmus student mobility network and to study the underlying mechanisms in network link formation by applying statistical modelling.

References

Barnett, G. A., Ke Jiang, M. L., Park, H. W. (2016). The flow of international students from a macro perspective: a network analysis. *Compare: A Journal of Comparative and International Education, 46*(4), 533–559.

Breznik, K., Skrbinjek, V, Law, K., Dakovic, G. (2013). On the erasmus student mobility for studies. In: Dermol, V., Trunk Sirca, N., Dakovic, G. (eds.). *Active citizenship by knowledge management & innovation: proceedings of the Management, Knowledge and Learning International Conference* (p. 13711377). Bangkok, Celje, Lublin: ToKnowPress.

Breznik, K., & Ragozini, G. (2015). Exploring the Italian erasmus agreements by a network analysis perspective. *IEEE/ACM International Conference on Advances in Social Networks Analysis and Mining (ASONAM), 2015*, 837–838.

Chadee, D., & Naidoo, V. (2009). Higher educational services exports: Sources of growth of Asian students in US and UK. *Servies Business, 3*(2), 173–187.

Chen, T. M., & Barnett, G. A. (2000). Research on international student flows from a macro perspective: A network analysis of 1985, 1989 and 1995. *Higher Education, 39*(4), 435–453.

De Benedictis, L., & Leoni, S. (2020). *Gender bias in the Erasmus students network.* arXiv preprint arXiv:2003.09167.

Derszi, A., Derszy, N., Kaptalan, E., & Neda, Z. (2011). Topology of the Erasmus student mobility network. *Physica A: Statistical Mechanics and its Applications, 390*(13), 2601–2610.

Doreian, P., Batagelj, V., & Ferligoj, A. (2005). *Generalized blockmodeling*. Cambridge University Press.

Fombona, J., Rodríguez, C., & Pascual Sevillano, M. A. (2013). The motivational factor of Erasmus students at the university. *International Educatio Studies, 6*(4), 1–9.

Freeman, L. C. (1979). Centrality in social networks conceptual clarification. *Social Networks, 1*(3), 215–239.

Jiang, K. (2014). International student flows between Asia, Australia, and Russia: A network analysis. *Journal of Contemporary Eastern Asia, 13*(1), 83–98.

King, R., & Ruiz-Gelices, E. (2003). International student migration and the European year abroad: Effects on European identity and subsequent migration behaviour. *International Journal of Population Geography, 9*(3), 229–252.

Kleinberg, J. M. (1999). Authoritative sources in a hyperlinked environment. *Journal of the ACM, 46*(5), 604–632.

Kondakci, Y., Bedenlier, S., & Zawacki-Richter, O. (2018). Social network analysis of international student mobility: Uncovering the rise of regional hubs. *Higher Education, 75*(3), 517–535.

Krackhardt, D., & Stern, R. N. (1988). Informal networks and organizational crises: An experimental simulation. *Social Psychology Quarterly, 51*(2), 123–140.

Macrander, A. (2017). Fractal inequality: A social network analysis of global and regional international student mobility. *Research in Comparative and International Education, 12*(2), 243–268.

Meara, P. (1994). The year abroad and its effects. *Language Learning Journal, 10*(1), 32–38.

Mitchell, K. (2012). Student mobility and European identity: Erasmus study as a civic experience? *Journal of Contemporary European Research, 8*(4), 490–512.

Molu, F. E., Başman, Eryiğit, D., Tunç, B., Yaman, G. (2014). Intercultural education and incoming-outcoming of Erasmus student study exchange: Marmara university students. *International Journal of 21st Century Education, 1*, 21–33.

Norris, E. M., & Gillespie, J. (2009). How study abroad shapes global careers: Evidence from the United States. *Journal of Studies in International Education, 13*, 392–397.

Papatsiba, V. (2006). Making higher education more European through student mobility? Revisiting EU initiatives in the context of the Bologna process. *Comparative Education, 42*(1), 93–111.

Parey, M., & Waldinger, F. (2011). Studying abroad and the effects on international labour market mobility: Evidence from the introduction of Erasmus. *The Economic Journal, 121*, 194–222.

Restaino, M., Vitale, M. P., & Primerano, I. (2020). Analysing international student mobility flows in higher education: A comparative study on European Countries. *Social Indicators Research, 149*, 947–965.

A Preference Index Design for Big Data

Venera Tomaselli and Giulio Giacomo Cantone

Abstract *TripAdvisor* is a business service that works as a reputation system to guarantee quality in tourism experience. This kind of new service is based on Big Data technologies and characterized by generating, managing and summarizing, even with rating indexes, a quantitative experimental size of information, representing a frontier issue for data analysis. These data are organized and offered to users by a filter system aimed at recommending consumer's choices. Through a methodological design oriented to reward competitive quality, this acts as a crowdsourced evaluation system. In this paper, we suppose that information provided through the website can be biased because past reviews and ratings can affect the process of data production. On the basis of an empirical research for approximately 26.000 scores on *TripAdvisor* multipoint scale organized into 8-years time series and harvested by *R* software, we observe non-linear dynamics and skewed distribution among values of the scale. In our study, we observed that the main goal of crowd rating platforms is to extensively rank subsets of a population of units. This is achieved through the systematic employment of estimation techniques of evaluative measures. We propose a design of rating indexes that reflects the original missions of crowd rating: to pragmatically decrease the risk of a bad experience for the customer, to coherently benchmark, and to reliably rank a list of competing units.

V. Tomaselli (✉)
Department of Political and Social Sciences, University of Catania, 8, Vitt. Emanuele II - 95131, Catania, Italy
e-mail: venera.tomaselli@unict.it

G. G. Cantone
Department of Physics and Astronomy, University of Catania, 64, S. Sofia, 95123, Catania, Italy
e-mail: prgcan@gmail.com

P. Mariani and M. Zenga (eds.), *Data Science and Social Research II*,
Studies in Classification, Data Analysis, and Knowledge Organization,
https://doi.org/10.1007/978-3-030-51222-4_27

1 Evaluation from Crowd Rating

Crowdsourcing is a data gathering process to collect opinions on a topic. A common application of crowd rating in tourism is for evaluation of perceived quality: platforms for online reviews (*TripAdvisor*, *Trustpilot*, *Google*, etc.) and businesses relying on recommender systems (*Facebook*, *Amazon Group*, etc.) are common examples.

Although a contributing factor of enthusiasm for crowd rating is the generally low cost to achieve acquisition of large structured datasets (Geiger et al. 2012), we identified four further relevant reasons that present a rationale for adoption of crowd rating in any activity organized around people's opinion:

- To build trust in digital communities (e.g., eBay).
- To display to the public a massive flux of information (e.g., to rearrange Big Data into sorted rankings).
- To develop matching algorithms for recommender systems.
- To lock-in and select users, as after they 'scored' a desirable reputation in a platform, it is less likely that they will leave the platform for a competitor, so as not to lose their previous 'score' (Dellarocas 2011).

May be the case for the adoption of a crowd rating system in order to not only, display but also to measure latent features in services or intangible goods. Nowadays, this is an established practice in customer satisfaction (Pizam et al. 2009) but we can relate the methodology of numerical estimation from crowd's opinions to historical Galton's experiment (Galton 1907).

The British polymath showed that, challenged 787 totally unknown people to estimate the weight of an ox, the difference between the median of crowd's opinions and the exact value was lesser than 1%. We summarized the differences between Galton's experiment and crowd rating (Table 1).

Table 1 Differences between experimental design of research and implemented rating systems

Controlled research design	Implemented design on websites
An exact value exists and it is approximate by a metric	Supplies the lack of unit of measures for features like 'taste'
The experiment has a fixed end, and until then, other's people opinion is secret	Public crowd rating websites run with no end times and no secrecy of what is trending
No competition among subjects of measurement	Enables a competition to get better positions in future rankings, or to influence recommender systems

More in details, the following features highlight structural complexity in data production in crowd rating:

- Lack of exact measure: while Galton asked people to estimate weights, crowd rating often aims to estimate features like quality or satisfaction. An established method to evaluate an inter-subjective value in perceptions is the use of ordinal multipoint scales, and this method is commonly observed in online rating systems.
- Secrecy of opinions: an experiment is structured to have a start and an end, and generally intermediary results are kept in secrecy to ensure control over variables and biases. While open platforms vigorously enforce secrecy on how their algorithms are 'hardcoded', their business model is still based on making public the monitored data that includes reviews and ratings.
- Enabled competition: it is common knowledge in digital marketing that when a new technology enables to rank products under a common criterion of interrogation ('query'), to keep a ranking position online becomes a primary target for any of those products, and in particular to be in the first visualized webpage of any related *query* on search engines (Varian 2016).
- To lock-in and select users, as after they 'scored' a desirable reputation in a platform, it's less likely that they will leave the platform for a competitor, so as not to lose their previous 'score' (Dellarocas 2011).

Jeacle and Carter (Jeacle 2011) define tourism online rating systems as micro-social systems based on trust. From this definition, a quantitative evaluation of satisfaction should not ignore the following statistical biases commonly associated to non-experimental studies on public opinions:

- Non-independency of observations: earlier ratings influence late ratings. Experimental studies (Salganik et al. 2006) and empirical findings (Lee 2015) on crowd rating suggest that, in the absence of secrecy of trends, judgements over products converge towards a strong modal class of answers ('herding'). Research on platforms Amazon and Yelp (Bai et al. 2009) confirmed the hypothesis of the existence of a social mechanism of herding ensuring that earlier ratings are more likely to influence future 'popularity' of products than later ones.
- Survivorship bias: competition of subjects actually reflects competition for survival in a market (Farmer 2011). By this struggle for survival, some subjects may disappear from the market, and others may show up. Not only do subjects in the same *query* or list have different lifespans, but also their data can be retroactively censored by platforms, for the reason that the platforms do not desire to host an inactive or misleading subject in their online rating service. This could be a misleading factor in analysis because it censors those subjects where it is more likely that 'unpopularity' and *weaknesses* will be observed. More generally, it skews the distribution of ratings into higher numerical values (Mangel and Samaniego 2009).
- Frauds and optimization strategies: platforms monitor data which are voluntarily submitted, and sometimes they lack clear procedures to confirm the general *sincerity* of the submitted data. While technologies to improve *fake detection* are

constantly in development,[1] frauds are usually a consistent factor of skewness in reviews (Ott et al. 2012, Li et al. 2014). A further reflection is necessary: while a subject who actually manipulates a ranking by the submission of *fakes* may be held responsible of crime under a variety of legislation, *TripAdvisor* states that 'optimization' and anything that does not involve a 'payment' to fake a review is not against its Terms of Service.[2] We could conclude that 'asking gently' to submit a max-scored rating should be considered a legitimate strategy of optimization of reputation and awareness, but it is made clear that material incentives in exchange for max-scored ratings are inadmissible behaviour under ToS.[3] Thus, those ratings will be subjected to censor, introducing another bias in observed results between ratings already revised and those not.

In addition, it is of relevance that people are free to submit or not reviews for an experienced unit. As a consequence, the amount of reviews for a unit is not fixed and nor the units are reviewed by the same amount of users.

These differences in amounts can be extremely relevant for benchmarking, in particular when we record a raising trend of inequalities in amount of reviews over the time. In presence of this trend, we cannot expect to measure a low variance, even if we record large amounts of reviews.

2 Web-Scraped Data

We sampled a list of 60 web pages of active restaurants on *TripAdvisor.com*, the first operating since October 10th, 2009. Restaurants are considered 'inactive units' until the time-point they receive the first review.

We define 'time-point' the standard interval of time within which the recorded review was submitted ('day', 'week', 'month', etc.). When is not stated otherwise, we adopt 'day' as time-point.

A restaurant with at least one recorded review at a time-point is labelled an 'active unit since that time-point' (from here on, 'active unit'), e.g., if we observe a review recorded in January, no reviews in February, and reviews again in March, the unit is labelled 'active' since January.

For all the restaurants, the sampling criteria were:

- Addressed in the tourism city of Catania, IT
- Not less than 20 reviews at August 5th, 2018, from a total of 3204 days of activity
- 'Pizza' in the menu

With a web scraping script in R framework, we collected metadata from the reviews in the sample ($N = 26.888$), in particular we recorded only the following variables of metadata from reviews:

[1] https://www.TripAdvisor.com/TripAdvisorInsights/w3703.

[2] *ibidem*.

[3] https://www.TripAdvisor.com/TripAdvisorInsights/w591.

- Day of submission, in the range of 3204 days, as t timing
- Uniquely associated ID of subject restaurant on *TripAdvisor*
- Recorded class of scores of the review, within the ordinal scale of 1 through 5

While the number of active subjects grows linearly (Fig. 1), the number of collected reviews does not (Fig. 2).

Even taking into account survivorship bias, which obscures data from subjects active in the past but inactive at August 5th, 2018, this does not explain the difference between the two growth ratios. The maximal divergence between the two growth ratios of (i) active subjects and (ii) collected reviews is reached on Day 1513th

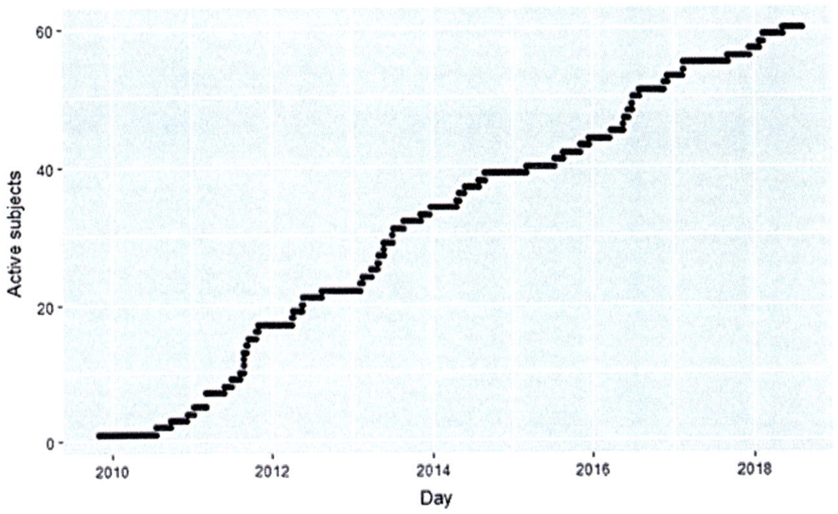

Fig. 1 Number of active subjects ('items') by day

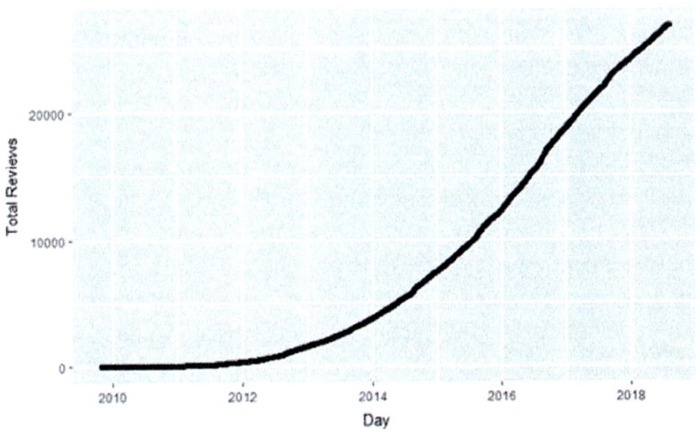

Fig. 2 Number of collected reviews by day

(December 18th, 2013) of 3204 (47%), when 34 of 60 subjects (57%) were already active.

2.1 Skewness of Frequency Distribution in Scores' Classes

Daily relative cumulative (until the last day) frequencies of classes of scores $F(\xi) = n_\xi/N$ were stable most of the time (Fig. 3).

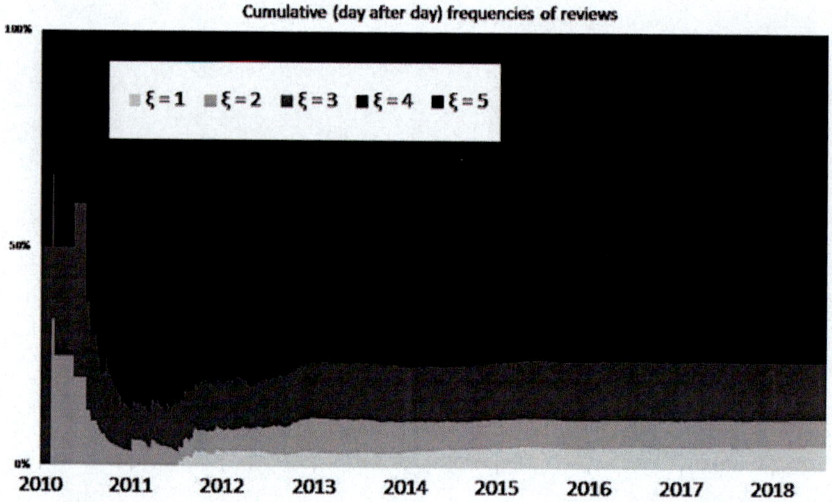

Fig. 3 Frequencies of time-cumulated reviews by day. The symbol ξ is used for the class of score

The modal class, indeed, was always $\xi = 5$, floating around a median frequency of .441. Data are consistent with results from previous studies on Italian cities on *TripAdvisor* (Baccianella et al. 2009). As the frequencies were stable and $\xi = 5$ scored almost half of total reviews, we supposed that weekly $\xi = 5$ should have been distributed around a central value of 0.445 and therefore $\xi \neq 5$ around 0.555.

After we aggregated daily data-points into weekly data-points by summing all the reviews with 7 days between a Sunday and its subsequent Saturday, starting from August 7th, 2011 and ending August 4th, 2018, for a total of 365 weeks in 7 years, we found the aforementioned hypothesis to be coherent with our data: *weekly* weekly $\xi = 5$ had a geometric mean of 0.4372, an average of 0.4487, a median of 0.4487, with a standard deviation of 0.0983, confirming the stable value.

We framed the time series from the starting week (August 7th, 2011) because this is the first week that satisfies this condition: every subsequent week had at least one $\xi = 5$ and one $\xi \neq 5$ reviews.

Another noteworthy property of this starting week is that at least 10 subjects were already active on that day. By doing this frame, we excluded 0.005 of total recorded reviews ('N') and 0.202 of total recorded time-points ('days'), as is shown in Fig. 2, above.

3 Ratings Estimation in a Ranking System

Although the debate between mean and median as estimators of the central value of records from multipoint scales is an open controversy (Lewis and Sauro 2016, Velleman et al. 1993), we will argue that for low amounts of classes of score, the mean must be adopted.

We noticed that, in ordinal scales, the robustness towards the extreme values of the median as estimator of central value is of no utility because the values are enclosed in a finite domain. Its lack of sensitivity towards small differences, on the other hand, is a disadvantage in cases where these differences, even the smallest, are decisive when the estimated parameter is argument of a rank-function in a benchmark analysis, like those proposed in customer satisfaction. This incongruence is exacerbated in a longitudinal context: the median as an argument of a rank-function is sensible to factors such as skewness in frequencies of classes of values, as in our case.

In particular, we observe that a minimal increase of the median of scores of an item after t causes a big 'jump' (permutation of rank; on the topic (Corain et al. 2016) of the item in the ranking towards the first positions. This property seems undesirable because, under ideal conditions, every permutation of ranks after t should be imputable much to a mutation of the measured performance, less to random or structural error in the model.

More specifically, the sum of amounts of $\xi = 4$ and $\xi = 5$ was always over 0.7 of the total, both in our data and in another study (Baccianella et al. 2009). Hence, to rank subjects by the median always produces a binary classification, which is of no use for ranking purposes for the aforementioned reasons. To estimate rating of subjects, we came to conclusion that a normalized average

$$[[(\sum n_{\xi} \times \xi)/N] - min(\xi)]/[max(\xi) - min(\xi)] \quad (1)$$

may be the viable solution when subjects are sorted in a ranking.

For those situations where we can be confident to detect strong skewedness towards the highest ('max') class of scores, we noticed that the simpler non-parametric ratio

$$f(\xi = max) = n_{max(\xi)}/N \quad (2)$$

will probably lead into a more robust, hence stable over time, ranking, without renouncing to give an information for a rate. We do not assume that the feature of lower variability over time of a ranking is valuable by default, but it can be in some cases.

A variant that helped further comparison of ratings between Baccianella et al. (2009) and Bai et al. (2009) in our case study reckons on adding the 4th class of score (the second highest) to the numerator of the ratio

$$f(\xi = 5) + f(\xi = 4) \tag{3}$$

4 Conclusions

The result of the normalization in (1) is to enclose estimations in the dominion (0,1). This is valuable for practical uses because allows for developing further arguments (i.e., on stability) about ranking techniques by parametric and non-parametric estimators of rating. Enclosing the final result in (0,1), it also gives the option to integrate different analyses of satisfaction based on scales that employ different number of points.

The feature related to the dominion (0,1) is useful to compare the results achieved through parametric measures with non-parametric rates. Indeed, in the (2) and (3) the ratio as non-parametric tool is employed to avoid the skewedness issue adding stability and robustness to rankings and summing at the 5 max point, the number of 4 points supports more properly further comparisons.

The issue of developing an established method of evaluation of performance for estimations in customer satisfaction, in particular if the mission is to provide a ranking, is still open. When dealing with extensive rankings our suggestion is to take in consideration the mathematical structure of a ranking, which is a sequence of natural number, where the distance among ranks is linear.

Therefore, the more the estimated ratings associated to ranks fit the assumption of linearity, the more that estimator fits the purpose of ranking the list.

References

Baccianella, S., Esuli, A., & Sebastiani, F. (2009). Multi-facet rating of product reviews. In: *Proceedings of 31th European Conference on IR Research on Advances in Information Retrieval* (pp. 461–472).

Bai, T., Zhao, X., He, Y., Nie, J.-Y., & Wen, J. R. (2018). Characterizing and predicting early reviewers for effective product marketing on E-Commerce websites. *IEEE Transactions on Knowledge and Data Engineering*, *30*(12), 1–14.

Corain, L., Arboretti, R., & Bonnini, S. (2016). *Ranking of multivariate populations: A permutation approach with applications*. Boca Raton, FL: CRC Press.

Dellarocas, C. (2011). Designing reputation systems for the social web. In: Masum, H., & Tovey, M. (Eds.) *The reputation society* (pp. 3–12). Cambridge, MA: MIT Press.

Farmer, R. (2011). Web reputation systems and the real world. In: Masum, H., & Tovey, M. (Eds.) *The reputation society* (pp. 13–24). Cambridge, MA: MIT Press.

Galton, F. (1907). Vox Populi. *Nature*, *75*, 450–451.
Geiger, D., Schader, M., Rosemann, M., & Fielt, E. (2012). Crowdsourcing information systems—definition, typology, and design. *Proceeding of International Conference on Information Systems*, 9–11.
Jeacle, I., & Carter, C. (2011). In tipadvisor we trust: Rankings, calculative regimes and abstract systems. *Accounting, Organizations and Society*, *36*(4/5), 293–309.
Lee, Y. J., Hosanagar, K., & Tan, Y. (2015). Do I follow my friends or the crowd? Information cascades in online movie ratings. *Management Science*, *61*(9), 2241–2258. https://doi.org/10.1287/mnsc.2014.2082.
Lewis, J. R. (1993). Multipoint scales: Mean and median differences and observed significance levels. *International Journal Human-Computer Interaction*, *5*, 382–392.
Lewis, J. R., & Sauro, J. (2016). *Quantifying the user experience: Practical statistics for user research*. Cambridge, MA: Morgan Kaufmann.
Li, J., Ott, M., Cardie, C., & Hovy, E. (2014). Towards a general rule for identifying deceptive opinion spam. In: *Proceedings of 52nd annual meeting of the association for computational linguistics* (pp. 1566–1576). Baltimore, MD.
Mangel, M., & Samaniego, F. (1984). Abraham Wald's work on aircraft survivability. *Journal of the American Statistical Association*, *79*, 259–267.
Ott, M., Cardie, C., & Hancock, J. (2012). Estimating the prevalence of deception in online review communities. In: *Proceedings of the 21st international conference on World Wide Web* (pp. 201–210).
Pizam, A., Shapoval, V., & Ellis, T. (2016). Customer satisfaction and its measurement in hospitality enterprises: A revisit and update. *International Journal of Contemporary Hospitality Management*, *28*(1), 2–35.
Salganik, M. J., Dodds, P. S., & Watts, D. J. (2006). Experimental study of inequality and unpredictability in an artificial cultural market. *Science*, *311*, 854–856.
Varian, H. R. (2016). The economics of Internet search. In: Bauer, J., Latzer, M. (Eds.) *Handbook on the economics of the Internet* (pp. 385–394). Cheltenham, UK: Edward Elgar Publishing.
Velleman, P. F., & Wilkinson, L. (1993). Nominal, ordinal, interval, and ratio typologies are misleading. *American Statistician*, *47*(1), 65–72.

Construction of an Immigrant Integration Composite Indicator through the Partial Least Squares Structural Equation Model K-Means

Venera Tomaselli, Mario Fordellone, and Maurizio Vichi

Abstract Integration is a multidimensional process, which can take place in different ways and at different times in relation to each of the single economic, social, cultural, and political dimensions. Hence, examining every single dimension is important as well as building composite indexes simultaneously inclusive of all dimensions in order to obtain a complete description of a complex phenomenon and to convey a coherent set of information. In this paper, we aim at building an immigrant integration composite indicator (IICI), able to measure the different aspects related to integration such as employment, education, social inclusion, active citizenship, and on the basis of which to simultaneously classify territorial areas such as European regions. For this application, the data collected in 274 European regions from the European Social Survey (ESS), Round 8, on immigration have been used.

1 Introduction

The immigrants' integration is a multidimensional process implying many economic, social, cultural, and political issues. This process is carried out according to several steps and in different conditions determining continuous redefinition of accomplishment outcomes. In fact, each single dimension, diachronically positioned over time, generates different integration levels. Hence, examining each single dimension is important as well as building composite indexes simultaneously comprehensive of all dimensions in order to obtain a complete description of a complex phenomenon and to convey a suitable set of information.

V. Tomaselli (✉)
Department of Political and Social Sciences, University of Catania, 8, Vitt. Emanuele II -95131, Catania, Italy
e-mail: venera.tomaselli@unict.it

M. Fordellone · M. Vichi
Department of Statistical Sciences, La Sapienza, University of Rome, Rome, Italy
e-mail: mario.fordellone@uniroma1.it
e-mail: maurizio.vichi@uniroma1.it

P. Mariani and M. Zenga (eds.), *Data Science and Social Research II*,
Studies in Classification, Data Analysis, and Knowledge Organization,
https://doi.org/10.1007/978-3-030-51222-4_28

According to the literature Entzinger (2000), Entzinger and Biezeveld (2003), the concept of integration can be broken down into different dimensions. Firstly, the socioeconomic dimension refers to housing conditions, work conditions, and income. The legal-political dimension takes into account the theme of citizenship and the rights of political participation, from the freedom of association to the voting right, which in some countries can be used at local government elections even without having achieved the citizenship status of the host country. Finally, the cultural and social dimension considers several elements, among which are knowledge of the language (Vermeulen 2004), free times activities, and access to information.

Due to the multidimensional nature of the integration concept, many studies underline the difficulty to identify core indicators (Ager and Eyber 2002; Strang et al. 2003) able to measure the integration level taking into account each dimension and subdimension of the integration concept (Cesareo and Blangiardo 2009). The factors more strictly connected to the host country approach toward migrants and also those related to country's socioeconomic conditions affect migrant integration (Di Bartolomeo et al. 2015) both at the local and regional levels (OECD 2018).

In this paper, we aim at providing a methodological proposal to build an immigrant integration composite indicator (IICI), able to measure the different aspects related to integration such as employment, education, social inclusion, and active citizenship and by which simultaneously to classify territorial areas (OECD 2008). With this in mind, we analyze the data collected in 274 European regions from European Social Survey (ESS), Round 8, by the structural equation modeling estimated via partial least squares (PLS-SEM) approach introduced by Lohmoller (1989) and developed by Tenenhaus et al. (2005).

In particular, we perform a simultaneous nonhierarchical clustering and partial least squares modeling, named partial least squares structural equation model *k*-means (PLS-SEM-KM), recently proposed by Fordellone and Vichi (2018), in order to obtain an immigrant integration composite indicator (IICI) and a clustering of the European regions.

Differently from the PLS-SEM methods, PLS-SEM-KM mainly focuses on the homogeneity between and within clusters of regions derived by a unique structural measurement model on immigrant integration. Thus, this study aims at both segmenting the immigrant population and simultaneously identifying the structural (i.e., the latent dimensions explaining the immigrants' integration) and measurement relations (i.e., the observed variables employed to build the latent dimensions) which have produced the segmentation among European regions grouped for immigrants' integration level.

The paper is structured as follows: in Sect. 2, a brief background on the PLS-SEM notation is provided. In Sect. 3, the PLS-SEM-KM model is presented; in Sect. 4, using the ESS data, the results obtained by IICI construction are shown.

2 Background Methods

2.1 Notation

Partial Least Squares (PLS) methodologies are algorithmic tools with analytic properties aiming at solving problems connected with stringent assumptions on data, e.g., distributional assumptions that are hard to meet in real life (Tenenhaus et al. 2005). Tenenhaus et al. try to better clarify the terminology used in the PLS field through a relevant review of the literature, focusing the attention on the Structural Equation Models standpoint.

Before showing the modeling details, the notation and terminology used in this paper are here presented to allow the reader to easily follow the subsequent formalizations and algebraic elaborations:

n, J	# of:	Observations, MVs
H, L, P	# of:	Exogenous LVs, endogenous LVs, LVs ($P = H + L$)
K	# of:	Clusters
Ξ	$n \times H$	Exogenous LVs matrix
$\mathbf{H}$	$n \times L$	Endogenous LVs matrix
$\mathbf{Y}$	$n \times P$	Scores matrix ($\mathbf{Y} = [\Xi, \mathbf{H}]$)
Γ	$L \times H$	Path coefficients matrix of the exogenous LVs
$\mathbf{B}$	$L \times L$	Path coefficients matrix of the endogenous LVs
$\mathbf{Z}$	$n \times L$	Errors matrix of the endogenous LVs
$\mathbf{X}$	$n \times J$	Data matrix
$\mathbf{E}$	$n \times J$	Errors matrix of the data
Λ_H	$J \times H$	Loadings matrix of the exogenous LVs
Λ_L	$J \times L$	Loadings matrix of the endogenous LVs
Λ	$J \times P$	Loadings matrix ($\Lambda = [\Lambda_H, \Lambda_L]$)
$\mathbf{T}$	$n \times H$	Errors matrix of the exogenous LVs
Δ	$n \times L$	Errors matrix of the endogenous LVs
$\mathbf{U}$	$n \times K$	Membership matrix (binary and row stochastic)

Usually, a PLS-SEM (called also PLS-PM, i.e., PLS path model) consists in a combination of two models:

- a structural model (or inner model) that specifies the relationships among latent variables (LVs). In this context, an LV is an unobservable variable (i.e., connected with a theoretical construct) indirectly described by a block of observable variables which are called manifest variables (MVs);
- a measurement model (or outer model) that relates the MVs to their LVs.

2.2 Structural Model

Let **X** be an $n \times J$ data matrix, with P endogenous and exogenous latent variables ($P \leq J$), let **H** be the $n \times L$ matrix of the endogenous LVs with generic element $\eta_{i,l}$, and let Ξ be the $n \times H$ matrix of the exogenous LVs with generic element $\xi_{i,h}$; the structural model is a causality model that relates the P LVs to each other through a set of linear equations (Vinzi et al. 2010). In matrix form:

$$\mathbf{H} = \mathbf{H}\mathbf{B}^T + \Xi\Gamma^T + \mathbf{Z} \tag{1}$$

where **B** is the $L \times L$ matrix of the path coefficients $\beta_{l,l}$ associated with the endogenous latent variables; Γ is the $L \times H$ matrix of the path coefficients $\gamma_{l,h}$ associated with the exogenous latent variables; **Z** is the $n \times L$ matrix of the residual terms $\zeta_{i,l}$.

Example 1 An example of structural model is shown in Fig. 1.

2.3 Measurement Model

In PLS-SEM, unlike the traditional SEM approach, there are two ways to relate MVs to their LVs: reflective and formative ways (Diamantopoulos and Winklhofer 2001; Tenenhaus et al. 2005). In the reflective way, it is supposed that each MV reflects its LV, i.e., the observed variables are considered as the effect of the latent construct; a reflective measurement model can be written in matrix form as

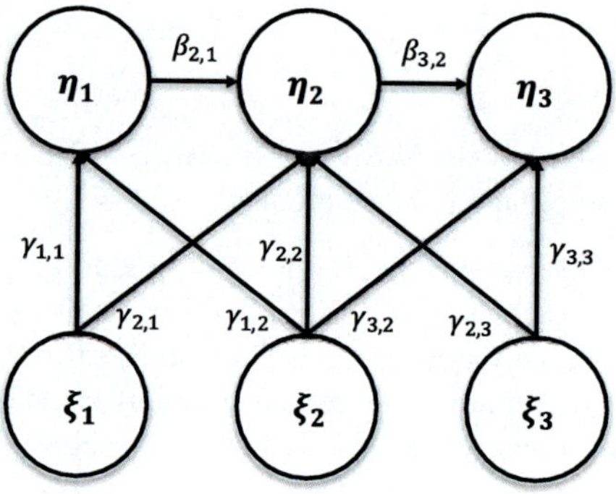

Fig. 1 Example of structural model with three endogenous LVs and three exogenous LVs

$$\begin{aligned}\mathbf{X} &= \mathbf{Y}\Lambda^T + \mathbf{E} \\ &= \begin{bmatrix}\Xi\ \mathbf{H}\end{bmatrix}\begin{bmatrix}\Lambda_H^T \\ \Lambda_L^T\end{bmatrix} + \mathbf{E} \\ &= \Xi\Lambda_H^T + \mathbf{H}\Lambda_L^T + \mathbf{E}\end{aligned} \tag{2}$$

where Λ_H is the $J \times H$ loadings matrix of the exogenous latent constructs with generic element $\lambda_{j,h}$; Λ_L is the $J \times L$ loadings matrix of the endogenous latent constructs with generic element $\lambda_{j,l}$; $\mathbf{E}$ is the $n \times J$ residuals matrix with element $\epsilon_{i,j}$, which have zero mean and are uncorrelated with $\xi_{i,h}$ and $\eta_{i,l}$. Then, the reflective way implies that each MV is related to its LV by a set of simple regression models with coefficients $\lambda_{j,l}$.

Conversely, in the formative way each MV is supposed to be *forming* its LV, i.e., the observed variables are considered as the cause of the latent construct. Formally, for an exogenous latent construct, the model can be written as

$$\Xi = \mathbf{X}\Lambda_H + \mathbf{T} \tag{3}$$

whereas, for endogenous latent construct the model can be written as

$$\mathbf{H} = \mathbf{X}\Lambda_L + \Delta \tag{4}$$

where $\mathbf{T}$ and Δ are, respectively, the $n \times H$ and $n \times L$ errors matrices with elements $\tau_{i,h}$ and $\delta_{i,l}$, which have zero mean and are uncorrelated with $x_{i,j}$. Then, the formative way implies that each MV is related to its LV by a multiple regression model with coefficients λs.

Example 2 In Fig. 2, two examples of PLS-SEM with three latent constructs (η_1, ξ_1, and ξ_2) and six observed variables (x_1, x_2, x_3, x_4, x_5, and x_6) are shown. In particular, there are two exogenous LVs (ξ_1 and ξ_2) and one endogenous LV (η_1). The MVs are related to their LVs in reflective way (left plot) and formative way (right plot).

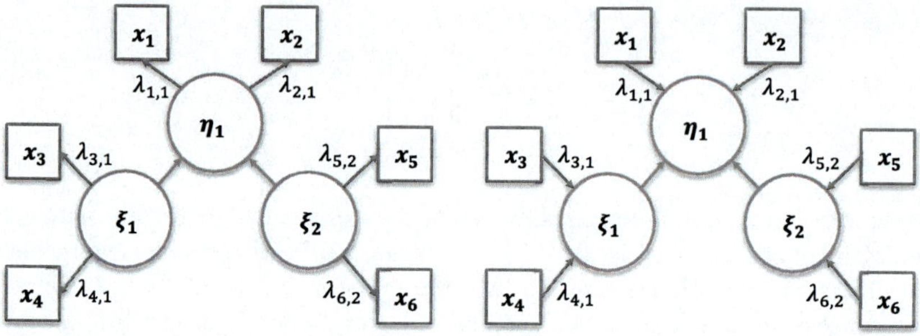

Fig. 2 Two examples of PLS path model with three LVs and six MVs: reflective measurement models (left) and formative measurement models (right)

3 Partial Least Squares *K*-Means

Given the $n \times J$ data matrix $\mathbf{X}$, the $n \times K$ membership matrix $\mathbf{U}$, the $K \times J$ centroids matrix $\mathbf{C}$, the $J \times P$ loadings matrix $\Lambda = [\Lambda_H, \Lambda_L]$, and the errors matrices $\mathbf{Z}$ $(n \times L)$ and $\mathbf{E}$ $(n \times J)$, the partial least squares structural equation model k-means (PLS-SEM-KM) model can be written as follows (Fordellone and Vichi 2018):

$$\begin{aligned}
&\mathbf{H} = \mathbf{H}\mathbf{B}^T + \Xi\Gamma^T + \mathbf{Z} \\
&\mathbf{X} = \mathbf{Y}\Lambda^T + \mathbf{E} = \Xi\Lambda_H^T + \mathbf{H}\Lambda_L^T + \mathbf{E} \\
&\mathbf{X} = \mathbf{UC}\Lambda\Lambda^T + \mathbf{E} = \mathbf{UC}\Lambda_H\Lambda_H^T + \mathbf{UC}\Lambda_L\Lambda_L^T + \mathbf{E},
\end{aligned} \tag{5}$$

subject to constraints: (i) $\Lambda^T\Lambda = \mathbf{I}$; and (ii) $\mathbf{U} \in \{0, 1\}$, $\mathbf{U}\mathbf{1}_K = \mathbf{1}_n$. Thus, the PLS-SEM-KM model includes the PLS and the clustering equations (i.e., $\mathbf{X} = \mathbf{UC}$ and then, $\mathbf{Y} = \mathbf{X}\Lambda$ becomes $\mathbf{Y} = \mathbf{UC}\Lambda$). The PLS-SEM-KM algorithm is composed by the following steps:

Algorithm 1 PLS-SEM-KM algorithm

1: Initialize $\Lambda = \mathbf{D}_{\Lambda}$;
Choose K through the *gap method* applied on scores matrix $\mathbf{Y} = \mathbf{X}\Lambda$;
$\omega = 10^{-12}$, iter=0, maxiter=300;
2: Random generate the memberships matrix $\mathbf{U}$;
Compute centers matrix $\mathbf{C} = (\mathbf{U}^T\mathbf{U})^{-1}\mathbf{U}^T\mathbf{X}$;
Compute latent scores matrix $\mathbf{Y} = \mathbf{UC}\Lambda$;
3: iter=iter+1;

Inner approximation

4: Estimate covariance matrix $\Sigma_Y = n^{-1}\mathbf{Y}^T\mathbf{JY}$ (with $\mathbf{J} = \mathbf{I}n^{-1}\mathbf{11}^T$);
5: Compute inner weights $\mathbf{W} = \mathbf{D}_B \otimes \Sigma_Y$;
6: Estimate new scores $\mathbf{Y}_W = \mathbf{YW}$;

Outer approximation

7: Update $\Lambda \rightarrow \Lambda_n = \mathbf{C}^T\mathbf{U}^T\mathbf{Y}_W(\mathbf{Y}_W^T\mathbf{Y}_W)^{-1}$; (Reflective way)
$\rightarrow \Lambda_n = (\mathbf{C}^T\mathbf{U}^T\mathbf{UC})^{-1}\mathbf{C}^T\mathbf{U}^T\mathbf{Y}_W$; (Formative way)
8: Update $\mathbf{U} \rightarrow \underset{\mathbf{U}}{\text{argmin}} \left\| \mathbf{X} - \mathbf{UC}\Lambda_n\Lambda_n^T \right\|^2$,
subject to $\Lambda_n^T\Lambda_n = \mathbf{1}_P, \mathbf{U} = \{0, 1\}, \mathbf{U1}_K = \mathbf{1}_n$;
9: Compute new centers $\mathbf{C}_n = (\mathbf{U}^T\mathbf{U})^{-1}\mathbf{U}^T\mathbf{X}$;

Stopping rule

10: Update $K \rightarrow Kn$ through the *gap method* applied on scores matrix $\mathbf{Y} = \mathbf{UC_n\Lambda_n}$
11: **if** $K_n \neq K$
go to step 2
12: **else**
13: **if** $\|\mathbf{C}\Lambda - \mathbf{C}_n\Lambda_n\|^2 > \omega$ & $\text{iter}<\text{maxiter}$, $\mathbf{C} = \mathbf{C}_n$, $\Lambda = \Lambda_n$;
repeat step 3-12;
14: **else**
exit loop 3-12;
15: **end if**
16: **end if**

Path coefficients estimation

17: **for** $l = 1$ to L **do**
18: **for** $h = 1$ to H **do**
19: Compute $\mathbf{Y}_h = \mathbf{X}\Lambda_h$
20: Compute $\mathbf{Y}_l = \mathbf{X}\Lambda_l$
21: Compute $\Gamma = (\mathbf{Y}_{h*}^T\mathbf{Y}_{h*})^{-1}\mathbf{Y}_{h*}^T\mathbf{Y}_l$
22: Compute $\mathbf{B} = (\mathbf{Y}_{l*}^T\mathbf{Y}_{l*})^{-1}\mathbf{Y}_{l*}^T\mathbf{Y}_l$
23: **end for**
24: **end for**

PLS-SEM-KM algorithm is based on the simultaneous optimization of PLS-SEM and reduced k-means (De Soete and Carroll 1994), where centroids of clusters are located in the reduced space of the LVs, thus, ensuring the optimal partition of the statistical units on the best latent hyperplane defined by the structural/measurement relations estimated by the prespecified model. The input parameters are the $n \times J$ standardized data matrix $\mathbf{X}$; the $J \times P$ design matrix of the measurement model $\mathbf{D}_{\Lambda}$, with binary elements equal to 1 if an MV is associated with an LV and 0 otherwise; the $P \times P$ path design matrix of the structural model $\mathbf{D}_B$, with binary elements equal to 1 if a latent exogenous or endogenous variable explains a latent endogenous variable and 0 otherwise. Matrix $\mathbf{D}_B$ is symmetrized.

Moreover, a different approach to select the optimal number of segments K is provided. In fact, PLS-SEM-KM algorithm includes the optimal K selection through the gap statistics proposed by Tibshirani et al. (2001). This statistics is embedded in the algorithm for estimating simultaneously the number of clusters together with PLS-SEM. In fact, the *gap method* may be applicable to any model-based clustering approach without restrictive assumptions on the scores distribution and therefore, is a valid method to be included in our methodology.

$\mathbf{Y}_h$ is the hth exogenous latent score and $\mathbf{Y}_l$ is the lth endogenous latent score; the symbol $\otimes$ indicates here the element-wise product of two matrices, while $*$ indicates the adjacent latent scores matrix, i.e., the set of latent scores that are related to the $\mathbf{Y}_h$ or $\mathbf{Y}_l$. The PLS-SEM-KM algorithm is a development of the Wold's original algorithm used to the PLS-SEM estimate in Lohmoller (1989). As you can see from the step 7 of the algorithm (i.e., in the loadings estimation), the method is performed for both reflective measurement models and formative measurement models. **U** matrix is optimized row by row solving an assignment problem through the objective function in the step 8 of the algorithm.

Therefore, the algorithm produces a matrix **U** of the segments assignment and a matrix **C** of centroids with a unique common measurement and structural model coefficients. However, researchers that wish determining segment specific measurement and structural model coefficients can apply group-specific PLS-SEM analysis. The unique measurement and structural model coefficients are interpreted as a consensus of the segment-specific coefficients.

The proposed methodology shows some important advantages with respect to the other proposed approaches for both cluster analysis and composite indicator construction: firstly, it is a simultaneous approach that identifies the best homogenous partition of the objects represented by the best causal relationships among latent and observed variables. Then, unlike a sequential approach, the identified partition is dependent on the prespecified composite-based (i.e., causal) relationships; moreover, distributional assumptions are not requested for the PLS-SEM-KM application (Fordellone and Vichi 2018), this because it uses a partial least squares (PLS) methodology that, unlike the covariance structure approach (CSA), is insensitive to the data distributional assumptions.

4 From Data to Results for IICI

The data used for the construction of the immigrant integration composite indicator (IICI) construction derive from the eighth iteration of the survey for ESS. Until now are available 18 of the 24 countries, which undertook fieldwork in 2016. Table 1 shows the principal topics included in ESS data.

Table 1 Topics and items of ESS survey

Items	Topic
Core A1–A6	Media use; internet use; social trust
Core B1–B43	Politics, including political interest, trust, electoral and other forms of participation, party allegiance, sociopolitical orientations immigration
Core C1–C44	Subjective well-being, social exclusion, crime, religion, perceived discrimination, national and ethnic identity, test questions (Sect. I), refugees
Core D1–D32	Climate change and energy, including attitudes, perceptions module and policy preferences
Core E1–E42	Welfare, including attitudes toward welfare provision, size of module claimant groups, attitudes toward service delivery and likely future dependence on welfare, vote intention in EU referendum
Core F1–F61	Sociodemographic profile, including household composition, sex, age, marital status, type of area, education and occupation, partner, parents, union membership, income and ancestry
Core Section H	Human values scale
Core Section 1	Test questions

After data aggregation, our data set is composed of 274 regions of the 18 countries and 64 Likert scale variables, defining the following 9 dimensions, i.e., *politics* with 19 MVs, *economics* with 2 MVs, *social* with 2 MVs, *cultural* with 2 MVs, *crime* with 2 MVs, *religion* with 2 MVs, *structural* with 11 MVs, *household* with 9 MVs, and *employment* with 15 MVs.

The application of the PLS-SEM-KM model has detected a number of clusters $K = 5$ obtaining the estimates of the path coefficients shown in Table 2.

The estimates reported in Table 2 show an overall good performance of the model both in terms of path coefficients (i.e., all the estimated coefficients are statistically significant) and in terms of explained deviance (i.e., high R^2 values). Observing the single coefficients, we can see that more remarkable significant effect on IICI is given by the *politics* (0.875) and *cultural* (−0.383) constructs. In contrast, a very low impact on IICI is given by the *structural* dimension (−0.046), which includes important demographic features of the respondents, followed by *household* (−0.154) and *religion* (−0.185) constructs together with *economics* (−0.215), *social* (0.211), *crime perception* (0.204), and *employment* (0.216) dimensions.

Figure 3 shows the loading estimates obtained for each latent dimension.

Table 2 Path coefficients estimated by PLS-SEM-KM

	Estimate	Std. error	t-value	Pr(> $\|t\|$)
(Intercept)	0.149	0.029	5.148	0.000
Politics	0.875	0.016	56.445	0.000
Economics	−0.215	0.029	−7.420	0.000
Social	0.211	0.022	9.385	0.000
Cultural	−0.383	0.022	−17.524	0.000
Crime	0.204	0.030	6.827	0.000
Religion	−0.185	0.019	−9.687	0.000
Structural	−0.046	0.012	−3.736	0.000
Household	−0.154	0.016	−9.743	0.000
Employment	0.216	0.013	16.211	0.000

F-statistic: 3756 on 9 and 264 DF (p-value = 0.000) $R^2 = 0.8823$, $R^2_{adj} = 0.882$

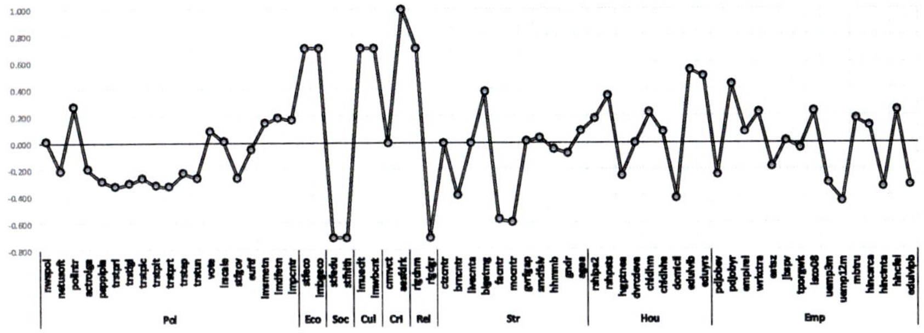

Fig. 3 Loadings estimates for each latent dimension

Figure 4 shows the cluster distributions on the 10 estimated latent scores (i.e., including also the composite indicator), while in Fig. 5 a geographical representation of the obtained clusters is shown. Note that the size of the 5 clusters comprising the identified partitions are 52, 46, 74, 39, and 63, respectively. In the representation of the loadings, we have used the official labels of the 64 MVs which we have selected for the definition of latent dimensions.[1]

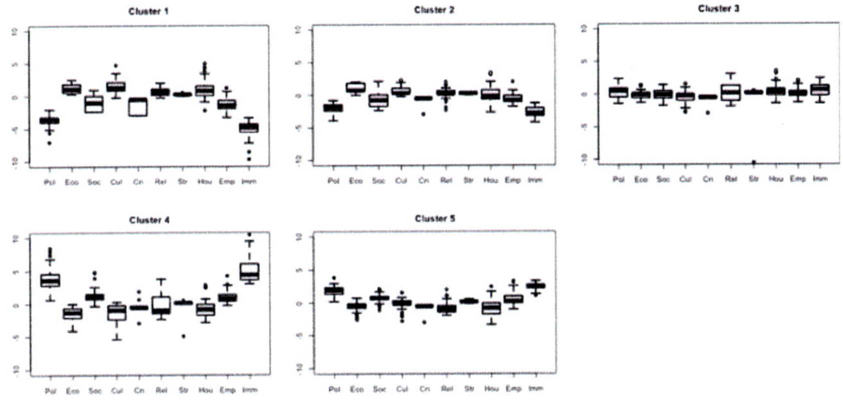

Fig. 4 Clusters distribution on the all latent constructs

[1]For more details on the selected MVs, you can see the official ESS website: http://www.europeansocialsurvey.org/about/news/essnews0038.html..

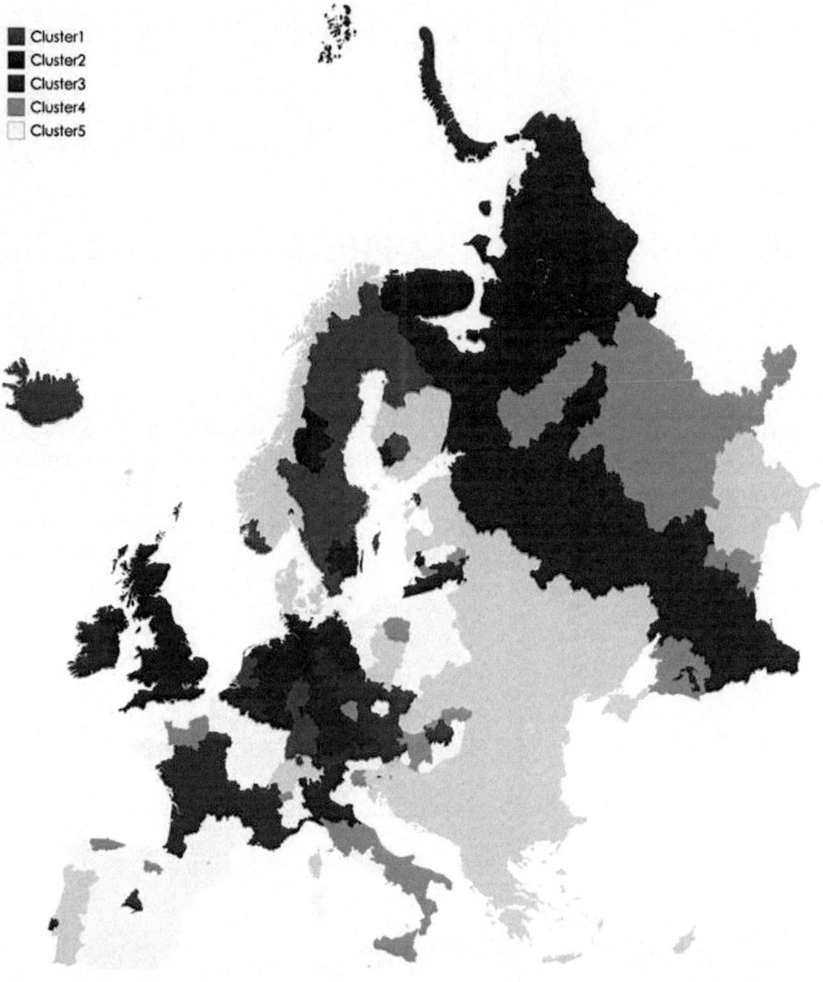

Fig. 5 Geographical representation of the clusters

From the cluster distributions in Fig. 4, we can note that the first and the fourth cluster are very discriminant of the immigrant integration level, because the IICI values are very low and very high, respectively. Moreover, we can also note that the political dimension has a very hard impact on the immigrant integration. So, the lower the IICI, the lower the political factor level is in the cluster 1. On the contrary, in the cluster 4 a high level of political dimension is related to a high level of the composite indicator.

The results obtained by employing the PLS-SEM-KM method show a reliable classification structure of 274 regions of 18 European countries where the level of immigrant integration is different for 4 clusters of regions.

The more discriminant ability of the 9 exogenous latent variables and also of the composite indicator (IICI) allows efficiently mapping overall the more northern regions in a cluster where the lowest values of the indicator represent a low level of immigrant integration while in the cluster 4, most of the southern and eastern regions are more discriminated on the basis of high values of the composite indicator for a higher level of immigrant integration. The effect of the political participation dimension is affecting the most both the classification of regions and the composite indicator building, assuming the same trend, thus, in each cluster: the higher/lower the level of political participation, the higher/lower the level of immigrant integration is in the European regions.

5 Conclusive Remarks

This work, employing the PLS-SEM methodology where SEM is estimated by PLS, is focused on the building of an integration composite indicator (IICI), in Europe. With this aim, we use a simultaneous PLS-SEM-KM approach introduced by Fordellone and Vichi (2018) (PLS-SEM-KM).

The results show a good performance of the global model, especially for the immigrant integration profile. Moreover, the conjoined clustering model defines partitions that add relevant information on the countries' features involved in the immigrant integration issue.

In our opinion, by employing composite indicators to measure a complex phenomenon like immigrant integration, an international comparative approach can help to focus and target nationally and locally immigrant integration policies.

References

Ager, A., & Eyber, C. (2002). Indicators of integration: A review of indicators of refugee integration. Report to the Home Office on Behalf of Michael Bell Associates.

Entzinger, H. (2000). The dynamics of integration policies: A multidimensional model. In R. Koopmans & P. Statham (Eds.), *Challenging immigration and ethnic relations politics* (pp. 97–118). Oxford: University Press.

Entzinger, H., & Biezeveld, R. (2003). *Benchmarking in immigrant integration*. Rotterdam: European Research Center on Migration and Ethnic Relations.

Cesareo, V., & Blangiardo, G. (2009). *Indici di integrazione*. Milano, IT: FrancoAngeli.

De Soete, G., & Carroll, J. D. (1994). K-means clustering in a low-dimensional Euclidean space. In E. Diday, Y. Lechevallier, M. Schader, P. Bertrand, & B. Burtschy (Eds.), *New approaches in classification and data analysis* (pp. 212–219). Berlin, Heidelberg: Springer.

Di Bartolomeo, A., Kalantaryan, S., & Bonfanti, S. (2015). Measuring integration of migrants: A multivariate approach. INTERACT RR 2015/01, Robert Schuman Centre for Advanced Studies, San Domenico di Fiesole, Firenze, IT: European University Institute.

Diamantopoulos, A., & Winklhofer, H. M. (2001). Index construction with formative indicators: An alternative to scale development. *Journal of Marketing Research, 38*(2), 269–277.

Fordellone, M., & Vichi, M. (2018). Structural equation modeling and simultaneous clustering through the partial least squares algorithm. arXiv:1810.07677.

Lohmoller, J. B. (1989). *Latent variable path analysis with partial least squares*. Heidelberg: Physica.

OECD. (2008). *Handbook on constructing composite indicators*. Methodology and user guide. Paris, FR: OECD Publishing. http://www.oecd.org/std/42495745.pdf.

OECD. (2018). *The integration of migrants in OECD regions: A first assessment*. Regional Development Working Papers 2018/01. Paris, FR: OECD Publishing. https://dx.doi.org/10.1787/fb089d9a-en.

Strang, A., Ager, A., & Brien, O. O. (2003). Indicators of integration: The experience of integration. Report to the Home Office on behalf of Michael Bell Associates.

Tenenhaus, M., Vinzi, E. V., Chatelin, Y. M., & Lauro, N. C. (2005). PLS path modeling. *Computational Statistics and Data Analysis*, *48*(1), 159–205.

Tibshirani, R., Walther, G., & Hastie, T. (2001). Estimating the number of clusters in a data set via the gap statistic. *Journal of the Royal Statistical Society: Series B*, *63*(2), 411–423.

Vermeulen, H. (2004). Models and modes of immigrant integration ...And where does Southern Europe Fit? In C. Inglessi, A. Lyberaki, H. Vermeulen & G. J. V. Wijngaarden (Eds.), *Immigration and integration in northern versus southern Europe* (pp. 11–26). Athens, EL: Netherlands Institute at Athens.

Vinzi, E. V., Trinchera, L., & Amato, S. (2010). PLS path modeling: From foundations to recent developments and open issues for model assessment and improvement. In V. Esposito Vinzi, W. Chin, J. Henseler & H. Wang (Eds.), *Handbook of partial least squares* (pp. 47–82). Berlin, Heidelberg: Springer.

Facebook Debate on Sea Watch 3 Case: Detecting Offensive Language Through Automatic Topic Mining Techniques

Alice Tontodimamma, Emiliano del Gobbo, Vanessa Russo, Annalina Sarra, and Lara Fontanella

Abstract Over the years, there has been growing concern about the disproportionate use of hate speech on social media platforms. In this paper, we present a text analysis for detecting abusive language in Italian messages on Facebook, surrounding the debate over the migrant-rescue ship, Sea Watch 3, and its captain Carola Rackete. The study data consists of more than 130,000 posts retrieved from two pages relating to Matteo Salvini, the leader of the Italian *Lega* political party, and from the official Facebook pages of five Italian newspapers. To explore the presence of offensive and hatred expressions in the corpus and to establish to what extent social users' language differs, depending on the type of Facebook pages analysed, we ran a topic model based on Latent Dirichlet Allocation. We have complemented this approach with tools from semantic network analysis.

1 Introduction

The Internet has drastically changed the ways in which people communicate, access news and establish relationships. It is important to note that, while, on the one hand, the Internet can be seen as a utopian platform for free speech and equality, on the

A. Tontodimamma · E. del Gobbo
Department of Neuroscience & Imaging, University G.d'Annunzio of Chieti-Pescara, Pescara, Italy
e-mail: alice.tontodimamma@unich.it

E. del Gobbo
e-mail: emiliano.delgobbo@unich.it

V. Russo · A. Sarra (✉) · L. Fontanella
Department of Legal and Social Sciences, University G.d'Annunzio of Chieti-Pescara, Pescara, Italy
e-mail: asarra@unich.it

V. Russo
e-mail: russov1983@gmail.com

L. Fontanella
e-mail: lara.fontanella@unich.it

P. Mariani and M. Zenga (eds.), *Data Science and Social Research II*, Studies in Classification, Data Analysis, and Knowledge Organization,
https://doi.org/10.1007/978-3-030-51222-4_29

other hand some of its distinctive features, such as ease of access, audience size and anonymity, also cause online spaces to become populated with content that others are likely to find degrading, abusive or otherwise harmful. In particular, hate speech has become prominent on social media platforms. While there is no universally accepted definition of hate speech, it is possible to affirm that this term covers all forms of hostile, bias-motivated and malicious expression directed against particular ethnic, religious, racial or sexually oriented groups or persons in society (Almagor 2011). The degree of intensity of hate speech can vary greatly and, in its extreme form, can be an incitement to violence. The perceived anonymity and lack of tangible consequences inherent to digital communication create the so-called *online disinhibition effect* (Suler 2004) and make the online setting a space where even unradicalised users are more likely to verbalise ideas which in other situations they would not express. Online hate speech is a highly topical research problem (see, Nobata et al. 2016; Schmidt and Wiegand 2017; Fortuna and Nunes 2018 and references therein) and an issue of public and political concern. International and European Union institutions are paying increasing attention to the spread of offensive comments in online forums and social media platforms. There are a number of factors which, in all European countries, cause and encourage the use of hate speech. These factors include: the resentment felt, by the indigenous population of those countries, of the austerity policies imposed on them by their governments; the provocative language used by political parties and by political movements when they take part in public debate; and the prejudiced manner used by some media when reporting matters that are relating to ethnic diversity or to minority groups. These ingredients, operating in conjunction with the large numbers of migrants and refugees arriving from different countries and with the difficulty of integrating them into the society, economy and culture of their host countries, act as catalysts of racist hate speech.

The plague of hateful and harassing online comments also affects Italy, which, according to Amnesty International, has become steeped in hostility, racism, xenophobia and unjustified fear of others. The latest Hate Barometer (Amnesty Internazional Italia 2019) highlights how one out of ten online contents released during the campaign for the 2019 European elections is offensive or discriminatory. Immigration, religious minorities and gender issues are the topics that most give rise to hate speech. As for the immigration issue, it predominates in online attacks on solidarity and those actors involved in immigration, ranging from those non-governmental organizations (NGOs) that still operate in the Mediterranean sea, to associations active in the territory of Italy and to the reception system in general.

Amnesty's reports (Amnesty International Italia 2018, 2019), analysing the relationships between language used by politicians and online hate speech, emphasise how some political forces have used stereotypes and hatred to make populist, identity and xenophobic sentiments their own, promoting the spread of incendiary, divisive, discriminatory language. On the same theme, some studies stress the leading role played by political ideology in determining rude and hateful comments on social media. Kurecic and Kuhar (2019) compare the rhetoric on illegal migration in the public speech of some European centre-right and right-wing populist parties. From their research it emerges that unlike the leaders of centrist and centre-right parties in

EU member states, whose rhetoric varies according to their countries of origin, right-wing populists appear to speak in a uniform fashion. Close examination of Matteo Salvini, leader of Italian *Lega* party, reveals that his rhetoric is harsh and explicit: well aware that the subject of migration arouses more public engagement than other topics, Salvini uses controversial policies and statements to keep the issue high on the political agenda. His stance regarding migrants is radical: equal treatment for all illegal migrants—with no recognition of possible "refugee status"—and repeated attacks on smugglers and NGOs are recurrent elements in his propaganda.

In this work, we focus on the analysis of Facebook debates centred on the case of Carola Rackete, captain of the Sea-Watch 3 migrant-rescue ship, who underwent investigation for breaking an Italian naval blockade that was trying to prevent her from docking the vessel on the island of Lampedusa, in Italian territory. To investigate the content and structure of online conversations on the Sea Watch 3 case, we applied methods from the fields of automatic topic mining and network analysis. The textual data, which captured the engagement metrics of thousand of posts retrieved from the official Facebook pages of five Italian newspapers and two pages linked to Matteo Salvini, have been explored to detect offensive and abusive comments and to determine to what extent the adopted language differs according to the Facebook analysed sources. The rest of this paper is organised as follows: in Sect. 2 we briefly outline the case study; in Sect. 3 data is presented, along with the results of the correspondence analysis performed on the corpus; the contribution of the paper in identifying actual topics of conversation on the Sea Watch 3 case and their structure is described in Sect. 4 whereas Sect. 5 gives a conclusion.

2 The Sea Watch 3 Case

In recent years, the particular political, economic and social conditions of countries in the Middle East and Africa have led to the development of substantial migratory flows towards Europe. The Central Mediterranean migratory route to Europe makes use of the proximity of the Libyan and Tunisian coasts to the Italian island of Lampedusa. In the persistent absence of both a legal safe passage and a large-scale search and rescue (SAR) operation, NGOs have filled the gap left by the restrictive policies of the European Union and its Member States, and between 2016 and 2017 they became the largest providers of SAR services off the coast of Libya, rescuing thousands of human lives at sea (Italian Coast Guard 2017). In spite of this extraordinary work, non-governmental rescue attracted a good deal of criticism at governmental level. After the 2018 Italian elections, Salvini, as Minister of the Interior, implemented a "closed-ports" policy, enabling his ministry to deny landing rights to ships carrying migrants. This policy has led to several cases in which migrant-carrying NGOs were forced to stay for days off the Italian coast before their ship was allowed to dock. In June 2019, the Sea Watch 3, belonging to a German NGO, rescued a group of migrants drifting in an inflatable raft off the coast of Libya. The Captain, Carola Rackete, declined to make them disembark in Tripoli, and instead set sail for Lampedusa, but was there denied access by the Ministry of the Interior under the provisions of the security decree. After a two-week standoff with the Italian authorities, Carola Rackete docked in the Italian port, and shortly afterwards was arrested, charged with resistance and violence against warships, and placed under investigation for aiding illegal immigration. The long duration of the case has given rise to enormous media exposure and considerable public debate, conducted for the most part on social networks.

3 Facebook Debate on Carola Rackete

Facebook, the most popular social network in Italy, according to the Global Digital Report 2019—We Are Social, was chosen as the source for data collection. The corpus was built by downloading the comments on Carola Rackete's posts. In particular, we decided to analyse the comments posted on the official page of Matteo Salvini and the closely linked page "Lega Salvini Premier". According to the Hate Barometer (Amnesty International Italia 2018, 2019), Salvini's page appears to be characterised by a high content of offensive and/or discriminatory comments, especially against immigrants, Roma people and women. We also gathered comments on posts retrieved from the official Facebook pages of five Italian newspapers. The comments were scraped using exportcomments.com and the analysed corpus consists of 131,403 posts (Salvini's pages: 34,305, Il Corriere della Sera: 22,306, Il Giornale: 10,843, La Repubblica: 28,998, Il Fatto quotidiano: 34,314, Libero: 637).

We developed a Python script to perform the screening and cleaning process of text documents in order to extract the relevant content and remove any unwanted stop words. Words with less than 75 occurrences have been pruned. The final vocabulary consists of 1467 terms and the ten most frequent ones are: *Salvini*, *Italia*, *casa*, *Carola*, *grande*, *Italiani*, *Paese*, *leggi*, *vai*, *Germania*. To explore the similarity in the language used in the analysed Facebook pages, we have firstly performed correspondence analysis on the corpus containing all the comments organised by sources. The analysis was carried out through the R package RcmdrPlugin.temis (Bouchet-Valat and Bastin 2013). The affinity can be visually detected from Fig. 1, where the Facebook pages are represented with respect to the first three dimensions which, overall, retain 75% of the total inertia. The projection on the first two dimensions clearly indicates the clustering of the sources into three groups: Salvini's pages; Libero and Il giornale; La Repubblica, Il Corriere della Sera and Il Fatto Quotidiano. The latter newspaper page shows, however, a detached position along the third dimension. This clustering is highly consistent with the political leanings of these newspapers.

Table 1 lists, for each of the identified clusters, the ten most meaningful specific terms, that is to say, the words whose observed frequency in the cluster is higher than expected given the length of the documents and the global distribution of the terms in the corpus.

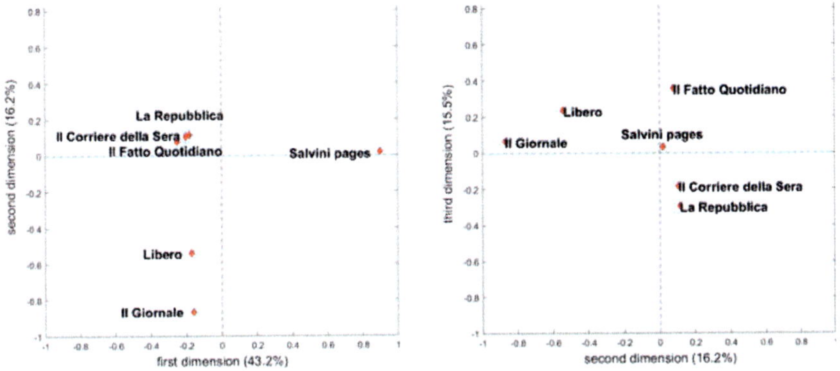

Fig. 1 Representation of the Facebook sources on the planes obtained through correspondence analysis

Table 1 Specific terms for the Facebook pages' groups

Il Corriere della Sera–La Repubblica	Il Giornale-Libero	Il Fatto Quotidiano	Salvini's pages
Nobel	Sicurezza	Stupro	Aiutali
Odiare	Ridicoli	Parigi	Educatamente
Linguaggio	Patria	Leoni	Bravissimo
Bravissima	Giustizia	Francesi	Affanculo
Zecca	Giudici	Medaglia	Chiudere
Vite	PD	Tastiera	Bravo
Umane	Merde	Premiano	Bianca
Santa	Buffoni	Querela	Cogliona
Umanità	Espulsione	Premiata	Grande
Schifo	Magistratura	Ipocrisia	Capitano

4 The Main Debated Topics

Automatic topic mining techniques are increasingly used by social scientists to detect hidden topics in online discussions. Topic modelling refers to a collection of methods and algorithms which aim to uncover the hidden thematic subjects in document collections by revealing recurring clusters of co-occurring words. While there are several different algorithms for performing topic modelling, the most common is Latent Dirichlet Allocation (LDA; Blei 2012), which assumes a probabilistic generative model where each document is represented as a random mixture over latent topics and each topic is characterised by a distribution over words. This generative process defines a joint probability distribution over both the observed and hidden random variables. The topic distribution in all documents shares a common Dirichlet prior, and the word distributions of topics shares a common Dirichlet prior as well.

Data analysis is performed by using the joint distribution and the priors to compute the posterior distribution of the hidden variables given the observed ones. Since the posterior distribution is intractable for exact inference, a wide variety of approximate inference algorithms, such as sampling-based algorithms (see, for example Steyvers and Griffiths 2006 for a detailed derivation of Gibbs-sampler for LDA) and variational algorithms (Blei et al. 2003) can be considered. In our research, to perform LDA we set the number K of desired topics, in turn, equal to 10, 12, 14 and we adopted the model estimated with $K = 12$, which guarantees the right trade-off between having enough words to disclose relevant information without making the topics cluttered. The analysis was performed using the fitlda MATLAB routine available in the Text Analytics Toolbox (MATLAB 2018).

In LDA, the topics are assumed to be latent variables, which need to be intuitively interpreted. This is usually achieved by examining the top keywords in each topic (Steyvers and Griffiths 2006). All the estimated topics, along with the most relevant among the top twenty terms, are listed in Table 2. Topics are sorted according to the estimated probability to be observed in the entire data set, and words are ordered according to their relevance obtained by normalising the posterior word probabilities per topic by the geometric mean of the posterior probabilities for the word across all topics.

Relevance measures suggest that these terms are highly coherent, clearly distinguish different topics and consequently facilitate their interpretation. The topics which seem to dominate the Facebook debate on Carola Rackete's case are strictly linked to Matteo Salvini. In particular, we can distinguish between a neutral general theme on his role as Minister of the Italian Republic (Topic 1), a sympathetic topic that groups enthusiastic and encouraging comments (Topic 2), and a theme related to unfavourable comments which includes a number of offensive and denigrative terms (Topic 3). In addition, Topic 6 focuses on media reports that the ruling Lega party sought to make millions of euros from a secret Russian oil deal. On the other hand, the two topics (4 and 11) that refer clearly to Carola Rackete contain both demeaning and insulting terms. The immigration issue characterises Topics 5, in a general perspective, and 12, where it is concentrated on the Sea Watch 3 case. The dynamic of the Sea Watch 3 incident is the main subject of Topic 10. The duty to respect Italian laws is clearly highlighted in Topic 8. Topic 7 contains comments on the decision of the City of Paris to award the Grand Vermeil Medal to Carola Rackete for saving migrants at sea. Finally, Topic 9 refers to the appalling story of Bibbiano, where the city mayor and a team of psychologists and social workers were arrested for their participation in a scheme to brainwash children into believing they had been abused and sell them. To get a broader characterisation of the defamatory topics (3, 4 and 11), we represent the most relevant words in Fig. 2.

LDA links each document to the detected topics with different probabilities and the documents that are most strongly linked to a topic are the ones that it best describes. We can, thus, assign each comment to a topic by considering its highest topic probability. Table 3 shows the topic distribution for all the comments in the corpus, along with the distributions for the different Facebook pages. The topics whose language seems to be most abusive (3, 4 and 11) show similar percentages of

Table 2 Top ten terms within the 12 topics sorted according to their relevance scores

Topic 1		Topic 2		Topic 3		Topic 4		Topic 5		Topic 6	
Word	*Score*	*Word*	*Score*	*Word*	*Score*	*Word*	*Score*	*Word*	*Score*	*Word*	*Score*
Italiano	0.302	Grande	1.947	Ministro	0.251	Galera	0.736	Persone	0.273	Salvini	0.611
Ministro	0.269	Matteo	1.336	Salvini	0.174	Carola	0.471	Umani	0.250	Lega	0.355
Popolo	0.263	Vai	0.999	Insulti	0.153	Casa	0.257	Esseri	0.211	Voti	0.160
Governo	0.215	Capitano	0.862	Basta	0.151	Santa	0.246	Problema	0.154	Milioni	0.156
Magistratura	0.186	Avanti	0.844	Merda	0.150	Torna	0.231	Ong	0.108	Russia	0.125
Salvini	0.169	Salvini	0.714	Cervello	0.131	Figlia	0.186	Poveri	0.083	Governo	0.110
Politica	0.164	Forza	0.517	Rubli	0.125	Delinquente	0.181	Traffico	0.082	Ministro	0.103
Repubblica	0.105	Bravo	0.395	Odio	0.122	Tedesca	0.170	Schiavi	0.077	Bacioni	0.083
Magistrati	0.092	Mollare	0.319	Schifo	0.114	Palle	0.122	Tratta	0.069	Putin	0.080
Giustizia	0.087	Ministro	0.286	Linguaggio	0.092	Rotto	0.112	Umanità	0.065	Storia	0.070
Topic 7		Topic 8		Topic 9		Topic 10		Topic 11		Topic 12	
Word	*Score*	*Word*	*Score*	*Word*	*Score*	*Word*	*Score*	*Word*	*Score*	*Word*	*Score*
Francia	0.464	Leggi	1.201	Milioni	0.525	Nave	1.232	Casa	2.516	Porto	0.510
Italia	0.362	Rispettare	0.275	Soldi	0.217	Comandante	0.272	Germania	0.890	Vite	0.494
Migranti	0.268	Italiane	0.228	Bibbiano	0.217	Finanza	0.229	Portali	0.807	Mare	0.393
Francesi	0.260	Regole	0.168	Bambini	0.203	Guardia	0.196	Ricca	0.732	Salvare	0.363
Africa	0.181	Galera	0.118	Processo	0.197	Blocco	0.196	Vai	0.582	Persone	0.278
Medaglia	0.164	Violato	0.113	Immunità	0.167	Bordo	0.180	Tedesca	0.428	Lampedusa	0.230
Clandestini	0.134	Giustizia	0.105	Giudice	0.159	Motovedetta	0.166	Stronza	0.303	Migranti	0.228
Europa	0.121	Diritto	0.102	Galera	0.127	Acque	0.157	Rompere	0.298	Umane	0.218
Immigrati	0.112	Civile	0.096	Sentenza	0.120	Ordine	0.148	Bianca	0.241	Tunisia	0.196
Parigi	0.086	Infranto	0.090	Sindaco	0.115	Seawatch	0.120	Coglioni	0.150	Seawatch	0.171

Fig. 2 Comparison of the most relevant words within the detected defamatory topics. The size of a word is proportional to its relevance, while colours represent topic membership

Table 3 Topic distributions for the entire corpus and for the analysed Facebook pages

	All comments	Salvini's pages	Newspapers					
			All	*Il Corriere della Sera*	*La Repubblica*	*Il Fatto quotidiano*	*Il Giornale*	*Libero*
Topic 1	13.2	9.6	14.4	13.4	13.5	13.9	20.3	24.5
Topic 2	14.0	30.6	8.1	7.7	11.4	6.3	5.3	6.1
Topic 3	10.1	7.0	11.1	10.8	12.3	10.4	11.0	11.6
Topic 4	11.1	9.5	11.6	13.0	11.4	10.3	13.8	9.4
Topic 5	8.2	4.9	9.4	9.3	9.7	9.6	8.1	10.4
Topic 6	7.4	6.9	7.6	7.6	7.4	8.3	6.1	8.4
Topic 7	8.0	4.7	9.1	7.7	7.9	11.6	7.7	6.5
Topic 8	6.4	3.5	7.5	7.3	6.7	7.9	8.6	7.6
Topic 9	6.1	2.8	7.2	9.1	5.4	7.6	7.1	7.8
Topic 10	5.6	2.4	6.8	6.6	6.4	7.1	6.9	3.7
Topic 11	7.0	17.1	3.4	3.5	3.8	3.1	3.5	1.8
Topic 12	2.9	0.8	3.7	3.9	4.1	3.8	1.7	2.2

comments across the pages of the different newspapers. On the other hand, offensive and denigrative terms addressed to Salvini show a lower representation on the pages linked to the Lega leader. It is worth noticing how in these pages there is a lower presence of comments containing terms specific to Topic 3 and a definitely higher

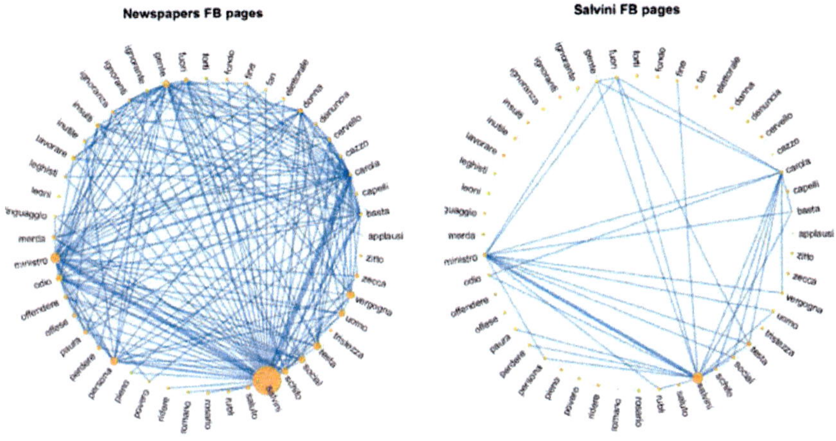

Fig. 3 Networks of the 50 most relevant words for Topic 3

association to Topic 11, whose percentage is equal to 17.1% compared to values between 1.8 and 3.8% for the comments retrieved from, the online pages of the newspapers.

To deepen the analysis of the use of abusive language in Salvini's Facebook pages and in those of the newspapers, we represent the co-occurrence matrix for the relevant words of each topic by means of semantic networks in Figs. 3, 4 and 5. Drawing a semantic network allows to detect what are the most frequent terms and how they relate to each other. For ease of interpretation, we constructed the networks by displaying the links corresponding to at least ten co-occurrences. It is worth noting that the sizing of nodes is proportional to word occurrences, while the edges are weighted according to the number of co-occurrences between terms. For each semantic network, we focus on *density* measure, defined as the ratio of present connections to the number of possible links. We find that the network density related to Topic 11 in Salvini's Facebook pages is higher than that observed in those of the newspapers (0.258 vs. 0.156). Conversely, we observe that the networks for Topics 3 and 4 on newspaper Facebook pages are considerably denser than those retrieved for Salvini's pages (Topic 3: 0.161 vs 0.031; Topic 4: 0.194 vs 0.082). To emphasise the extent to which the importance of a node depends on its neighbours, we also consider the *eigenvector centrality* (EC). This last metric makes possible to find keywords shaping the dominant discourse patterns within the topic. Paying particular attention to the densest network (Fig. 5), we find out the the words with the highest EC measure in Salvini's Facebook pages are "casa" (EC = 0.111),"portali" (EC = 0.009), "Germania" (EC = 0.082), "ricca" (EC = 0.075). In contrast, the narrative appearing in newspaper Facebook pages reveals that the corresponding semantic network contains as central concepts words like "Italia" (EC = 0.132), "paese" (EC = 0.098), "Germania" (EC = 0.095), "Olanda" (EC = 0.061).

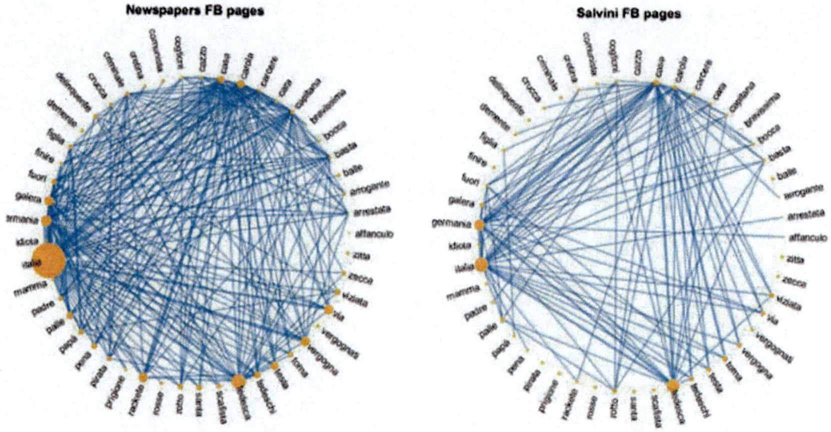

Fig. 4 Networks of the 50 most relevant words for Topic 4

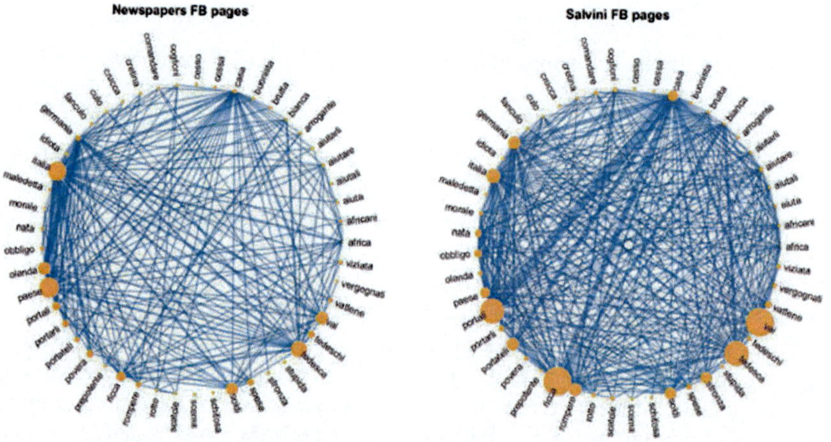

Fig. 5 Networks of the 50 most relevant words for Topic 11

5 Conclusions

The results of our research make it clear that within Facebook there are a number of specific languages that embrace peculiar groups of opinion. Specifically, the results of correspondence analysis, and the topics retrieved using the LDA algorithm describe characteristic niches of opinion. Accordingly, when studying the online public sphere, it is necessary to reason not in terms of the receiving public but in terms of receiving audiences (Suler 2004). In order to disentangle the complex phenomenon of hate on online spaces, certain considerations have to be observed. The first of these concerns the boundaries of online hate speech: there is a fuzzy area, including hatred contents and neutral expressions, which is difficult to define. In our case study, within the three topics identified as "hatred" (3, 4, 11), we can detect several expressions that while not be exactly offensive, cannot be considered totally neutral ("basta","galera", "delinquente" and "ricca"). The second issue concerns the sharing of specific hate languages: the incitement to hatred is not a phenomenon that extends only in intensity, it also derives from specific shared cultural backgrounds. From the LDA results, it emerges that out of the three Topics of incitement to hatred, Topic 11 strongly characterises the followers of Salvini's page. Topic 2 is also particularly shared by Salvini's followers. In this topic, we recovered contents of encouragement towards the Lega Leader ("grande", "vai", "capitano" and"mollare"). It is interesting to note that both these topics are composed of words of encouragement positioned on two positive/negative poles that together make up a single choral language. In conclusion, all the elements mentioned above, that is the sharing of certain ideological positions, the agreement with and support for the leader, and a choral and unifying language, identify the existence of a "rethoric" peculiar to the public of Matteo Salvini, part of a set of cultural insights, that can translate into forms of incitement to hatred (Wodak and Krzyżanowski 2017).

References

Almagor, R. C. (2011). Fighting hate and bigotry on the internet. *Policy and Internet*, *3*(3), 1–28. https://doi.org/10.2202/1944-2866.1059.

Amnesty International Italia. (2018). Conta fino a 10. Barometro dell'odio in campagna elettorale.

Amnesty International Italia. (2019). *Il barometro dell'odio—Elezioni Europee*.

Blei, D. M. (2012). Probabilistic topic models. *Communications of the ACM*, *55*(4), 77–84. https://doi.org/10.1145/2133806.2133826.

Blei, D. M., Ng, A. Y., & Jordan, M. I. (2003). Latent Dirichlet allocation. *Journal of Machine Learning Research*, *3*(1), 993–1022. https://doi.org/10.1162/jmlr.2003.3.4-5.993.

Bouchet-Valat, M., & Bastin, G. (2013). Rcmdrplugin.temis, a graphical integrated text mining solution in r. *The R Journal*, *5*(1), 188–196. https://doi.org/10.32614/RJ-2013-018.

Fortuna, P., & Nunes, S. (2018). A survey on automatic detection of hate speech in text. *ACM Computing Surveys*, *51*(85), 1–85. https://doi.org/10.1145/3232676.

Italian Coast Guard. (2017). SAR operations in the Mediterranean Sea.

Kurecic, P., & Kuhar, P. (2019). The rhetoric on illegal migration of centre-right parties and right-wing populist parties in "Old" and "New" EU Member States: A content analysis of leaders' speeches. In *Data Value Chains in Science & Territories 2019 International Conference Proceedings* (pp. 79–86).

MATLAB. (2018). *version 9.5.0.944444 (R2018b)*. Natick, Massachusetts: The MathWorks Inc.

Nobata, C., Tetreault, J., Thomas, A., Mehdad, Y., & Chang, Y. (2016). Abusive language detection in online user content. In: *Proceedings of the 25th International Conference on World Wide Web* (pp. 145–153). https://doi.org/10.1145/2872427.2883062.

Schmidt, A., & Wiegand, M. (2017). A survey on hate speech detection using natural language processing. In: *Proceedings of the fifth international workshop on natural language processing for social media* (pp. 1–10). Valencia, Spain: Association for Computational Linguistics. https://doi.org/10.18653/v1/W17-1101.

Steyvers, M., & Griffiths, T. (2006). Probabilistic topic models. In: Landauer, T., Dennis, M. S., & Kintsch, W. (Eds.) *Latent Semantic analysis: A road to meaning*. Lawrence Erlbaum.

Suler, J. (2004). The online disinhibition effect. *Cyberpsychology and Behavior*, *7*, 321–326. https://doi.org/10.1089/1094931041291295.

Walther, J. B. (2011). Theories of computer-mediated communication and interpersonal relations. In M. L. Knapp & J. A. Daly (Eds.), *The handbook of interpersonal communication* (pp. 443–479). Thousand Oaks, CA: Sage.

Wodak, R., & Krzyżanowski, M. (2017). Right-wing populism in Europe & USA: Contesting politics & discourse beyond 'Orbanism' and 'Trumpism'. *Journal of Language and Politics*, *16*(4), 471–484. https://doi.org/10.1075/jlp.17042.krz.

Martini's Index and Total Factor Productivity Calculation

Biancamaria Zavanella and Daniele Pirotta

Abstract The axiomatic property of consistency in aggregation is fundamental for example in the analysis of economic systems divided into sectors and sub-sectors, to ensure consistency of results at different levels of sectoral aggregation. The expenditure ratios index numbers satisfy this property by construction; however, they often do not satisfy other properties such as the factor reversibility. Martini's index satisfies both properties and in this work it is applied to the problem of calculating the Total Factor Productivity (TFP), which typically refers to a context in which consistency in aggregation is important. In the proposed applications the results obtained are very encouraging and stimulate interest in further study.

1 Introduction

Given the vastness of the available literature concerning the problem of calculating the Total Factor Productivity (TFP), it is important to clarify that this work does not aim to provide a new contribution to the topic or to identify an additional TFP measurement in competition with the one given by Istat and the OECD.

The goal to be achieved is the presentation of a completely original application of the index proposed by Marco Martini in his books (1992, 2001) to calculate TFP. The Martini Index had never been applied before to real phenomena since an iterative and complex procedure aimed at searching for its distinguishing parameter M needs to be defined to calculate it. The calculation tools available at the time of the index formalization would have required a significant investment of time and would have provided approximate results. Nowadays powerful and easy-to-use calculation tools, which make it possible to apply the Martini formula in real contests, are available and widely used.

The Martini Index is part of the axiomatic theory of index numbers, in the version among the many available in the literature proposed by Martini himself in the above-

B. Zavanella (✉) · D. Pirotta
Università degli Studi di Milano-Bicocca, Milan, Italy
e-mail: biancamaria.zavanella@unimib.it

P. Mariani and M. Zenga (eds.), *Data Science and Social Research II*, Studies in Classification, Data Analysis, and Knowledge Organization,
https://doi.org/10.1007/978-3-030-51222-4_30

mentioned books. By construction, the formula in question satisfies all the axiomatic properties proposed by the author. In particular, this work focuses on consistency in aggregation. Such a property, in fact, is extremely useful in various areas of economic analysis, in which the data of the entire system can be aggregated in the different business sectors, or, in any case, they can be aggregated according to specific variables of interest. A typical example of data having these characteristics is the one related to TFP which is mostly calculated by aggregating the data from both sectors of economic activity and capital and labour inputs.

The data used is the one of the EU_KLEMS (Jager 2017) project. It was not possible to use the corresponding Istat data as the capital and labour remuneration shares are published exclusively as an arithmetic average of the data relating to two consecutive years, that is, in a form suitable for the Törnqvist index, currently used by institutional bodies for the calculation of TFP, but not sufficient for the Martini one.

The work is organized as follows. The first paragraph briefly presents the methodology proposed by the OECD and followed by Istat for the calculation of TFP. In the second paragraph, the axiomatic theory of index numbers and their properties are briefly recalled. In the third paragraph, the Martini index with its properties is described. In the fourth one, the original application of the index is introduced and the reworked versions of the Martini formula, which are necessary to make it applicable to the context data, are described. Finally, in the fifth and sixth paragraphs the two applications of the index are described. They are constructed in such a way as to highlight two different types of consistency in the aggregation and the results obtained are presented and discussed.

2 The Calculation of the Total Factor Productivity (TFP)

The reference methodology, which is briefly presented here for the sole purpose of the following application, is that contained in the guidelines of the OECD (2001a, b). In particular, the notation defined by Istat in the Methodological Note (ISTAT 2018) is used, to which reference is made for a more detailed description of the procedure.

Starting from Solow (1957), the concept of Total Factor Productivity coincides with the measurement of the movements of the production function caused by technical progress. The production function used is the classic Cobb-Douglas one, that is,

$$Y_t = A_t * L_t^{\alpha_t} * K_t^{\beta_t}$$

where A_t, α_t and β_t are parameters, L_t and K_t are the labour and capital inputs, Y_t is the gross product expressed in real terms and represented by the Value Added at chain-linked values with a fixed reference year (see Statistical Glossary www.istat.it). In particular, A_t is the parameter representing the variations of the production function over time, which are attributed to technical progress. The variations of the production function interpret the variations of VA in real terms, which are not explained by corresponding variations in the factors of production and are identified with TFP:

$$TFP_t = A_t = Y_t/(L_t^{\alpha_t} * K_t^{\beta_t})$$

In logarithmical terms, the change in Total factor productivity is defined as

$$\ln(TFP_t/TFP_{t-1}) = \ln(Y_t/Y_{t-1}) - \ln(I_t/I_{t-1})$$

where (Y_t/Y_{t-1}) is the measure of the variation of VA in real terms, i.e. it is a quantity index, and (I_t/I_{t-1}) is a synthetic index of the total quantity of inputs used to produce it. The choice of the formula for calculating the synthetic index of the variations of total input is crucial for the calculation of the TFP variation and the literature on the subject is very large. For reasons of space, only very few of the most famous examples are quoted here: Star and Hall (1976), Diewert (1976), Jorgenson and Griliches (1967). The formula adopted by international statistical bodies (OCSE) is the one by Tornquist (1936) about quantities which, in general, is expressed like this

$${}_{t-1}T_t = \prod_{i=1}^{n} \left(\frac{q_{it}}{q_{it-1}}\right)^{\left(\frac{v_{it} + v_{it-1}}{2}\right)}$$

where vit and vit-1 are the shares of the value of the single asset on the total of the aggregate. In particular, volume index of inputs is calculated, in logarithmic form, as follows:

$$\log(I_t/I_{t-1}) = 0.5 * (svl_t/svl_{t-1}) * \log(L_t/L_{t-1}) + 0.5 * (svk_t/svk_{t-1}) * \log(K_t/K_{t-1})$$

where svk_t and svl_t are the remuneration shares for labour and capital on value added.

$$svl_t = \frac{w_t * L_t}{p_t * Y_t} \qquad svk_t = \frac{u_t * K_t}{p_t * Y_t}$$

where u_t, w_t and p_t are the prices of capital, labour and of the output respectively, and the following ratio is valid:

$$u_t * K_t + w_t * L_t = p_t * Y_t$$

so that $p_t * Y_t$ is the value added at current prices.

It is obvious that the choice of the formula has very important consequences on the results, especially taking into account that the comparison does not take place only between two times t and $t - 1$, but even considering long periods of time by using the chain index formula:

$$ {}_0T_n = \prod_{t=1}^{N} {}_{t-1}T_t $$

which makes the effects of choosing the formula more evident.

3 Elements of Axiomatic Theory of Index Numbers

To evaluate the effect of choosing the formula on the results obtained, reference can be made to axiomatic theory of index numbers. Also considering this matter the bibliography is endless, and even here only three of the most important authors are mentioned Balk (1995), Diewert (1976), Eichorn and Voeller (1976), Martini (1992, 2001). In this work we refer to the last author. The characteristics to be studied in depth regarding the choice of the Törnqvist index would be different if it is studied in the light of the axiomatic theory. However, here we limit ourselves to analyse the property of consistency in aggregation which the Törnqvist index does not satisfy. The consistency in aggregation is appropriate when we have a set of goods (or sectors) $A = \{a_1, a_2, \ldots, a_n\}$ which is constituted by the union of disjoint sets A_k, namely,

$$ A = \bigcup_{k=1}^{K} A_k $$

where the intersection among the different subsets A_k is empty. If the index is calculated from the subsets, which are then aggregated using the same formula, or from the individual assets directly, the consistency in aggregation applies if the two results are equal. This property is very important in the context of productivity analysis. Indeed, the value added is calculated for the single sectors of economic activity and for all the aggregated activities, therefore it would be very appropriate that even the index calculated on the single sectors and then aggregated for the whole economic system would coincide with the one calculated directly for all goods considered jointly. So, here we come to the purpose of this work: to present an alternative which can be applied in some specific circumstances, that is the index proposed by Marco Martini in the two books cited.

4 The Martini Index

Martini (1992, 2001) defines the indispensable (axiomatic) properties of Proportionality, Commensurability, linear Homogeneity and Monotonicity, and the desired properties of Reversibility of the Base and the Factors and Consistency in the Aggregation. Martini proposes in his works even a new index which satisfies all the axiomatic and desirable properties.

To introduce the Martini index, two concepts must be first defined: the cofactor and the factor antithesis. If the prices, quantities and the value index are known, it is natural to think that the following ratio between the dynamics of prices, quantities and value exists:

$$_bV_t = {_bQ_t} * {_bP_t}$$

which links the value index $_bV_t$ to the product of the price index ($_bPV_t$) multiplied by a quantity index ($_bQ_t$). The concept of cofactor $_bQ_t$ of the price index is here introduced. If $_bV_t$ and $_bP_t$ are known, it is always calculated

$$_bQ_t =_b V_t / {_bP_t}$$

If the cofactor $_bQ_t$ is equal to the quantity index calculated with the same formula as $_bP_t$, then it is said that the latter satisfies the reversibility of the factors. The factor antithesis of the price index is instead defined as

$$_b\underline{P}_t =_b V_t / {_bQ_t}$$

that is the ratio of the value index to the quantity index calculated with the same formula used for the price index. The formulas which respect the consistency in the aggregation are those belonging to the class of expenditure ratio indices represented for the quantities in the following way:

$$_bQ_t(x) = \frac{\sum_{h=1}^{n} q_{th}(p_{th}^{x} p_{bh}^{1-x})}{\sum_{h=1}^{n} q_{bh}(p_{th}^{x} p_{bh}^{1-x})}$$

with $(h = 0, 1, 2, \ldots, n)$ and $0 \leq x \leq 1$.

It is shown (Martini 1992) that if the sign of the covariance between the elementary indices of price and quantity, weighted by values at time b (base time), is positive (negative) the index $_bP_t(x)$ is monotone increasing (decreasing) in relation to x. The opposite occurs by factor antithesis

$$_bP_t(x) = \frac{\sum_{h=1}^{n} p_{th}q_{th}}{\sum_{h=1}^{n} p_{bh}q_{bh}} \Big/ \frac{\sum_{h=1}^{n} q_{th}(p_{th}^{x} p_{bh}^{1-x})}{\sum_{h=1}^{n} q_{bh}(p_{th}^{x} p_{bh}^{1-x})}$$

monotone decreasing (increasing) if the sign is positive (negative).

Since the two curves obtained by drawing the index and its antithesis as a function of x are monotone in the opposite direction, the intersection occurs at a single abscissa point: $x = M$; the two diagonals are not symmetrical, thus $M \neq 1/2$.

Therefore the value $x = M$ is the only one for which

$$_bP_t(M) =_b \underline{P}_t(M)$$

occurs (Fig. 1). The Martini index $_bP_a(M)$ is an expenditure ratio:

$$ {}_bP_t(M) = \Sigma_h\, p_{th} q_h^M / \Sigma_h\, p_{bh} q_h^M $$

weighted by geometric mean of quantities: $q_h^M = q_{ah}^M q_{bh}^{1-M}$ $(h = 0, 1, 2, \ldots, n)$.

The value M depends on the prices and quantities of the situations being compared and can be calculated by iteration, looking for the value assumed by M at the intersection of the two curves,

Given the characteristics of the Martini index, it appears to be very suitable for the application to the calculation of TFP. In the following pages an original application of the index in relation to productivity data is presented. An original application in all senses as the Martini index had never been applied before for a series of reasons which will not be explained here due to lack of space.

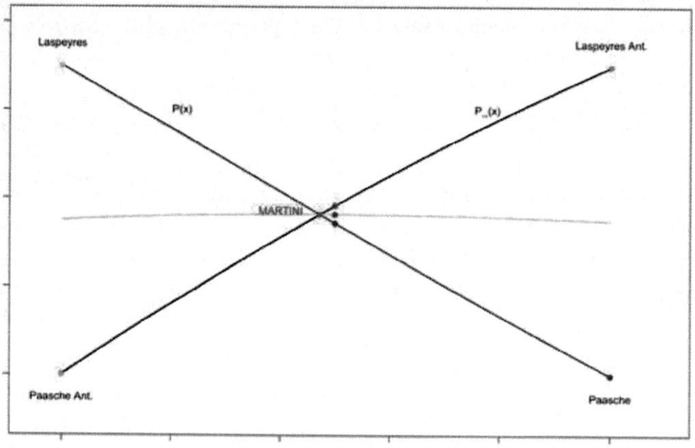

Fig. 1 The index ${}_bP_t(x)$ and the factor antithesis ${}_b\underline{P}_t(x)$

5 Application to the Calculation of the Total Factor Productivity (TFP)

To perform this application, data are needed which make it possible to calculate the Martini index for the purpose of evaluating productivity and these ones were found within the EU KLEMS project.

This project aims to create a database on measures of economic growth, productivity, employment creation, capital formation and technological change at the industry level for all European Union member states from 1970 onwards. The database should facilitate the production of high quality statistics by using national accounts and input-output analysis methodologies, with the aim to improve the international comparability. Input measure includes various capital, labour, energy, material and service categories (K-L-E-M-S). 15 organizations from across the EU take part in the realization of this project. They are a mix of academic institutions and research institutions of national economic policy and are supported by a number of statistical offices, and by OCSE as well.

In this work we will focus on data about Italy and refer to the latest published version (EU KLEMS Growth and Productivity Accounts, 2017 Release, Statistical Module). The variables are divided into values, prices, volumes and additional variables. The concepts and methodologies to calculate the various growth and productivity variables have been adapted to the new European System of National Accounts (ESA 2010). The coverage of the time period is 1995–2015. The data on production, value added and employment, as well as those relating to gross fixed investments, prices and capital stocks are consistent with Eurostat at the corresponding sector levels. The data are based on the NACE 2 classification distinguishing 34 sectors and 8 macro-sectors.

The variables used have been chosen according to the calculation of the Martini index.

VA is the value added at current prices; it is expressed in millions of euros. This variable is used to calculate the value index. **VA-QI** is the measure of volume of value added at 2010 prices. This measure will be used for the calculation of TFP. **LAB** is the remuneration of the labour factor, expressed in millions of euros. **CAP** is the remuneration of the capital factor, expressed in millions of euros. CAP equals value added (VA) minus the labour remuneration (LAB). **CAP-QI** is a volume index based on 2010 = 100. It is the capital input in terms of quantity. It will be used to calculate the volume index of the capital input **QK**. **H-EMP** is the number of hours worked by the total number of employees and freelancers, and it is expressed in thousands of units. This variable is used to calculate the labour input index **QL**. The following two additional variables can then be calculated using EU KLEMS data. **PL** Labour price index; it is obtained by the ratio between the remuneration of the **LAB** work and the number of **H-EMP** hours worked, and then making the ratio of this new measure obtained by the times $t/t-1$. **PK** Capital price index. It is obtained by calculating the $t/t-1$ index of the remuneration of **CAP** capital, and then making the ratio of this one to the capital volume index **CAP-QI**. Given the available variables, it is necessary to rewrite the Martini index in a different way, while keeping its properties intact.

$$
{}_{t-1}P_t^M = \frac{\sum_{h=1}^{n} p_{th}(q_{th}^M q_{t-1h}^{1-M})}{\sum_{h=1}^{n} p_{t-1h}(q_{th}^M q_{t-1h}^{1-M})} =
$$

$$
= \frac{p_{tL}(q_{tL}^M q_{t-1L}^{1-M}) + p_{tk}(q_{tk}^M q_{t-1k}^{1-M})}{p_{t-1L}(q_{tL}^M q_{t-1L}^{1-M}) + p_{t-1k}(q_{tk}^M q_{t-1k}^{1-M})} =
$$

$$
= \frac{\left(\frac{p_{tL}}{p_{t-1L}}\right)(p_{t-1L}\cdot q_{t-1L})\left(\frac{q_{tL}}{q_{t-1L}}\right)^M + \left(\frac{p_{tk}}{p_{t-1K}}\right)(p_{t-1K}\cdot q_{t-1K})\left(\frac{q_{tK}}{p_{t-1K}}\right)^M}{(p_{t-1L}\cdot q_{t-1L})\left(\frac{q_{t-1L}}{q_{t-1l}}\right)^M + (p_{t-1K}\cdot q_{t-1K})\left(\frac{q_{tK}}{q_{t-1K}}\right)^M} =
$$

While the Martini index of quantities becomes

$$
{}_{t-1}Q_t^M = \frac{\sum_{h=1}^{n} q_{th}(p_{th}^M p_{t-1h}^{1-M})}{\sum_{h=1}^{n} q_{t-1h}(p_{th}^M p_{t-1h}^{1-M})} =
$$

$$
= \frac{q_{tL}(p_{tL}^M p_{t-1L}^{1-M}) + q_{tk}(p_{tk}^M p_{t-1k}^{1-M})}{q_{t-1L}(p_{tL}^M p_{t-1L}^{1-M}) + p_{t-1k}(p_{tL}^M q_{t-1L}^{1-M})} =
$$

$$
= \frac{\left(\frac{q_{tL}}{p_{t-1L}}\right)(q_{t-1L}\cdot p_{t-1L})\left(\frac{p_{tL}}{p_{t-1L}}\right)^M + \left(\frac{q_{tK}}{q_{t-1K}}\right)(q_{t-1K}\cdot p_{t-1K})\left(\frac{p_{tK}}{p_{t-1K}}\right)^M}{(q_{t-1L}\cdot p_{t-1L})\left(\frac{p_{tL}}{p_{t-1L}}\right)^M + (q_{t-1K}\cdot p_{t-1K})\left(\frac{p_{tK}}{p_{t-1K}}\right)^M} =
$$

The results of the application of the Martini index are reported below and they are compared with those of the Törnqvist index, which has been recalculated as well using EU KLEMS data in order to make the comparison possible.

6 Productivity Sectors

The following list shows the sectors into which the total economy is divided. Capital letters indicate the sectors, while numbers the sub-sectors. In the EU KLEMS publications there are also 8 macro-sectors which have not been considered in this work. The consistency in aggregation of the indices was researched by directly bringing together the 17 sectors indicated with capital letters, just excluding sector U, for obvious reasons, and sector T which, by its nature, does not have the capital data (Table 1).

Table 1 Productivity sectors

Code	Productivity sector
A	Agriculture, forestry and fishing
B	Mining and quarrying
C	Total manufacturing
10–12	Food products, beverages and tobacco
13–15	Textiles, wearing apparel, leather and related products
16–18	Wood and paper products; printing and reproduction of recorded media
19	Coke and refined petroleum products
20–21	Chemicals and chemical products
22–23	Rubber and plastics products, and other non-metallic mineral products
24–25	Basic metals and fabricated metal products, except machinery and equipment
26–27	Electrical and optical equipment
28	Machinery and equipment n.e.c.
29–30	Transport equipment
31–33	Other manufacturing; repair and installation of machinery and equipment
D-E	Electricity, gas and water supply
F	Construction
G	Wholesale and retail trade; repair of motor vehicles and motorcycles
45	Wholesale and retail trade and repair of motor vehicles and motorcycles
46	Wholesale trade, except motor vehicles and motorcycles
47	Retail trade, except motor vehicles and motorcycles
H	Transportation and storage
49–52	Transport and storage
53	Postal and courier activities
I	Accommodation and food service activities
J	Information and communication
58–60	Publishing, audiovisual and broadcasting activities
61	Telecommunications
62–63	IT and other information services
K	Financial and insurance activities
L	Real estate activities
M-N	Professional, scientific, technical, administrative and support service activities
O	Public administration and defence; compulsory social security
P	Education
Q	Health and social work
R	Arts, entertainment and recreation
S	Other service activities
T	Activities of households as employers; undifferentiated goods- and services-producing activities of households for own use
U	Activities of extraterritorial organizations and bodies

In this regard, it must be highlighted that the data used here as elementary price and quantity indices are actually data already aggregated according to the Törnqvist formula. This is one of the reasons why comparisons between the results obtained in this work and the official published results will never be made. For the aggregation of sectoral indices, which have been calculated according to the Martini formula in the general aggregate, the following weights are used:

$$W_{ti} = \frac{(q_{t-11\,Li} \cdot p_{t-1Li})\left(\frac{p_{tLi}}{P_{t-1Li}}\right)^M + (q_{t-1}K_i \cdot p_{t-1Ki})\left(\frac{p_{tKi}}{P_{t-1Ki}}\right)^M}{\sum_{i=1}^{17}(q_{t-11\,Li} \cdot p_{t-1Li})\left(\frac{p_{tLi}}{P_{t-1Li}}\right)^M + \sum_{i=1}^{17}(q_{t-11\,Ki} \cdot p_{t-1Ki})\left(\frac{p_{tKi}}{P_{t-1Ki}}\right)^M}$$

With $i = 1, \ldots, 17$.

For the calculation of the Törnqvist index, the classic formula is used in which the weights are obtained as an arithmetic mean of the remuneration shares for capital and labour. Table 2 shows results of applying the two indices.

Table 2 Martini and Törnqvist on the total and on aggregated sectors

Year	Martini		Törnqvist	
	Total	Aggregated	Total	Aggregated
1995	1,0000	1,0000	1,0000	1,0000
1996	1,0159	1,0150	1,0159	1,0152
1997	1,0254	1,0257	1,0253	1,0268
1998	1,0473	1,0491	1,0473	1,0515
1999	1,0649	1,0688	1,0655	1,0738
2000	1,0859	1,0896	1,0866	1,0981
2001	1,1084	1,1106	1,1100	1,1227
2002	1,1294	1,1311	1,1317	1,1469
2003	1,1455	1,1446	1,1477	1,1633
2004	1,1599	1,1580	1,1616	1,1789
2005	1,1716	1,1698	1,1738	1,1940
2006	1,1960	1,1919	1,1975	1,2182
2007	1,2190	1,2140	1,2202	1,2417
2008	1,2242	1,2207	1,2258	1,2508
2009	1,1995	1,1957	1,2009	1,2308
2010	1,1975	1,1904	1,1961	1,2298
2011	1,2011	1,1930	1,1993	1,2339
2012	1,1804	1,1696	1,1790	1,2136
2013	1,1599	1,1478	1,1582	1,1926
2014	1,1568	1,1426	1,1559	1,1897

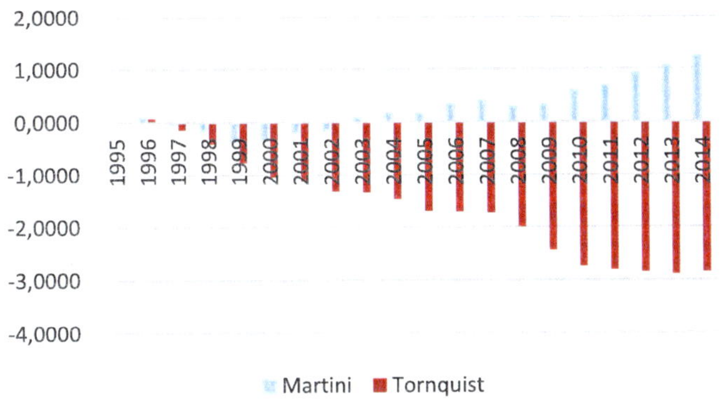

Fig. 2 Percentage variations between the Martini index and the Törnqvist index calculated on the total production and on aggregated sectors

The graph of percentage variations between the index calculated on the total of the economic system and on the aggregation of indices calculated for single production sectors is also inserted here to make the comparison more effective (Fig. 2).

By observing the graph, it is immediately noticed that the variations relating to the Martini index are definitely minor compared to the ones of the Törnqvist index. Percentage variations of both indices increase as you move away from the period chosen as the basis. It must be remembered that the indices are calculated with the chain formula and this leads to making the changes increasingly evident. However, it is particularly noticeable that the Martini index presents small percentage variations, perhaps due to the fact that the starting indices are not elementary indices but aggregations generated with the Törnqvist formula. The data for the complete calculation of the Martini index are not available.

6.1 Aggregation in Relation to Labour and Capital

We now proceed to verify the property of the aggregation by following another procedure, with the aim of obtaining the same results, that is respecting the property. A labour index and a capital index are calculated separately according to the Martini formula, and by aggregating separately the input measurements of the two factors for all 17 sectors. In this way, a general labour index and a capital index are obtained and then they are combined into an overall index, still according to the Martini formula. The same steps are plainly carried out by applying the Törnqvist formula.

The formula for the labour index $t/t-1$ which aggregates the 17 sectors taken into consideration is the following one:

$$_{t-1}Q_{Lt} = \frac{\sum_{i=1}^{17}\left(\frac{q_{tLi}}{q_{t-1Li}}\right)(q_{t-1Li}\cdot p_{t-1Li})\left(\frac{p_{tLi}}{P_{t-1Li}}\right)^{M}}{\sum_{i=1}^{17}(q_{t-11\,Li}\cdot p_{t-1Li})\left(\frac{p_{tLi}}{P_{t-1Li}}\right)^{M}}$$

The chain index is then built. The same formula is applied for the calculation of the capital index by replacing labour data with capital data. Moreover, labour and capital price indices are obtained by exchanging prices and quantities. The aggregation of the two factors then takes place in the following way:

$$_{t-1}Q_t = \frac{_{t-1}Q_{Lt}(q_{t-1Li}\cdot p_{t-1Li})_{t-1}P^{M}_{Lt} + {}_{t-1}Q_{K1}(q_{t-1Ki}\cdot p_{t-1Ki})_{t-1}P^{M}_{Kt}}{(q_{t-1L}\cdot p_{t-1L})_{t-1}P^{M}_{Lt} + (q_{t-1K}\cdot p_{t-1K})_{t-1}{P_{Kt}}^{M}}$$

The results of this application are shown in the following table and graph (Table 3 and Fig. 3). As before, in order to better evaluate the results, the graph of the percentage variations and absolute differences between the Martini and Törnqvist indices are reported. They have been calculated on the total production and on the labour and capital indices calculated on the 17 sectors. As can be seen in this case, the different behaviour of the two indices is much more evident than in the other comparison.

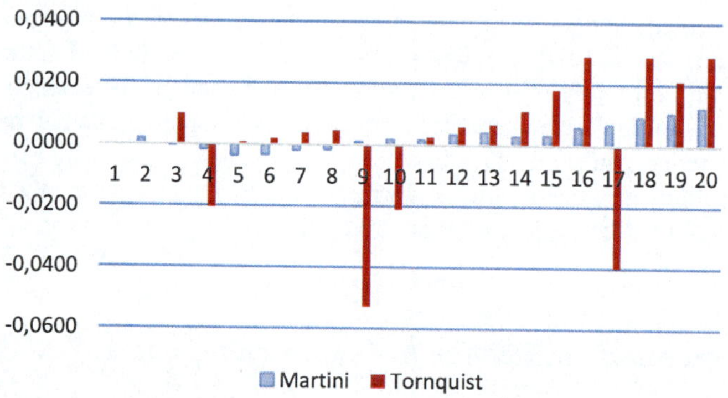

Fig. 3 Percentage variations between the index calculated on the total system and the aggregated index of capital and labour

Table 3 Martini and Törnqvist indices on the whole system and on capital and labour separately

Year	Martini		Törnqvist	
	Total	Aggregated	Total	Aggregated
1995	1,0000	1,0000	1,0000	1,0000
1996	1,0159	1,0141	1,0159	1,0159
1997	1,0254	1,0257	1,0253	1,0153
1998	1,0473	1,0491	1,0473	1,0685
1999	1,0649	1,0688	1,0655	1,0649
2000	1,0859	1,0896	1,0866	1,0845
2001	1,1084	1,1106	1,1100	1,1058
2002	1,1294	1,1311	1,1317	1,1264
2003	1,1455	1,1446	1,1477	1,2082
2004	1,1599	1,1580	1,1616	1,1859
2005	1,1716	1,1698	1,1738	1,1708
2006	1,1960	1,1919	1,1975	1,1904
2007	1,2190	1,2140	1,2202	1,2121
2008	1,2242	1,2207	1,2258	1,2122
2009	1,1995	1,1957	1,2009	1,1792
2010	1,1975	1,1904	1,1961	1,1612
2011	1,2011	1,1930	1,1993	1,2471
2012	1,1804	1,1696	1,1790	1,1448
2013	1,1599	1,1478	1,1582	1,1338
2014	1,1568	1,1426	1,1559	1,1222

7 Conclusions

The goal of this work was to apply for the first time the index proposed by Martini in 1992 and 2001 to the calculation of the Total Factor Productivity, by using it as a measure volume of input (labour and capital) instead of the Törnqvist index applied in the official statistics. The difficulty in applying the Martini Index consisted in the presence of an exponent M which assumes different values every year as it is the result of the intersection of the Price Index curve with the one of its Factor Antithesis. Moreover, no elementary data were available but only the data from the EU KLEMS project which had already been previously processed. So, the data used as elementary indices are actually indices already calculated with the Törnqvist formula that is the one used by national statistical institutes. In order to use the Martini Index, some reworked versions of the data have been carried out to make its application possible. In particular, given that the main goal was the study of the property of consistency in aggregation, two different approaches were followed: the aggregation of labour, capital input data for each sector of economic activity and then the aggregation through ad hoc weights, which appear in the work, of the sectors to the entire economic system. If the index owns the property of the aggregation by calculating the total aggregate index starting from the data of the economic system as a whole and the index calculated by aggregating the sectoral data, the same result should be achieved.

The aggregation for all sectors of the global labour input index and the global capital input index according to the Martini formula, by calculating M each time.

The results obtained are very encouraging, despite the data problems already highlighted. They were compared with the results obtained by applying the Törnqvist formula to the same data and calculating the percentage variations. It also occurs that as the length of the time period considered for the Martini index increases, the effects of the distortions are much less and in some cases almost irrelevant.

Much remains to be done, above all, the properties of the index in question, that can be verified through simulations, remain to be investigated, given the difficulty of obtaining real data with the necessary features.

References

Balk, B. M. (1995). Axiomatic price index theory: A survey. *International Statistical Review, 63*, 69–93.

Balk, B. M. (1998). *Industrial price, quantity and productivity indices: The micro-economic theory and an application*. Boston/Dordrecht/London: Kluwer Academic Publishers.

Diewert, W. E. (1976). Exact and superlative index numbers. *Journal of Econometrics, 4*(2), 115–145.

Eichhorn, W., & Voeller, J. (1976). *Theory of the price index: Fisher's test approach and generalisations*. Lecture notes in economics and mathematical systems, Berlin: Springer.

EU KLEMS PROJECT. (2017). Growth and productivity account, release 2017.

Evans, W. D., & Siegel, I. H. (1942). The meaning of productivity indexes. *Journal of the American Statistical Association, 37*(217), 103–111.

ISTAT. (2017). Productivity measures, November 2017. https://www.istati.it.

ISTAT. (2018). Productivity measures, November 2018. https://www.istat.it.

Jager, K. (2017). EU KLEMS Growth and Productivity Accounts 2017 Release, Statistical Module. Description of Methodology and country motes for Italy, September 2017. http://www.euklems.net.

Jorgenson, D. W., & Griliches, Z. (1967). The Explanation of Productivity Change. *Review of Economic Studies*.

Martini, M. (1992). *I numeri indice in un approccio assiomatico*. Milan: Giuffrè.

Martini, M. (2001). *Numeri indice per il confronto nel tempo e nello spazio*. CUSL.

OECD. (2001a). Measurement of aggregate and industry level productivity growth. *Technical Report, 59–102*, 125–158.

OECD. (2001b). Measuring productivity. OECD productivity manual: A guide to the measurement of industry level and aggregate productivity growth, Paris.

Solow, R. (1957). Technological change and the aggregate production function. *The Review of Economics and Statistics*, 312–320.

Star, S., & Hall, R. E. (1976). An approximate divisia index of total factor productivity. *Econometrica: Journal of the Econometric Society*, 257–263.

Tornquist, L. (1936). The bank of Finland's consumption price index. *Bank of Finland Monthly Bulletin 10*, 1–8.

Author Index

P. Mariani and M. Zenga (eds.), *Data Science and Social Research II*,
Studies in Classification, Data Analysis, and Knowledge Organization,
https://doi.org/10.1007/978-3-030-51222-4